Fundamentals of
Statistics & Probability Theory

A Tutorial Approach

Vol. 1 Probability Theory

Howard Dachslager, Ph.D.
Irvine Valley College

PATHWAYS TO CLEAR LEARNING

Learning Step by Step

FUNDAMENTALS OF
Statistics
&
Probability Theory

A *Tutorial Approach*

Vol. 1 Probability Theory

Howard Dachslager, Ph.D.

Published by
Path Ways To Clear Learning
Telephone (949) 375-1675
Web Site: PathwaysToClearLearning.com
E-Mail address: info@pathwaystoclearlearning.com

ISBN: 978-1492245100

To My Dearest Friends

Frankie Besch & Albert Murtz

Table Of Contents

VOLUME I PROBABILITY THEORY

Descriptive Statistics

Set Theory

Probability Theory

VOLUME II STATISTICS
Inference Theory

Complete solutions to all unsolved problems and supplementary problems are in
the web site: http://www.PathwaysToClearLearning.com

Preface

How to Use this Book

This book is designed to tutor the reader in a first course study of statistics and probability theory. It covers all the major topics. Each topic is covered by several lessons. For each lesson, the reader is presented with exercises as examples and problems. For each example, its solution is worked out step by step in detail. Following the examples, solved problems that mimic the examples are also worked out in similar detail. Finally to test the reader's understanding, unsolved problems that mimic the examples as well as the solved problems are presented with answers. At the end of each lesson, a set of supplementary problems is given. These problems are more challenging then those in the main body of the tutorial. Detailed solutions of the unsolved problems and supplementary problems can be found in the solution manual.

To see how the tutorial works, let us look at a section of this book. In the book, turn to page 47, Lesson 4, section 1. Let us look at 4.1 - Example 1 (copied below). We see that this example comes with a solution.

Next go to the bottom of page 47 for 4.1 - Solved problem 1 (copied below). We see that this example comes with a solution (top of page 48).

Note that 4.1- Example 1 and 4.1 - Solved problem 1 are same type of problems and have the same type of solutions.

Next go to page 48 for 4.1 - Problem 1 (copied below). Here only the answer is given. This problem is also similar to 4.1 - Example 1 and 4.1 - Solved Problem 1. The line

Refer back to **4.1 - Example 1 & 4.1 - Solved Problem 1.**

reminds the reader to go back to 4.1 - Example 1 and Solved Problem 1 to help in solving the unsolved problem.

Throughout the entire book we use this tutorial method for each example and associated problems.

Why Study Statistics?

Statistics is used to bring meaningful relationships between data and events. The subject can be divided into three areas of study:

- Descriptive Statistics

- Probability Theory

- Inference Theory

We begin this discussion, using examples, to show what statistics can do and finish with examples to show what statistics cannot do.

What Statistics Can Do.

Descriptive Statistics

The application of descriptive statistics to data is extensively used in data collection and applications.

Example: Ms Smith teaches a class in Ancient Greek history at a local senior center. To get a better understanding of her students, she has them fill out a questionnaire about their personal and academic background. Questions such as the student's age, gender, income, and academic background are asked. This information will be organized in a way to give Ms. Smith a better understanding of her students.

Example: Mr. Fuente is the track coach at a local high school. At the beginning of each academic year, he has several male and female students try out for the varsity teams. To qualify, the students first need to complete a background questionnaire of themselves and make several runs on the track to determine their speed and endurance at different distances. All the information about the students is collected, organized and evaluated. From this data, students will be selected to participate on the varsity team.

Probability Theory

The tools of probability theory can be used to show and interpret interesting relationships between several events.

Example: On the evening weather news, it was reported that there is a 60 percent chance that it will rain over the weekend. Several questions can be raised about this forecast:

1. How did this number come about?

2. What exactly does 60 percent mean?

Probability theory will help in answering these types of questions.

Example: Mr. Jones is playing a card game. His hand has one king. To win he needs to draw two more kings. What are his chances?

Questions like this are not hard to answer when tools of probability theory are properly applied.

Example: According to a recent study, 7.0% of the population has a lung disease. Of those people that have lung disease, 90% smoke. Suppose a person who smokes is selected at random from the population. Can we conclude that there is a 90% chance that he or she will have a lung disease?

Most people would conclude there is a 90% chance or at least a very high chance that the person selected will have a lung disease. Proper use of probability theory will show this not to be the case. In fact one can show the chance is significantly less than 90%.

__Inference__ __Theory__

 Inference theory is nothing more than drawing certain conclusions about a large population from samples taken from this population. Perhaps proper application of inference theory is the most important discipline of statistics.

Example: The president of a local chamber of commerce wants to find out the opinions of Orange county residents on the building of a new shopping center. Since asking all the residents is not possible, he takes a sample of 200 residents at random and finds that 52% would support the new shopping center. From this poll several questions could be raised:

1. Can he conclude from the results of the sample that 52% of all residents are in support of a new shopping center?

2. Can he use the figure 52% to estimate the true percentage of residents that support a new shopping center?

3. If he decides to use this figure of 52% as an argument that the majority of the residents support a new shopping center, how far off can his claim be? In fact, what is the chance that a majority do

not support a new shopping center?

The application of inference theory, will answer many of these types of questions.

Example: A manufacturer has recently developed a new gasoline additive. It claims that, when added to the fuel tank, the average car will experience a 15% increase in fuel mileage. Assume a consumer group decides to challenge the company's claim. They run a test on ten automobiles, where five have the additive in their tanks and the others do not. From this test they discover that the cars with the additive only did 10% better then those cars tested that did not have the additive. Several conclusions can occur:

　1. The claim of the manufacturer is false.

　2. The claim of the manufacturer is not necessarily false due to uncontrolled circumstances that could have
　　　occurred in the test.

　3. The manufacturer's claim is true, since 10% is "close" to 15%.

Whichever of these three conclusions are decided upon, there is always a possibility that an error is made in that

the conclusion is not true. For example, assume from the data collected the consumer group rejects the claim of the manufacturer. Even so, there is still a chance that the claim of the manufacturer is valid.

Such questions as these are dealt with using inference theory.

What Statistics Cannot Do.

Over the years statistics has been successful in solving many of the above questions. Using statistics to establish relationships between several events is a very important part of statistics. However, one serious flaw in the application of statistics is concluding that there are causal relationships between events. Such a use of statistics is highly questionable.

Example: Some recent studies showed that, with all else being equal, married men live longer then single men. Can we conclude that if a man marries, he has a better chance of living longer then if he remains single? If we answer yes to this question, then we assume there is a causal relationship between being married and living longer. The following could be presented to support such a conclusion:

　1. Since a married man eats at home his nutrition is better. Therefore, a married man lives a healthier life style.

　2. Because of his responsibilities to his family, he is motivated to live a healthier life style.

3. To help with the family responsibility, his wife will pay close attention to his well being.

Can one present a non causal relationship? Consider the following argument:

When selecting a husband, most single women will prefer a single man who is healthy and lives a healthy life style over a man that is not healthy or lives a life style not acceptable to a woman. Therefore, this selection process assures that a higher percentage of men selected for marriages are already healthy and therefore will have a higher chance of living longer over single men.

Which is correct? This is easier said then done. The one thing we can safely conclude is that the collection of data and statistical analysis does not give us a clear answer.

Example: A result of a study was release by a large Eastern university. The longevity of one thousand adults that exercised vigorously each week was compared to the longevity of one thousand adults that were sedentary in their life style. The conclusion from this study was that a life style of vigorous exercise was important in living longer.

One is tempted to draw a causal relationship between vigorous exercise and longevity. However, is it possible that this study does not warrant such a conclusion? The following is a simple explanation that points out the flaw in concluding that vigorous exercise causes one to live longer: Adults that are born with good health have a greater interest in exercising, over those that are born with poor health. For example, if one is born with asthma, there is less chance that this person will do much vigorous exercise. In fact, perhaps it can be argued that people with asthma simply do not live as long.

One can certainly argue both cases. However, it is difficult to see how statistics can determine such causality. As in beauty, causality lies in the eyes of the beholden.

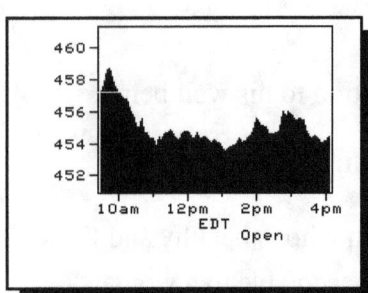

Descriptive Statistics
Lesson 1
Frequency Distributions

1.1 - What is a Frequency Distribution Table and Histogram?

It is difficult to interpret most data in its raw form. One effective way to interpret raw data is to construct a distribution table and histogram where the data is tabulated according to classes. There are two types of classes:
(1) individual numeric values and (2) fixed intervals.

A histogram is a graphic representation of a distribution table made up of rectangles where the base of the rectangles represents the numeric or interval classes and the height of the rectangles measure the frequency of the values.

Example: The following set of raw data is listed:

    6   5   4   9   8

    8   3   9   7   6

    4   1   2   4   2

    2   10 5   2   1

    1   4   6   6   6

From this data, we form a distribution table and histogram where the classes are individual values:

Data Class	Frequency of Data Occurring	Histogram
1	3	
2	4	
3	1	
4	4	
5	2	
6	5	
7	1	
8	2	
9	2	
10	1	

	Total	25	

Example: The following set of data is listed:

4.15	9.06	8.41	7.72	2.57	8.73
7.39	5.62	3.74	1.13	1.82	4.28
1.01	6.07	2.74	8.28	6.27	5.32
2.33	2.49	9.90	5.01	2.07	1.04
1.08	4.40	5.78	6.14	6.42	6.46

From this data, we form a distribution table and histogram where the classes are the following intervals[1]

[0.5,1.5), [1.5,2.5), [2.5,3.5), [3.5,4.5), [4.5,5.5),
[5.5,6.5), [6.5,7.5), [7.5,8.5), [8.5,9.5), [9.5,10.5).

The interval [0.5,1.5) includes all values, from the table, greater than or equal to 0.5 but **less** than 1.5. The interval [1.5,2.5) includes all values, from the table, greater than or equal to 1.5 but less than 2.5, etc.

Data Class	Frequency of Data Occurring	Histogram
[0.5,1.5)	4	
[1.5,2.5)	4	
[2.5,3.5)	2	
[3.5,4.5)	4	
[4.5,5.5)	2	
[5.5,6.5)	7	
[6.5,7.5)	1	
[7.5,8.5)	3	
[8.5,9.5)	2	
[9.5,10.5)	1	
Total	30	

To construct a frequency distribution, use the following rules:

---

[1]Selection of class types and values depends on the application needed.

1. From the raw data, find the minimum and maximum values. This gives the range of data.

2. Decide on one of two types of classes: single numeric values or class intervals of a fixed interval size and the class values.

1.1 - Example 1: A sample of 30 families in New York City was recently taken. The following data represents the number of children per family:

4	7	3	2	3	4
2	5	4	2	2	2
3	4	5	2	1	5
2	6	6	3	5	3
5	2	4	2	0	2

(a). Construct a frequency distribution table for single value classes.

(b). Construct the appropriate histogram.

Solutions:

➤ **(a).**
Step 1: Scanning these values, we find the smallest value is 0 and the largest value is 7.

Step 2: From the data, we see that no children occur in one family, one child occurs in one family, two children occur in ten families, three children occur in five families, four children occur in five families, five children occur in five families, six children occur in two families and seven children occur in only one family.

# Children Per Family	Number of Families	➤ (b). Histogram
0	1	
1	1	
2	10	
3	5	
4	5	
5	5	
6	2	
7	1	
	Total 30	

1.1 - Example 2: A survey of hourly wages of fifty employees at a local fast food restaurant resulted in the following data:

$6.37, $5.44, $5.29, $6.21, $6.35, $5.86, $5.62, $8.43, $6.85, $7.89

$6.93, $9.27, $4.63, $5.15, $6.50, $5.14, $7.35, $6.21, $5.34, $7.10
$6.77, $5.62, $4.08, $7.10, $6.33, $6.58, $5.98, $5.86, $6.84, $4.06
$5.04, $8.40, $5.93, $4.63, $6.45, $5.20, $5.93, $4.81, $5.99, $4.29
$5.87, $5.11, $6.83, $4.46, $5.34, $6.00, $6.71, $5.09, $5.27, $6.70

(a). Construct a frequency distribution table for the following classes:

[3.5,4.5), [4.5,5.5), [5.5,6.5), [6.5,7.5), [7.5,8.5), [8.5,9.5).

(b). Construct the appropriate histogram.

Solutions:

➤ **(a).**
Scanning the data, there are four employees that earn between $3.50 and $4.49, fourteen employees that earn between $4.50 and $5.49, sixteen employees that earn between $5.50 and $6.49, twelve employees that earn between $6.50 and $7.49, three employees that earn between $7.50 and $8.49 and 1 employees earns between $8.50 and $9.49.

Hourly wage Classes	Number of Employees	➤ (b). Histogram
[$3.50,$4.50)	4	
[$4.50,$5.50)	14	
[$5.50,$6.50)	16	
[$6.50,$7.50)	12	
[$7.50,$8.50)	3	
[$8.50,$9.50)	1	
Total 50		

Solved Problems

1.1 - Solved Problem 1: Ms. Jones recently gave a final examination in Spanish. Twenty five students in her class were surveyed as to the number of hours they studied for the final. The following table is the results of this survey:

8	3	5	6	7
2	3	3	8	2
4	5	7	10	9

7	9	9	1	6
9	10	10	7	5

(a). Construct a frequency distribution table for single value classes.

(b). Construct the appropriate histogram.

Solutions:

➤ **(a).**
Step 1: Scanning these values, we find the smallest value is 1 and the largest value is 10.

Step 2: From the data, we see that one student studied one hour, two students studied two hours, three students studied three hours, one student studied four hours, three students studied five hours, two students studied six hours, four students studied seven hours, two students studied 8 hours, four students studied nine hours, and three students studied ten hours.

# Of Hours Studied	# Of Students	➤ (b). Histogram
1	1	
2	2	
3	3	
4	1	
5	3	
6	2	
7	4	
8	2	
9	4	
10	3	
	Total 25	

1.1 - Solved Problem 2: The Frozen Foods Company recently developed a new pizza. To check the retail prices the supermarkets are charging for this pizza, it takes a survey of the prices that are charged by 20 supermarkets. The following data was collected:

$5.37, $6.11, $4.88, $5.33, $5.54
$5.80, $4.71, $4.85, $5.01, $6.15
$4.78, $5.08, $5.47, $5.89, $6.47
$6.32, $5.77, $6.21, $6.17, $4.52

(a). Construct a frequency distribution table for the following classes:

[$4.50,$4.70), [$4.70,$4.90), [$4.90,$5.10), [$5.10,$5.30), [$5.30,$5.50), [$5.50,$5.70), [$5.70,$5.90), [$5.90,$6,10), [$6.10,$6.30), [$6.30,$6.50).

(b). Construct the appropriate histogram.

Solutions:

➤ **(a).**

Scanning the data, there is one supermarket that charges between $4.50 and $4.69, four supermarkets that charges between $4.70 and $4.89, two supermarkets between $4.90 and $5.09, no supermarkets between $5.10 and $5.29, three supermarkets between $5.30 and $5.49, one supermarket between $5.50 and $5.69, three supermarkets between $5.70 and $5.89, no supermarkets between $5.90 and $6.09, four between $6.10 and $6.29 and two between $6.30 and $6.49.

Price per Pizza	Number of supermarkets	➤ (b). Histogram
[$4.50,$4.70)	1	
[$4.70,$4.90)	4	
[$4.90,$5.10)	2	
[$5.10,$5.30)	0	
[$5.30,$5.50)	3	
[$5.50,$5.70)	1	
[$5.70,$5.90)	3	
[$5.90,$6,10)	0	
[$6.10,$6.30)	4	
[$6.30,$6.50)	2	
Total 20		

Unsolved Problems with Answers

1.1 - Problem 1: A die is a six sided cube, where each side is marked with the numbers 1 through 6. A pair of these dice are tossed 35 times where the sum of the dice each time is recorded. The following is the numbers recorded:

7	3	7	4	5	5	5
8	8	4	10	7	8	7
4	8	9	5	8	11	9
2	7	8	8	6	8	8
5	4	7	9	9	11	7

(a). Construct a frequency distribution table for single value classes.

(b). Construct the appropriate histogram.

Answers:

➤ (a).

# Possible Outcomes	Frequency of Occurrences	➤ (b). Histogram
2	1	
3	1	
4	4	
5	5	
6	1	
7	7	
8	9	
9	4	
10	1	
11	2	
12	0	
Total	**30**	

⇑ *Refer back to* **1.1 - Example 1 & 1.1 - Solved Problem 1**

1.1 - Problem 2: Twenty students at a local college tried out for the track team. Each student had to run the one hundred yard dash. The following is the running speed, in seconds for each student:

9.66, 9.41, 9.42, 10.23, 10.57

10.66, 10.72, 10.74, 9.77, 10.86

10.39, 10.17, 9.53, 10.74, 11.12

11.17, 10.54, 10.06, 11.33, 11.00.

(a). Construct a frequency distribution table for the following classes:

[9.3,9.5), [9.5,9.7), [9.7,9.9), [9.9,10.1), [10.1,10.3),

[10.3,10.5), [10.5,10.7), [10.7,10.9), [10.9,11.1),

[11.1,11.3), [11.3,11.5).

(b). Construct the appropriate histogram.

Answers:
➤ **(a).**

Racing speed in Seconds	Number of Racers
[9.3,9.5)	2
[9.5,9.7)	2
[9.7,9.9)	1
[9.9,10.1)	1
[10.1,10.3)	2
[10.3,10.5)	1
[10.5,10.7)	3
[10.7,10.9)	4
[10.9,11.1)	1
[11.1,11.3)	2
[11.3,11.5)	1
Total	20

➤ **(b). Histogram**

⇑ *Refer back to* **1.1 - Example 2** & **1.1 - Solved Problem 2**

1.2 - Relative-Frequency Distribution

The relative-frequency distribution of a class of data is the frequency of the data divided by the total frequency.

1.2 - Example 1: Using the frequency distribution table in 1.1 - Example 1,

(a). Construct a relative-frequency distribution.

(b). Interpret the meaning of the relative-frequency distribution.

Solutions:
➤ **(a).**
Step 1: The frequency distribution is

# Children per Family	Number of Families
0	1
1	1
2	10
3	5
4	5
5	5
6	2
7	1
Total	30

Step 2: In the above table, divide each number in the second column by 30:

# Children per Family	Relative Frequency (rounded)
0	0.03
1	0.03
2	0.33
3	0.17
4	0.17
5	0.17
6	0.07
7	0.03
Total	1

➤ **(b).**

3% of the families have no children (0.03 = 3%).

3% of the families have one child.

33% of the families have two children.

17% of the families have three children.

17% of the families have four children.

17% of the families have five children.

7% of the families have 6 children.

3% of the families have 7 children.

Solved Problems

1.2 - Solved Problem 1: Using the distribution in the 1.1 - Solved Problem 1

(a). Construct a relative-frequency distribution.

(b). Interpret the meaning of the relative-frequency

Solutions:

➤ **(a).**

Step 1: The frequency distribution is

#Of hours studied	# of students
1	1
2	2
3	3
4	1
5	3
6	2
7	4
8	2
9	4
10	3
	Total 25

Step 2: Construct the relative-frequency distribution by dividing each of the value in the second column by 25:

# Of Hours Studied	Relative Frequency of Students
1	0.04
2	0.08
3	0.12
4	0.04
5	0.12
6	0.08
7	0.16
8	0.08
9	0.16
10	0.12
Total	1

➤ **(b).**

4% of the students studied one hour.

8% of the students studied two hours.

12% of the students studied three hours.

4% of the students studied four hours.

12% of the students studied five hours.

8% of the students studied six hours.

16% of the students studied seven hours.

8% of the students studied eight hours.

16% of the students studied nine hours.

12% of the students studied ten hours.

Unsolved Problems with Answers

1.2 - Problem 1: Using the distribution in 1.1 - Example 2,

(a). construct a relative-frequency distribution table.

(b). Interpret the meaning of the relative-frequency distribution

Answers:

➤(a).

Hourly Wage Classes	Number of Employees
[$3.5,$4.5)	0.08
[$4.5,$5.5)	0.28
[$5.5,$6.5)	0.32
[$6.5,$7.5)	0.24
[$7.5,$8.5)	0.06
[$8.5,$9.5)	0.02
Total	1

➤ (b).

8% of the employees earn between $3.50 and $4.49 an hour.

28% of the employees earn between $4.50 and $5.49 an hour.

32% of the employees earn between $5.50 and $6.49 an hour.

24% of the employees earn between $6.50 and $7.49 an hour.

6% of the employees earn between $7.50 and $8.49 an hour.

2% of the employees earn between $8.50 and $9.49 an hour.

 Refer back to **1.2 - Example 1** & **1.2 - Solved Problem 1**

1.3 - Cumulative-Relative-Distribution

The Cumulative-Relative-Distribution is the sum of all relative frequencies at and above each line of the relative-distribution table.

1.3 - Example 1: From 1.1 - Example 1,

(a). Construct a cumulative relative-distribution table.

(b). Interpret this table.
Solutions:

➤ **(a).**
Step 1: The distribution table for this example is

Number Children per Family	Number of Families
0	1
1	1
2	10
3	5
4	5
5	5
6	2
7	1

Step 2: The relative-frequency distribution table is

# Children per Family	Relative Frequency (rounded)
0	0.03
1	0.03
2	0.33
3	0.17
4	0.17
5	0.17
6	0.07
7	0.03
Total	1

Step 3: For each line, sum the numbers at and above in the relative frequency table:

# Children per family	Cumulative Frequency (rounded)
0	0.03
1	0.06
2	0.39
3	0.56
4	0.73
5	0.90
6	0.97
7	1.00

➤ **(b).**

Three percent of the families have no children.

Six percent have at most one child.

Thirty nine percent have at most two children.

Fifty six percent have at most three children.

Seventy three percent have at most four children.

Ninety percent have at most five children.

Ninety seven percent have at most six children.

One hundred percent have at most 7 children.

Solved Problems

1.3 - Solved Problem 1: Using the distribution in Solved Problem 1.1,

(a). Construct a cumulative-relative-frequency distribution table.

(b). Interpret the meaning of the cumulative relative-frequency distribution table.

Solutions:

➤ **(a).**

Step 1: The frequency distribution is

# of Hours Studied	# of Students
1	1
2	2
3	3
4	1
5	3
6	2
7	4
8	2
9	4
10	3
	Total 25

Step 2: The relative-frequency distribution table is:

# of Hours Studied	Relative Frequency of Students
1	0.04
2	0.08
3	0.12
4	0.04
5	0.12
6	0.08
7	0.16
8	0.08
9	0.16
10	0.12
Total	1

Step 3: Sum the values at and above each line of the above table:

# Of Hours Studied	Cumulative Relative Frequency of Students
1	0.04
2	0.12
3	0.24
4	0.28
5	0.40
6	0.48
7	0.64
8	0.72
9	0.88
10	1.00

➤ **(b).**

Four percent of the students studied one hour.

Twelve percent of the students studied at most two hours.

Twenty four percent of the students studied at most three hours.

Twenty eight percent of the students studied at most four hours.

Forty percent of the students studied at most five hours.

Forty eight percent of the students studied at most six hours.

Sixty four percent of the students studied at most seven hours.

Seventy two percent of the students studied at most eight hours.

Eighty eight percent of the students studied at most nine hours.

One hundred percent of the students studied at most ten hours.

Unsolved Problems with Answers

1.3 - Problem 1: Using the relative distribution in 1.1 - Example 2,

(a). Construct a Cumulative-relative-frequency distribution table.

(b). Interpret the Cumulative-relative-frequency distribution table.

Answers:

➤ **(a).**

Hourly Wage Classes	Number of Employees
[$3.5,$4.5)	0.8
[$4.5,$5.5)	0.36
[$5.5,$6.5)	0.68
[$6.5,$7.5)	0.92
[$7.5,$8.5)	0.98
[$8.5,$9.5)	1.00

(b).
Eight percent earn less than $4.50.
Thirty six percent earn less than $5.50.
Sixty eight percent earn less than $6.50.
Ninety two percent earn less than $7.50.
Ninety eight percent earn less than $8.50.
One hundred percent earn less than $9.50.

⇑ *Refer back to* **1.3 - Example 1** & **1.3 - Solved Problem 1**

Supplementary Problems

1. A die is tossed 120 times. The following distribution occurred:

Single Class Outcomes	Frequency for Each Outcome
1	20
2	20
3	22
4	19
5	19
Total	120

a. Interpret the frequency table.

b. Construct the histogram.

c. Construct a relative frequency distribution table.

d. Construct a cumulative frequency distribution table.

e. Interpret the cumulative frequency distribution table.

2. A survey of ten children each in twenty cities as to whether they believe in Santa Claus was performed. The following is the distribution table resulting from this survey:

Number of Children that Believe in Santa Claus	Number of Cities
0	0
1	1
2	2
3	2
4	6
5	4
6	2
7	2
8	1
9	0
10	0
	Total 20

a. Interpret the distribution table.

b. Draw the histogram.

c. Construct a relative frequency distribution.

d. Interpret the relative frequency distribution.

3. From the grade records at a local college, the following cumulative relative distribution of all students was constructed:

Grade Point Average	Cumulative Relative Frequency Distribution
[0.00,1.00)	0.09
[1.00,1.50)	0.18
[1.50,2.00)	0.41
[2.00,2.50)	0.56
[2.50,3.00)	0.78
[3.00,3.50)	0.92
[3.50,4.00]	1.00

a. Interpret this distribution.

b. Construct a relative frequency distribution.

c. Interpret this distribution.

d. If the enrollment at this college is 10,000, construct a frequency distribution.

━━━━━━━━━━━━━━━━━━━━━━━━━━━━━━━

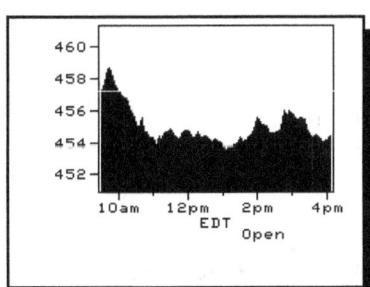

Since the collection of data usually consists of several numerical values, there is frequently a need for a single numeric value to represent this data. Such a value is called an average of the data. For most applications, there are three types of averages: mean, median and mode.

2.1 - How are the Mean, Median, and Mode values computed?

How to compute the mean value.

The mean value is the same as the arithmetic mean of the numeric data. To compute the mean value of data:

1. Add the values.

2. Divide this sum by the number of values.

The mean value is represent by $\overline{X}$.

Example: $\overline{X} = \dfrac{1 + 10 + 3 + 7 + 10 + 100 + 4 + 33}{8} = 21$

How to compute the median value.

Case 1:

The median value for an odd number of data is the middle value, where the data is arranged in ascending order (low to high).

To compute the median value for an odd number of data:

1. Rearrange the data in ascending order.

2. The median value is the middle value.

Example: 1, 10, 3, 7, 10, 100, 4.

Step 1: Arranging these values in ascending order: 1, 3, 4, 7, 10, 10, 100

Step 2: The number 7 is the middle value. Therefore, 7 is the median value.

Case 2:

The median value for an even number of data is the average of the middle two numbers, where the data is arranged in ascending order.

Example: 1, 10, 3, 7, 10, 100, 4, 33.

Step 1: Arranging these values in ascending order: 1, 3, 4, 7, 10, 10, 33, 100

Step 2: The two middle values are 7 and 10.

Step 3: The median value is $\dfrac{7 + 10}{2} = 8.5$.
Therefore, 8.5 is the median value.

How to compute the mode.

The mode of a set of data is the value that occurs the most frequently.

Case 1:

1, 10, 3, 7, 10, 100, 4, 33.

Here the number 10 appears twice and the other values occur only once. Therefore, the number 10 is the mode of the data.

Case 2:

1, 13, 3, 7, 10, 100, 4, 33

Here all numbers appear once. Therefore, there is no mode.

Case 3:

1, 10, 3, 7, 10, 100, 4, 33, 1.

Here the numbers 1 and 10 appear twice. The other numbers appear once. Therefore, the data has two modes: 1 and 10.

2.1 - Example 1: This past summer, Ms. Gardener read six books. The following data represents the number of pages in each of these books: 238, 132, 542, 601, 401, 225.

(a). Find the mean

(b). Find the median

(c). Find the mode.

Solutions:

➤(a).
Step 1: $238 + 132 + 542 + 601 + 401 + 225 = 2139$

Step 2: $\overline{X} = \dfrac{2139}{6} = 356.5$ pages.

➤(b).
Step 1: Arrange the numbers 238, 132, 542, 601, 401, 225 in ascending order: 132, 225, 238, 401, 542, 601.

Step 2: Since there are six values, average the middle two numbers

$$\dfrac{238 + 401}{2} = 319.5$$

➤(c).
Since all the numbers occur only once, there is no mode.

2.1 - Example 2: A sample of 30 families in New York City was recently taken. The following data represents the number of children per family:

4	7	3	2	3	4
2	5	4	2	2	2
3	4	5	2	1	5
2	6	6	3	5	3
5	2	4	2	0	2

(a). Find $\overline{X}$

(b). Find the median

(c). Find the mode.

(d). Construct the appropriate histogram and locate these three averages.

Solutions:

➤(a).
Step 1: $4 + 7 + 3 + 2 + 3 + 4 + 2 + 5 + 4 + 2 + 2 + 2 +$
$3 + 4 + 5 + 2 + 1 + 5 + 2 + 6 + 6 + 3 + 5 + 3 +$
$5 + 2 + 4 + 2 + 0 + 2 = 100$

Step 2: $\overline{X} = \dfrac{100}{30} = 33.33.$

➤**(b).**
Step 1: Arrange the numbers in ascending order:

0, 1, 2, 2, 2, 2, 2, 2, 2, 2, 2, 2,
3, 3, 3, 3, 3, 4, 4, 4, 4, 4, 5, 5,
5, 5, 5, 6, 6, 7

Step 2: Since there are an even number of values, average the two middle values which are 3 and 3.

Step 3: Therefore, the median is $\dfrac{3 + 3}{2} = 3.$

➤**(c).**
From the values arranged in ascending order in (b), we see that the number 2 appears ten times. Therefore, the number 2 is the mode.

➤**(d).**

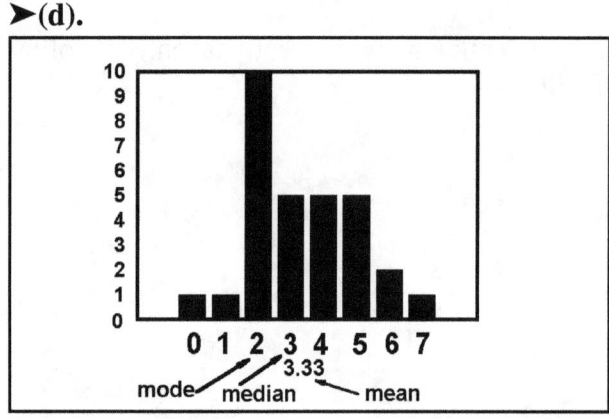

Solved Problems

2.1 - Solved Problem 1: At a local community college, nine students applied for a special science scholarship. The following are their grade point averages:

3.8, 3.2, 3.2, 3.1, 4.0, 3.7, 3.5, 3.0, 3.9.

(a). Find the mean

(b). Find the median

(c). Find the mode.

Solutions:

➤(a).
Step 1: $3.8 + 3.2 + 3.2 + 3.1 + 4.0 + 3.7 + 3.5 + 3.0 + 3.9 = 31.4$.

Step 2: $\overline{X} = \dfrac{31.4}{9} = 3.49$.

➤(b).
Step 1: Arrange the above numbers in ascending order:

3.0, 3.1, 3.2, 3.2, 3.5, 3.7, 3.8, 3.9, 4.0.

Step 2: Since there are nine values, the middle number is 3.5. Therefore, the median value is 3.5.

➤(c).
The mode is 3.2 since this number occurs twice.

2.1 - Solved Problem 2: Ms. Jones recently gave a final examination in Spanish. Twenty five students in her class were surveyed as to the number of hours they studied for the final. The following table is the results of this survey:

8	3	5	6	7
2	3	3	8	2
4	5	7	10	9
7	9	9	1	6
9	10	10	7	5

(a). Find $\overline{X}$

(b). Find the median

(c). Find the mode.

(d). Construct the appropriate histogram and locate these three averages.

Solutions:

➤(a).
Step 1: Add the numbers:

$8 + 3 + 5 + 6 + 7 +$
$2 + 3 + 3 + 8 + 2 +$
$4 + 5 + 7 + 10 + 9 +$
$7 + 9 + 9 + 1 + 6 +$
$9 + 10 + 10 + 7 + 5 = 155$

Step 2: Divide 155 by 25: $\dfrac{155}{25} = 6.2$.

Step 3: The mean $\overline{X} = 6.2$ hours.

➤**(b)..**
Step 1: Arrange the data in ascending order:

1, 2, 2, 3, 3, 3, 4, 5, 5, 5, 6, 6, 7, 7, 7, 7, 8, 8, 9, 9, 9, 9, 10, 10, 10.

Step 2: Since there is an odd number of values, the median value occurs at the thirteen number: 7. Therefore, the median value is 7.

➤**(c).**
The number 7 and 9 both occur four times. Therefore, both these numbers are modes.

➤**(d).**

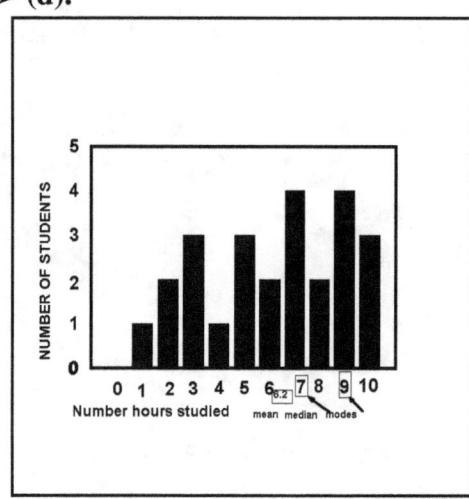

Unsolved Problems with Answers

2.1 - Problem 1: A consumer group recently tested eight new automobiles for their mileage per gallon. The following is the mileage attained:

32.7, 31.3, 32.7, 29.7, 21.9, 31.3, 38.1, 37.7

(a). Find the mean

(b). Find the median

(c). Find the mode(s).

Answers:

➤**(a).** $\overline{X} = 31.9$

➤**(b).** 32

➤**(c).** 32.7, 31.3

⇑ *Refer back to* **2.1 - Example 1** & **2.1 - Solved Problem 1**

2.1 - Problem 2: A die is a six sided cube, where each side is marked with the numbers 1 through 6. A pair of these dice is tossed 35 times where the sum of the dice is recorded each time. The following are the numbers recorded:

7	3	7	4	5	5	5
8	8	4	10	7	8	7
4	8	9	5	8	11	9
2	7	8	8	6	8	8
5	4	7	9	9	11	7

(a). Find $\overline{X}$

(b). Find the median

(c). Find the mode.

(d). Construct the appropriate histogram and locate these three averages.

Answers:

➤**(a).** $\overline{X}$ = 6.89

➤**(b).** 7

➤**(c).** 8

➤**(d).**

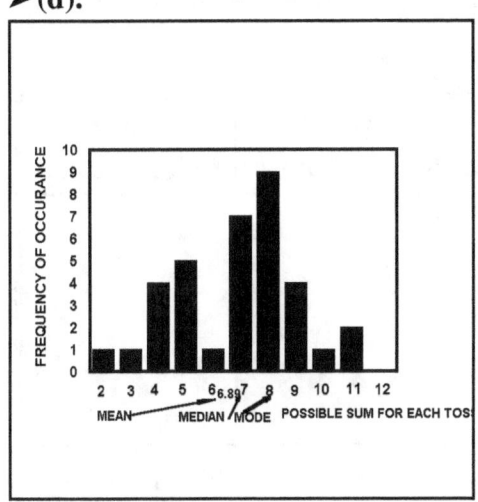

⇑ *Refer back to* **2.1 - Example 2** & **2.1 - Solved Problem 2**

2.2 - How are the Mean, Median, and Modal values for a frequency distribution computed?

The best way to show how to compute these averages is by example.

2.2 - Example 1: A survey of hourly wages of fifty employees at a local fast food restaurant resulted in the
following frequency distribution:

Hourly wage Classes	Number of Employees
[$3.50,$4.50)	4
[$4.50,$5.50)	14
[$5.50,$6.50)	16
[$6.50,$7.50)	12
[$7.50,$8.50)	3
[$8.50,$9.50)	1
Total 50	

(a). Compute the mean of the frequency distribution.

(b). Compute the median of the frequency distribution.

(c). Compute the mode of the frequency distribution.

Solutions:

➤**(a).**
To compute the mean of a frequency distribution we complete the following table:

Col 1 Hourly wage Classes	Col. 2 Number of Employees	Col. 3 Mid-Class Values	Col. 4 Col. 2 X col. 3
[$3.50,$4.50)	4		
[$4.50,$5.50)	14		
[$5.50,$6.50)	16		
[$6.50,$7.50)	12		
[$7.50,$8.50)	3		
[$8.50,$9.50)	1		
	Total 50		**Total**

1. The values in Col. 3 are the middle value for each interval.

For example, the mid-class value for [$3.50,$4.50) equals $\dfrac{\$3.50 + \$4.50}{2} = \$4$

2. The mean of the frequency distribution is $\overline{X} = \dfrac{\text{total of Column 4}}{50}$

Col 1 Hourly wage Classes	Col. 2 Number of Employees	Col. 3 Mid-Class Values	Col. 4 Col. 2 X col. 3
[$3.50,$4.50)	4	$4.00	$16.00
[$4.50,$5.50)	14	$5.00	$70.00
[$5.50,$6.50)	16	$6.00	$96.00
[$6.50,$7.50)	12	$7.00	$84.00
[$7.50,$8.50)	3	$8.00	$24.00
[$8.50,$9.50)	1	$9.00	$9.00
	Total 50		**Total** $299.00

Therefore, $\overline{X} = \dfrac{299}{50} = \$5.98$

➤**(b)..**
We start with the frequency distribution table:

Hourly wage Classes	Number of Employees
[$3.50,$4.50)	4
[$4.50,$5.50)	14
[$5.50,$6.50)	16
[$6.50,$7.50)	12
[$7.50,$8.50)	3
[$8.50,$9.50)	1
Total 50	

The following rules should be followed to compute the median value of the frequency distribution:

1. From the frequency column, we compute 50/2 = 25 employees.

2. The twenty-fifth employee occurs in the class [$5.50,$6.50).

3. The median value is given by the formula:

$$\frac{(25 - 18)}{25}(\$6.50 - \$5.50) + \$5.50 = \$5.78, \text{ where the } 18 = 4 + 14.$$

➤ (c). To compute the mode of the frequency distribution we use the formula

$$\text{Mode} = L_1 + \left(\frac{\Delta_1}{\Delta_1 + \Delta_2}\right)C,$$

where C is the width of the class intervals,

Δ_1 = the largest frequency value(s) of the second column minus the preceding frequency value,

Δ_2 = the largest frequency value(s) of the second column minus the following frequency value,

L_1 is the lowest value for the most frequently used class,

where C = $6.50 - $5.50 = $1.00.

$\Delta_1 = 16 - 14 = 2, \Delta_2 = 16 - 12 = 4$,

$L_1 = \$5.50.$

Therefore, mode = $L_1 + \left(\frac{\Delta_1}{\Delta_1 + \Delta_2}\right)C = \$5.50 + \left(\frac{2}{2 + 4}\right)(\$1.00) = \$5.83.$

Solved Problems

2.2 - Solved Problem 1: The Frozen Foods Company recently developed a new pizza. To check the retail prices the supermarkets are charging for this pizza, a survey of the prices that are charged by 20 supermarkets is taken. The following frequency distribution was collected:

Price per Pizza	Number of Supermarkets
[$4.50,$4.70)	1
[$4.70,$4.90)	4
[$4.90,$5.10)	2
[$5.10,$5.30)	0
[$5.30,$5.50)	3
[$5.50,$5.70)	1
[$5.70,$5.90)	3
[$5.90,$6.10)	0
[$6.10,$6.30)	4
[$6.30,$6.50)	2
Total	20

(a). Compute the mean of the frequency distribution.

(b). Compute the median of the frequency distribution.

(c). Find the mode of the frequency distribution.

Solutions:

➤(a).
To compute the mean of a frequency distribution, we complete the following table:

Price per Pizza	Number of Supermarkets	Col. 3 Mid-Class values	Col. 4 Col. 2 x Col. 3
[$4.50,$4.70)	1		
[$4.70,$4.90)	4		
[$4.90,$5.10)	2		
[$5.10,$5.30)	0		
[$5.30,$5.50)	3		
[$5.50,$5.70)	1		
[$5.70,$5.90)	3		
[$5.90,$6.10)	0		
[$6.10,$6.30)	4		
[$6.30,$6.50)	2		
Total	20		

1. The values in Col. 3 are the middle value for each interval.

For example, the mid-class value for [$4.50,$4.70] equals $\dfrac{\$4.50 + \$4.70}{2} = \$4.60$.

2. The mean of the frequency distribution is $\overline{X} = \dfrac{\text{total of Column 4}}{20}$.

The following is the completed table

Price per Pizza	Number of supermarkets	Col. 3 Mid-Class values	Col. 4 Col. 2 x Col. 3
[$4.50,$4.70)	1	$4.60	$ 4.60
[$4.70,$4.90)	4	$4.80	$19.20
[$4.90,$5.10)	2	$5.00	$10.00
[$5.10,$5.30)	0	$5.20	$ 0.00
[$5.30,$5.50)	3	$5.40	$16.20
[$5.50,$5.70)	1	$5.60	$ 5.60
[$5.70,$5.90)	3	$5.80	$17.40
[$5.90,$6.10)	0	$6.00	$ 0.00
[$6.10,$6.30)	4	$6.20	$24.80
[$6.30,$6.50)	2	$6.40	$12.80
Total 20			**Total** $110.60

Therefore, $\overline{X} = \dfrac{\$110.60}{20} = \$5.51$.

➤(b).

We start with the following table:

Price per Pizza	Number of Supermarkets
[$4.50,$4.70)	1
[$4.70,$4.90)	4
[$4.90,$5.10)	2
[$5.10,$5.30)	0
[$5.30,$5.50)	3
[$5.50,$5.70)	1
[$5.70,$5.90)	3
[$5.90,$6.10)	0
[$6.10,$6.30)	4
[$6.30,$6.50)	2
Total	20

The following rules should be followed to compute the median value of the frequency distribution:

1. From the frequency column, we compute 20/2 = 10 supermarkets.

2. The tenth supermarket occurs in the class [$5.30,$5.50).

3. The median value is given by the formula:

$\dfrac{(10 - 7)}{10}(\$5.50 - \$5.30) + \$5.30 = \$5.36$, where the $7 = 1 + 4 + 2$.

➤(c).

Here, we have two classes which occur the most frequently: [$4.70,$4.90) and [$6.10,$6.30). To compute the two modes of the frequency distribution, we use the formula

$$\text{Mode} = L_1 + \left(\dfrac{\Delta_1}{\Delta_1 + \Delta_2}\right)C$$

For the class [$4.70,$4.90) we have $C = \$4.90 - \$4.70 = \$0.20$

$\Delta_1 = 4 - 1 = 3$, $\Delta_2 = 4 - 2 = 2$,

$L_1 = \$4.70$.

Therefore, $\text{Mode}_1 = L_1 + (\dfrac{\Delta_1}{\Delta_1 + \Delta_2})C = \$4.70 + (\dfrac{3}{3 + 2})(\$0.20) = \$4.82.$

For the class [$6.10,$6.30) we have C = $6.30 - $6.10 = $0.20

$\Delta_1 = 4 - 0 = 4$, $\Delta_2 = 4 - 2 = 2$,
$L_1 = \$4.70$.

Therefore, $\text{Mode}_2 = L_1 + (\dfrac{\Delta_1}{\Delta_1 + \Delta_2})C = \$6.10 + (\dfrac{4}{4 + 2})(\$0.20) = \$6.23.$

Unsolved Problems with Answers

2.2 - Problem 1: Mrs. Clark is the manager of a weight reduction club for women. For the 294 members, she recorded their individual weights at the time they joined club. The following frequency distribution represents this data:

Weight Classes	Number of Members
[130,140)	15
[140,150)	45
[150,160)	55
[160,170)	42
[170,180)	32
[180,190)	48
[190,200)	25
[200,210)	12
[210,220)	9
[220,230)	9
[230,240)	2

(a). Compute the mean of the distribution.

(b). Compute the median of the distribution.

(c). Compute the mode of the distribution.

Answers:

➤(a). $\overline{X}$ = 170.92 pounds

➤(b). 162.18 pounds

➤(c). 154.35 pounds

⇑ *Refer back to* **2.2 - Example 1** & **2.2 - Solved Problem 1**

Supplementary Problems

1. Ms. Cary is a commodity trader. Over the past twenty weeks, she has recorded the end-of-the-week closing prices where prices are measured in cents per pound. To get a better understanding of the changes of these prices, she needs to compute a five week average of these prices. The second column in the following table is a list of these prices:

Week	Weekly Meat Prices (cents per pound)	5 Day Moving Average.
1	0.4524	
2	0.4222	
3	0.3987	
4	0.4123	0.4262
5	0.4454	0.4358
6	0.5002	0.4536
7	0.5112	
8	0.5321	
9	0.4986	
10	0.4541	
11	0.4199	
12	0.4041	
13	0.3989	
14	0.4000	
15	0.3910	
16	0.3876	
17	0.3818	
18	0.3870	
19	0.3987	
20	0.3597	

The rules for computing the 5 day moving average is as follows:

1. Add the first 5 numbers and record this average in the last column.

2. Drop the first number in the list of prices, compute the average of the next 5 numbers and record this average in the last column.

3. Continue this process to the end of the price data.

In the last column we have computed the first three 5 day moving averages. Complete the remainder of this column.

2. For the numbers 4, 12, 654, 132, -10, 13, 0, -125, 13, p, find the value p so that the average of these ten number is $\overline{X} = 1$.

3. Using the formula $1 + 2 + 3 + 4 + ... + n = \dfrac{n(n + 1)}{2}$, find the mean average $\overline{X}$ of the

numbers
5, 6, 7, 8, 9,..., 10,000.

4. Using the formula $1^2 + 2^2 + 3^2 + 4^2 + ... + n^2 = \dfrac{n(n + 1)(2n + 1)}{6}$, find the mean $\overline{X}$

of the numbers 9, 16, 25, 36, 49, 64, 81, 100, ..., 10,000.

5. Consider the list of data A: 1, 2, 3, 4, 5, 6, 7, 8, 9, 10. From this set of data, we list all sub-lists of numbers consisting of nine numbers:

A1: {2, 3, 4, 5, 6, 7, 8, 9, 10}

A2: {1, 3, 4, 5, 6, 7, 8, 9, 10},

A3: {1, 2, 4, 5, 6, 7, 8, 9, 10},

A4: {1, 2, 3, 5, 6, 7, 8, 9, 10},

A5: {1, 2, 3, 4, 6, 7, 8, 9, 10},

A6: {1, 2, 3, 4, 5, 7, 8, 9, 10},

A7: {1, 2, 3, 4, 5, 6, 8, 9, 10},

A8: {1, 2, 3, 4, 5, 6, 7, 9, 10},

A9: {1, 2, 3, 4, 5, 6, 7, 8, 10},

A10: {1, 2, 3, 4, 5, 6, 7, 8, 9 },

For the original list of data as well as all sub-lists, compute their mean values.

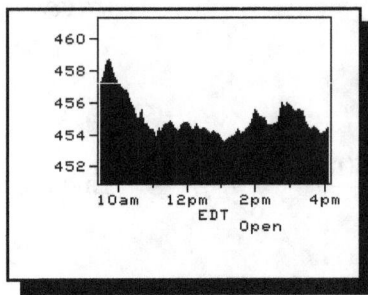

Descriptive Statistics
Lesson 3
Measuring Variation

3.1- What is Data Variation?

Data variation is a numeric value which measures the spread of data from the mean $\bar{x}$. For example, the two sets of numbers 20, 22, 25, 30 and 5, 22, 25, 45 both have the same mean $\bar{x} = 24.25$, but the spreads from 24.25 are different since the first group of data is not as varied as the second group of data. The following two histograms graphically demonstrate two sets of data both having a mean $\bar{x} = 100$ but different variations.

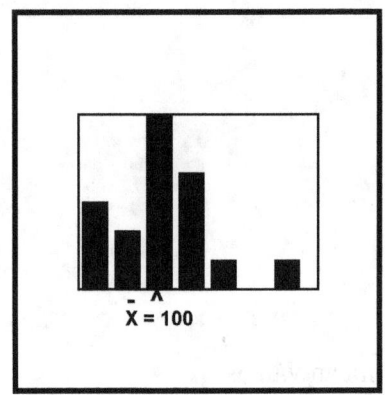

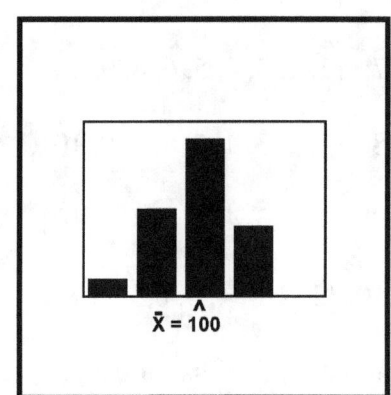

The following are three common methods for representing the variation of data.

The Range

The range of data is the difference between the largest and smallest numbers in the data.

Example: Assume the data is 1, 2, 3, 4, 5, 6, 7, 8, 9, 10. Since 10 and 1 are the largest and smallest numbers respectively, the range is 10 - 1 = 9.

The Absolute Mean Variation (AMV)

The following table shows how the absolute mean variation is computed:

Data x	$\lvert x - \bar{x} \rvert$
1	$\lvert 1 - 5.5 \rvert = 4.5$
2	$\lvert 2 - 5.5 \rvert = 3.5$
3	$\lvert 3 - 5.5 \rvert = 2.5$
4	$\lvert 4 - 5.5 \rvert = 1.5$
5	$\lvert 5 - 5.5 \rvert = 0.5$
6	$\lvert 6 - 5.5 \rvert = 0.5$
7	$\lvert 7 - 5.5 \rvert = 1.5$
8	$\lvert 8 - 5.5 \rvert = 2.5$
9	$\lvert 9 - 5.5 \rvert = 3.5$
10	$\lvert 10 - 5.5 \rvert = 4.5$
$\bar{x} = 5.5$	AMV $= 25/10 = 2.5$

The Standard Deviation (s)

The following table shows how the standard deviation is computed:

(1) Data x	(2) $x - \bar{x}$	(3) $(x - \bar{x})^2$
1	-4.5	$(-4.5)^2 = 20.25$
2	-3.5	$(-3.5)^2 = 12.25$
3	-2.5	$(-2.5)^2 = 6.25$
4	-1.5	$(-1.5)^2 = 2.25$
5	-0.5	$(-0.5)^2 = 0.25$
6	0.5	$(0.5)^2 = 0.25$
7	1.5	$(1.5)^2 = 2.25$
8	2.5	$(2.5)^2 = 6.25$
9	3.5	$(3.5)^2 = 12.25$
10	4.5	$(4.5)^2 = 20.25$
$\bar{x} = 5.5$		$s = \sqrt{\dfrac{82.5}{10}} = \sqrt{8.25} \approx 2.87$

The following are the rules for computing the standard deviation:

Rule 1: Compute $\bar{x}$ from the data in column 1.

Rule 2: The numbers in column 2 are computed using the formula $x - \bar{x}$.

Rule 3: The numbers in column 3 are computed using the formula $(x - \bar{x})^2$.

Rule 4: To compute the standard deviation:

a. sum the numbers in column 3. In the table, this sum is 82.5.

b. Divide this sum by the total number of values in column 3. This gives 8.25.

c. The standard deviation s is the square root of the value computed in b. This gives s = 2.87.

Variance

The variance of a set of data is defined as the square of the standard deviation s^2.

For the above example, the variance is $s^2 = 2.87^2 = 8.25$.

Of these methods used to compute the variation of a set of data, the variance and standard deviation are the most frequently used.

3.1 - Example 1: Ms. Jones teaches a Latin class at a Senior center. The following data is the ages of her students: 74, 67, 65, 74, 67, 81, 65, 85, 67, 80.

Find the

(a). range.

(b). absolute mean variation.

(c). standard deviation.

(d). variance.

Solutions:

➤**(a).**
The oldest and youngest ages are 85 and 65 respectively. Therefore, the range is 85 - 65 = 20 years old.

➤(b).

The following table shows how the absolute mean variation is computed.

Data x	$\lvert x - \bar{x} \rvert$
74	$\lvert 74 - 72.5 \rvert = 1.5$
67	$\lvert 67 - 72.5 \rvert = 5.5$
65	$\lvert 65 - 72.5 \rvert = 7.5$
74	$\lvert 74 - 72.5 \rvert = 1.5$
67	$\lvert 67 - 72.5 \rvert = 5.5$
81	$\lvert 81 - 72.5 \rvert = 8.5$
65	$\lvert 65 - 72.5 \rvert = 7.5$
85	$\lvert 85 - 72.5 \rvert = 12.5$
67	$\lvert 67 - 72.5 \rvert = 5.5$
80	$\lvert 80 - 72.5 \rvert = 7.5$
$\bar{x} = 725/10 = 72.5$	$AMV = 63/10 = 6.3$

➤(c).

The following table shows how the standard deviation is computed:

(1) Data x	(2) $x - \bar{x}$	(3) $(x - \bar{x})^2$
74	$74 - 72.5 = 1.5$	$(1.5)^2 = 2.25$
67	$67 - 72.5 = -5.5$	$(-5.5)^2 = 30.25$
65	$65 - 72.5 = -7.5$	$(-7.5)^2 = 56.25$
74	$74 - 72.5 = 1.5$	$(1.5)^2 = 2.25$
67	$67 - 72.5 = -5.5$	$(-5.5)^2 = 30.25$
81	$81 - 72.5 = 8.5$	$(8.5)^2 = 72.25$
65	$65 - 72.5 = -7.5$	$(-7.5)^2 = 56.25$
85	$85 - 72.5 = 12.5$	$(12.5)^2 = 156.25$
67	$67 - 72.5 = -5.5$	$(-5.5)^2 = 30.25$
80	$80 - 72.5 = 7.5$	$(7.5)^2 = 56.25$
$\bar{x} = 72.5$		$s = \sqrt{\dfrac{492.5}{10}} = \sqrt{49.25} \approx 7.02$

➤(d).

The variance is $s^2 = 7.02^2 \approx 49.25$.

Solved Problems

3.1 - Solved Problem 1: Rick is a member of the All Star Bowling Team. Last week he bowled the following scores: 187, 167, 201, 185, 167, 210, 205, 167.

Find the
(a). range.
(b). absolute mean variation.
(c). standard deviation.
(d). variance.

Solutions:

➤(a).
The highest and lowest scores are 210 and 167 respectively. Therefore the range is 210 - 167 = 43 points.

➤(b).
The following table shows how the absolute mean variation is computed.

Data x	$\|x - \bar{x}\|$
187	$\|187 - 186.125\| = 0.875$
167	$\|167 - 186.125\| = 19.125$
201	$\|201 - 186.125\| = 14.875$
185	$\|185 - 186.125\| = 1.125$
167	$\|167 - 186.125\| = 19.125$
210	$\|210 - 186.125\| = 23.875$
205	$\|205 - 186.125\| = 18.875$
167	$\|167 - 186.125\| = 19.125$
$\bar{x} = 1489/8 = 186.125$	**AMV** $= 117/8 = 14.625$

➤(c).

The following table shows how the standard deviation is computed:

(1) **Data x**	(2) $x - \bar{x}$	(3) $(x - \bar{x})^2$
187	187 - 186.125 = 0.875	0.77
167	167 - 186.125 = -19.125	365.77
201	201 - 186.125 = 14.875	221.27
185	185 - 186.125 = -1.125	1.27
167	167 - 186.125 = -19.125	365.77
210	210 - 186.125 = 23.875	570.02
205	205 - 186.125 = 18.875	356.27
167	167 - 186.125 = -19.125	365.77
$\bar{x} = 186.13$		$s = \sqrt{\dfrac{2246.91}{8}} \approx 16.76$

➤(d).

The variance is $s^2 = 16.76^2 \approx 280.90$.

Unsolved Problems with Answers

3.1 - Problem 1: A die is tossed 20 times with the following outcomes:
4, 5, 6, 4, 2, 4, 2, 2, 4, 4, 2, 4, 1, 1, 3, 4, 5, 3, 3, 4.

Find the

(a). range.

(b). absolute mean variation.

(c). standard deviation.

(d). variance.

Answers:

➤(a). 5

➤(b). 1.12

➤(c). 1.31

➤(d).1.73

⇑ *Refer back to* **3.1 - Example 1** & **3.1 - Solved Problem 1.**

3.2 - Computing the Variance and Standard Deviation for Frequency Distributions.

The following example of a frequency distribution demonstrates how to compute its standard deviation:

Class	(1) Mid-Value	(2) Frequency	(3) (1)x(2)	(4) $[(1) - \bar{x}]^2$	(5) (2)x(4)
[25,35)	30	5	150	$(30 - 57.5)^2 = 757.25$	3781.25
[35,45)	40	7	280	$(40 - 57.5)^2 = 306.25$	2143.75
[45,55)	50	10	500	$(50 - 57.5)^2 = 56.25$	562.50
[55,65)	60	6	360	$(60 - 57.5)^2 = 6.25$	37.50
[65,75)	70	4	280	$(70 - 57.5)^2 = 156.25$	625.50
[75,85)	80	12	960	$(80 - 57.5)^2 = 506.25$	6075.50
Total		44	2530		13225

$\bar{x} = 2530/44 = 57.50$

$s^2 = 13225/44 = 300.57$

$s = \sqrt{300.57} \approx 17.34$

3.2 - Example 1: A survey of hourly wages of fifty employees at a local fast food restaurant resulted in the following frequency distribution:

Hourly wage Classes	Number of Employees
[$3.50,$4.50)	4
[$4.50,$5.50)	14
[$5.50,$6.50)	16
[$6.50,$7.50)	12
[$7.50,$8.50)	3
[$8.50,$9.50)	1
Total 50	

Compute the variance and standard deviation.

Solution:

The following table computes the standard deviation and variance:

Classes	(1) Mid-values	(2) frequency	(3) (1)x(2)	(4) $[(1) - \bar{x}]^2$	(5) (2)x(4)
[$3.50,$4.50)	$4.00	4	$16.00	$(4 - 5.98)^2 = 3.92$	15.68
[$4.50,$5.50)	$5.00	14	$70.00	$(5 - 5.98)^2 = 0.96$	13.44
[$5.50,$6.50)	$6.00	16	$96.00	$(6 - 5.98)^2 = 0.0004$	0.0064
[$6.50,$7.50)	$7.00	12	$84.00	$(7 - 5.98)^2 = 1.04$	12.48
[$7.50,$8.50)	$8.00	3	$24.00	$(8 - 5.98)^2 = 4.08$	12.24
[$8.50,$9.50)	$9.00	1	$ 9.00	$(9 - 5.98)^2 = 9.12$	9.12
Total		50	$299.00		62.97

$\bar{x} = \$299/50 = \$5.98$

$s^2 = 62.97/50 \approx \$1.26$

$s = \sqrt{1.26} \approx \$1.12$

Solved Problems

3.2 - Solved Problem 1: The Frozen Foods Company recently developed a new pizza. To check the retail prices the supermarkets are charging for this pizza, a survey of the prices that are charged by 20 supermarkets is taken. The following frequency distribution was collected:

Price per Pizza	Number of Supermarkets
[$4.50,$4.70)	1
[$4.70,$4.90)	4
[$4.90,$5.10)	2
[$5.10,$5.30)	0
[$5.30,$5.50)	3
[$5.50,$5.70)	1
[$5.70,$5.90)	3
[$5.90,$6.10)	0
[$6.10,$6.30)	4
[$6.30,$6.50)	2
Total	20

Compute the variance and standard deviation.

Solution:

The following table computes the standard deviation and variance:

Classes	(1) Mid-values	(2) frequency	(3) (1)x(2)	(4) $[(1)-\bar{x}]^2$	(5) (2)x(4)
[$4.50,$4.70)	$4.60	1	4.60	0.86	0.86
[$4.70,$4.90)	$4.80	4	19.20	0.53	2.12
[$4.90,$5.10)	$5.00	2	10.00	0.28	0.56
[$5.10,$5.30)	$5.20	0	0.00	0.11	0.00
[$5.30,$5.50)	$5.40	3	16.20	0.02	0.06
[$5.50,$5.70)	$5.60	1	5.60	0.01	0.01
[$5.70,$5.90)	$5.80	3	17.40	0.07	0.21
[$5.90,$6.10)	$6.00	0	0.00	0.22	0.00
[$6.10,$6.30)	$6.20	4	24.80	0.45	1.80
[$6.30,$6.50)	$6.40	2	12.80	0.76	1.52
Total		20	$110.60		$7.14

$\bar{x} = \$110.60/20 = \$5.53$

$s^2 = 7.14/20 \approx 0.36$

$s = \sqrt{0.36} \approx \$0.60$

Unsolved Problems with Answers

3.2 - Problem 1: Ms. Clark is the manager of a weight reduction club of 294 women. For the members, she recorded their individual weights at the time they joined the club. The following frequency distribution represents this data:

Weight classes	Number of Members
[130,140)	15
[140,150)	45
[150,160)	55
[160,170)	42
[170,180)	32
[180,190)	48
[190,200)	25
[200,210)	12
[210,220)	9
[220,230)	9
[230,240)	2
Total	294

Find the variance and standard deviation.

Answers:

$s^2 = 528.92$

$s = 23.00$

⇑ *Refer back to* **3.2 - Example 1 & 3.2 - Solved Problem 1.**

3.3 - An Application for the Standard Deviation

In Statistics, we frequently are interested in the data that fall within a given number of standard deviations from the mean $\bar{x}$.

3.3 - Example 1: A sample of 30 families in New York City was recently taken. The following data, listed in ascending order, represents the number of children per family:

0, 1, 2, 2, 2, 2, 2, 2, 2, 2,
2, 2, 3, 3, 3, 3, 3, 4, 4, 4,
4, 4, 5, 5, 5, 5, 5, 6, 6, 7

Find

(a). the mean $\bar{x}$.

(b). the standard deviation.

(c). the numbers that are within one standard deviation of $\bar{x}$.

(d). the numbers that are within two standard deviations of $\bar{x}$.

(e). the numbers that are within three standard deviations of $\bar{x}$.

(f). the percent of numbers that are within two standard deviations of $\bar{x}$.

Solutions:

➤(a).
To find the mean, add all the above numbers and divide by 30. This gives $\bar{x} \approx 3.33$.

➤(b).
Following the rules in section 1 of this lesson, we find the standard deviation $s = 1.65$.

➤(c).
To find the numbers that are within one standard deviation of $\bar{x}$, we select those numbers that are

between

3.33 - 1.65 = 1.68 and 3.33 + 1.65 = 4.98. This would include all numbers between 2 and 4 children:

2, 2, 2, 2, 2, 2, 2, 2, 2, 2,
3, 3, 3, 3, 3, 4, 4, 4, 4, 4.

➤(d).
To find the numbers that are within two standard deviations of $\bar{x}$, we select those numbers that are between

3.33 - 2(1.65) = 0.03 and 3.33 + 2(1.65) = 6.63. This would include all numbers between 1 and 6 children:

1, 2, 2, 2, 2, 2, 2, 2, 2, 2, 2,
3, 3, 3, 3, 3, 4, 4, 4, 4, 4, 5,
 5, 5, 5, 5, 6, 6.

➤(e).
To find the numbers that are within three standard deviations of $\bar{x}$, we select those numbers that are between 3.33 - 3(1.65) = -1.62 and 3.33 + 3(1.65) = 8.28. This would include all 30 numbers.

➤(f).
 Since 28 out of 30 numbers are within 2 standard deviations of $\bar{x}$, the percent is 28/30 ≈ 93%.

Solved Problems

3.3 - Solved Problem 1: A computer generates the following 25 random numbers ranging from 1 to 10:

1, 2, 2, 3, 3, 4, 5, 5, 5, 6,
6, 6, 6, 6, 6, 6, 7, 7, 7, 8,
9, 9, 9, 10 ,10.

Find

(a). the mean $\bar{x}$.

(b). the standard deviation.

(c). the numbers that are within one standard deviation of $\bar{x}$.

(d). the numbers that are within two standard deviations of $\bar{x}$.

(e). the numbers that are within three standard deviations of $\bar{x}$.

(f). the percent of numbers that are within two standard deviations of $\bar{x}$.

Solutions:

➤**(a).**
To find the mean, add all the above numbers and divide by 25. This gives $\bar{x} \approx 5.92$

➤**(b).**
Following the rules in section 1 of this lesson, we find the standard deviation $s \approx 2.48$.

➤**(c).**
To find the numbers that are within one standard deviation of $\bar{x}$, we select those numbers that are between
5.92 - 2.48 = 3.44 and 5.92 + 2.48 = 8.40. This would include all numbers between 4 and 8:

4, 5, 5, 5, 6,
6, 6, 6, 6, 6,
6, 7, 7, 7, 8.

➤**(d).**
To find the numbers that are within two standard deviations of $\bar{x}$, we select those numbers that are between
5.92 - 2(2.48) = 0.96 and 5.92 + 2(2.48) = 10.88. This would include all the numbers.

➤**(e).**
To find the numbers that are within three standard deviations of $\bar{x}$, we select those numbers that are between 5.92 - 3(2.48) = -1.52 and 5.92 + 3(2.48) = 13.36. This would include all numbers 25 numbers.

➤**(f).**
Since all of the 25 numbers are within 2 standard deviations of $\bar{x}$, the percent of these numbers is 100%.

Unsolved Problems with Answers

3.3 - Problem 1: Ms. Jones recently gave a final examination in Spanish. The thirty students in her class received the follow grades:

57.47, 60.05, 60.83, 61.78, 62.70,
62.73, 62.80, 63.16, 63.24, 63.27,
63.31, 63.94, 64.18, 64.68, 64.83,
64.97, 65.18, 65.26, 65.31, 65.51,
65.60, 65.88, 66.31, 66.46, 66.56,
67.21, 67.57, 67.64, 68.44, 69.61.

Find the

(a). mean $\bar{x}$.

(b). standard deviation.

(c). numbers that are within one standard deviation of $\bar{x}$.

(d). numbers that are within two standard deviations of $\bar{x}$.

(e). numbers that are within three standard deviations of $\bar{x}$.

(f). the percent of numbers that are within two standard deviations of $\bar{x}$.
Answers:

➤(a). $\bar{x}$ = 64.55

➤(b). s = 2.57

➤(c).
62.70,62.73, 62.80, 63.16, 63.24,
63.27,63.31, 63.94, 64.18, 64.68,
64.83,64.97, 65.18, 65.26, 65.31,
65.51,65.60, 65.88, 66.31, 66.46,
66.56

➤(d).
60.05, 60.83, 61.78, 62.70, 62.73,
62.80, 63.16, 63.24, 63.27, 63.31,
63.94, 64.18, 64.68, 64.83, 64.97,
65.18, 65.26, 65.31, 65.51, 65.60,
65.88, 66.31, 66.46, 66.56, 67.21,
67.57, 67.64, 68.44, 69.61

➤(e). All the numbers.

➤(f). 96.67%

⇑ *Refer back to* **3.3 - Example 1 & 3.3 - Solved Problem 1.**

Supplementary Problems

1. For the set of numbers 0,1,2,3,4,5,6,...,100, use the formulas

$$1 + 2 + 3 + 4 + ... + n = \frac{n(n + 1)}{2} ,$$

$$1^2 + 2^2 + 3^2 + 4^2 + \dots + n^2 = \frac{n(n + 1)(2n + 1)}{6}.$$

a. to find the mean $\bar{x}$.

b. to find its standard deviation.

c. List all numbers that fall within one standard deviation of the mean.

d. List all numbers that fall within two standard deviations of the mean.

2. A statistician was hired by a professional basketball team to do a study on the performance of the team. The following is a list, in numeric order, of the final scores of the team's last 50 games:

85	85	85	86	87	89	90	91	91	92	92	92	93
93	94	97	97	99	99	100	100	103	103	103	104	106
106	106	107	108	108	108	109	109	109	112	113	114	115
116	117	117	118	119	119	120	121	122	123	125.		

a. Find the standard deviation.

b. For this data complete the following frequency distribution table:

Classes	Frequency
[80,90)	
[90,100)	
[100,110)	
[110,120)	
[120,130)	

c. For the above distribution, compute the standard deviation.

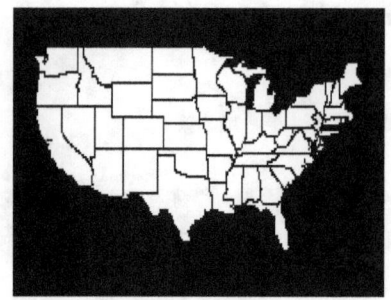

Set Theory
Lesson 4
Basic Concepts

In order to understand probability theory and statistics, we shall begin with a study of sets.

4.1 - What is a Set?

A set is a collection of items (elements) with the following rules:

1. Each item in the set is distinct.

2. The order of the items is not important.

4.1 - Example 1: **A** = {Mary, Jane, Harry, Bill} is a set?

Solution:

This collection of items is a set since all the elements are distinct.

4.1 - Example 2: F = {100, -6, 10, 100, 76.5} is a set?

Solution:

This collection of items is not a set since the element 100 is repeated twice. **F** = {-6,10,100,76.5 } is now a set since the elements are distinct.

4.1 - Example 3: D = {(a,b), (c,f), (e,k), (a,g)} is a set?

Solution:

This collection of items is a set since all the elements are distinct. In this set, the elements are pairs of letters. Note that the letter a is not a member of this set.

Solved Problems

4.1 - Solved Problem 1: A = {a, 65, 3, Harry}.is a set?

Solution:

This is a set even though the elements are of different types. Note that each element is distinct.

4.1 - Solved Problem 2: B = {a, c, harry, a} is a set?

Solution:

This expression is not a set since the element a is repeated twice.

B can be made a set by writing **B** ={a, c, harry}.

4.1 - Solved Problem 3: D = {(2,5), (3,5), (7,2)} is a set?

Solution:

This expression is a set consisting of 3 distinct pair of elements: (2,5), (3,5),(7,2). Since the elements of this set are only pairs, the individual numbers are not members of the set. For example, the number 2 is not a member of this set. The fact that 2 appears in two separate pairs does not violate the rule that the elements have to be distinct.

Unsolved Problems With Answers

4.1 Problem 1: A = {a, b, c, d, ff} is a set?

Answer:

Yes

⇑ *Refer back to* **4.1 - Example 1 & 4.1 - Solved Problem 1.**

4.1 - Problem 2: A = {1, 5, 2, 1} is a set?

Answer:

No

⇑*Refer back to* **4.1 - Example 2 & 4.1 - Solved Problem 2.**

4.1 - Problem 3: A = {(h,h), (h,t), (t,h), (t,t)} is a set?

Answer:

Yes

⇑ *Refer back to* **4.1 - Example 3 & 4.1- Solved Problem 3.**

4.2 - When are Two Sets Equal?

Two sets **A = B** are said to be equal if they both contain the same identical elements.

4.2 - Example 1: The set {1, 3, 5, 7} = {5, 7, 1, 3}?

Solution:

Even though the order is different, these two sets have the same elements. Therefore they are equal.

4.2 - Example 2: The set {a, John, c} = (a, c, John, z}?

Solution:

These two sets are not equal since one set has the element z and the other does not.

Solved Problems

4.2 - Solved Problem 1: The set {10, 3, 5, 7} = {5, 7, 10, 3}?

Solution:

Even though the order is different, these two sets are equal since they contain exactly the same elements.

4.2 - Solved Problem 2: The set {1, 2, 3, 4, 5} = {0, 1, 2, 3, 4, 5}?

Solution:

These two sets are not equal since one set has the element 0 and the other does not.

Unsolved Problems with answers.

4.2 - Problem 1: The set {2, 7, -3} = {2, -3, 7}?

Answer:

Yes.

⇑ *Refer back to* **4.2 - Example 1 & 4.2 - Solved Problem 1.**

4.2 - Problem 2: The set {(a,b), (c,d), (a,a)} = {(b,a), (c,d), (a,a)}?

Answer:

No.   ⇑*Refer back to* **4.2 - Example 2 & 4.2 - Solved Problem 2**

4.3 - What is an Empty Set?

An empty set is a set that does not contain any elements and is indicated by the null symbol φ.

4.3 - Example 1: If **A** = {all students taking statistics that are over 20 feet tall}, then **A** = φ ?

Solution:

A is empty since there are no people over 20 feet tall. Therefore, **A** = φ.

4.3 - Example 2: If **B** = {all numbers that are both greater than zero and less than five} then **B** = φ?

Solution:

B is not empty since there are numbers between zero and less than five.

Solved Problems

4.3 - Solved Problem 1: If **A** = {all numbers that are both greater than 5 and less than 2} then **A** is empty?

Solution:

The set **A** is empty since there are no numbers that are both greater than 5 and less than 2 .

4.3 - Solved Problem 2: If **A** = {all students that are studying math and English} then **A** is empty?

Solution:

The set **A** is not empty since there are many students that are studying math and English.

Unsolved Problems with answers.

4.3 - Problem 1: Is the set **C** = {all people born both in England and Mexico} empty?

Answer:

Yes.

⇑*Refer back to* **4.3 - Example 1 & 4.3 - Solved Problem 1.**

4.3 - Problem 2: Is the set **D** = {all birds that cannot fly} empty?

Answer:

No.   ⇑ *Refer back to* **4.3 - Example 2 & 4.3 - Solved Problem 2.**

Supplementary Problems

1. Is {1, 3, {1,3}} a set?

2. Is{ф} a set?

3. How many elements does the set {1,2,{1,2,3}} have?

4. Construct a set consisting of four elements, where three of the elements are themselves sets and the fourth is the empty set.

5. Are the sets {1, 2, {3, 4, 5}}, {1, 2, {5, 4, 3}} equal?

6. For the following sets, indicate which ones are empty:

a. All numbers greater than 3 or less than 0.
b. All numbers greater than 3 and less than 0
c. All number greater than 3 and less than 4.
d. All students majoring in English and studying math
e. All people born in China and Japan.
f. All people born in China or Japan.

7. For the following sets **A,** find the number of elements in **A**:

a. **A** = {a,b,c,{a}}   b. **A** = {{a,b,c}}

c. **A** = {a,b,{a},{a,b},{a,b,c},c}   d. **A** = {{a,{a},b,{b,c},ab}}

e. **A** = {{ф}}   f. **A** = {{{ф}, ф},ф}

8. Are the following sets:

a. {a {a}}

b. {a,b,c,{a,b},{a,b,c}}

c. {a,b,c,{a,b,c,d}, {a,c,b,d}}?

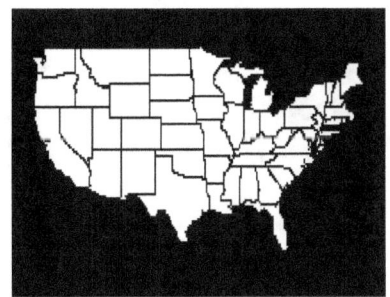
In order to work with numbers, we use certain operators, such as addition, subtraction, multiplication and division. The same holds true for sets. For sets there are three operators: union, intersection, and complement. With these operators we will develop new sets by forming the union of two or more sets, forming the intersection of two or more sets and taking the complement of a set. This lesson discusses these operators in detail.

5.1 - What is the Union of Two or More Sets?

The union of two or more sets is the collection of all the distinct elements from these sets. This union results in a new set.

The symbol for the union operator is $\cup$.

5.1 - Example 1: If **A** = {Mary, Jane, Harry, Bill} and **B** = {Jack, Bill, Frankie}, find **A**$\cup$**B**.

Solution:

A$\cup$**B** = {Mary, Jane, Harry, Bill, Jack, Frankie}. The union of **A** and **B** is the collection of both sets. Note that the element Bill is in both **A** and **B**. However, for the union to be a set, the element Bill can only appear once.

5.1 - Example 2: If **F** = {99,- 6, 10, 100, 76 } and **D** = {1, 2, 3}, find **F**$\cup$**D**.

Solution:

F$\cup$**D** = {99, -6, 10, 100, 76, 1, 2, 3}. Again the union is the collection of the sets **F** and **D**.

5.1 - Example 3: If **F** = {(a,b), (c,d), (e,f)}, **D** = {(b,a), (e,f)} find **F**$\cup$**D**.

Solution:

F$\cup$**D** = {(a,b), (c,d), (e,f), (b,a)}. Note that (a,b) and (b,a) are distinct elements. Again, the elements of **F** and **D** are collected together to form the union.

5.1 - Example 4: If **T** = {a, e, i, o, u}, **V** = {a, e, t, v}, find **T**$\cup$**V**$\cup$**T**.

Solution:

$T \cup V \cup T$ ={a, e, i, o, u, t, v}. The union **T** with itself is equal to **T** since each element of **T** can only appear once. **5.1 - Example 5:** If **K** = {a, b, c, d}, **L** = {a, d}, find **K**∪**L**.

Solution:

K∪**L** = **K** = {a,b,c,d}. All elements of **L** are also elements of **K**.

5.1 - Example 6: If **G** = {Billy, Mary, Harry}, find **G**∪ϕ.

Solution:

G∪ϕ = {Billy, Mary, Harry}∪ϕ = {Billy, Mary, Harry} = **G**

Solved Problems

5.1 - Solved Problem 1: If **G** = {cats, dogs, cattle, horses, sheep} and B = {horses, sheep, mules}, find **G**∪**B**.

Solution:

G∪**B** = {cats, dogs, cattle, horses, sheep, mules}. Remember, that duplication of elements is not allowed.

5.1 - Solved Problem 2: If **F** = {1, 3, 5, a, b} and **D** = {5, a, c, 0}, find **F**∪**D**.

Solution:

F∪**D** = {1,3,5,a,b,c,0}. Sets can consist of different types of elements.

5.1 - Solved Problem 3: If **K** = {a, t, mm}, **D** = {t, rd, m} and **R**= {1, 2, 3, 4, 5}, find **K**∪**D**∪**R**.

Solution:

K∪**D**∪**R**= {1, 2, 3, 4, 5, t, rd, m, a, mm}. Here we have a collection of all three sets.

5.1 - Solved Problem 4: If **T** = {7, 8, 9, 10}, **P** = {a, e, t, v}, find **T**∪**P**∪**T**.

Solution:

T∪**P**∪**T** = **T**∪**P** = {7, 8, 9, 10, a, e, t, v}. Again repetition is not allowed.

5.1 - Solved Problem 5: If **W** = {(H,H), (H,T), (T,H), (T,T)}, **X** = {(H,H), (T,T)}, find **W**∪**X** .

Solution:

W∪**X** = {(H,H), (H,T), (T,H), (T,T)}. **W** has 4 elements where each element is a pair of letters.

For example the pair (H,H) is a single element.

5.1 - Solved Problem 6: Simplify **E** = {(1,2), (2,1), (3,3)}∪ϕ.

Solution:

E = {(1,2), (2,1), (3,3)}∪ϕ= **E** = {(1,2), (2,1), (3,3)}. The empty set has no elements. So the union is just the collection of elements of **E**.

Unsolved Problems with Answers.

5.1 - Problem 1: If **G** = {airplanes, automobiles, skates} and **D** = {horses, skates, tricycles, rockets}, find
G ∪**D**.

Answer:

G ∪**D** = {airplanes, automobiles, skates, horses, tricycles, rockets}.

⇑ *Refer back to* **5.1 - Example 1 & 5.1 - Solved Problem 1**.

5.1 - Problem 2: If **F** = {California, Washington, a, b, Howard} and **D** = {1, 2, 3, 4, 5}, find **F** ∪**D**.

Answer:

F ∪**D** = {California, Washington, a, b, Howard, 1, 2, 3, 4, 5}.

⇑ *Refer back to* **5.1 - Example 2 & 5.1 - Solved Problem 2**.

5.1 - Problem 3: If **K** = {a, b, c}, **D** = {c, d, e} and **R**= {e, f, g}, find **K**∪**D**∪**R**.

Answer:

K ∪**D** ∪**R** = {a, b, c, d, e, f, g}.

⇑ *Refer back to* **5.1 - Example 3 & 5.1 - Solved Problem 3**.

5.1 - Problem 4: If **T** = {(a,b,c,d)}, **P** = {(a,b,d,c), (a,c,b,d), (a,d,c,b)}, find **T**∪**P**∪**T**.

Answer:

T∪**P**∪**T** = **T**∪**P** = {(a,b,c,d), (a,b,d,c), (a,c,b,d), (a,d,c,b)}.

⇑ *Refer back to* **5.1 - Example 4 & 5.1 - Solved Problem 4**.

5.1 - Problem 5: If **W** = {(1,1), (1,2) ,(2,1), (2,2)}, **X** = {(1,1), (2,2)}, find **W**∪**X**.

Answer:

$\mathbf{W} \cup \mathbf{X}$ = {(1,1), (1,2), (2,1), (2,2)}

⇑ *Refer back to* **5.1 - Example 5 & 5.1 - Solved Problem 5**.

5.1 - Problem 6: Simplify $\mathbf{E}$ = {1, 2, 3, 4,...} ∪ φ.

Answer:

$\mathbf{E}$ = {1, 2, 3, 4,...} ∪ φ = $\mathbf{E}$ = {1, 2, 3, 4,...}.

⇑ *Refer back to* **5.1 - Example 6 & 5.1 - Solved Problem 6**.

5.2 - What is the Intersection of Two or More Sets?

The intersection of two or more sets is the collection of all the distinct elements that are in common to these sets. This intersection results in a new set. The symbol for the intersection operator is ∩.

5.2 - Example 1: If $\mathbf{A}$ = {Mary, Jane, Harry, Bill, Howard} and $\mathbf{B}$ = {Jack, Bill, Frankie, Howard}, find $\mathbf{A} \cap \mathbf{B}$.

Solution:

$\mathbf{A} \cap \mathbf{B}$ = {Bill, Howard}. The elements Bill and Howard are the only elements that are in both sets.

5.2 - Example 2: If $\mathbf{A}$ = {a, b, c}, $\mathbf{B}$ = {a, b, c, d, e}, $\mathbf{C}$ = {b, c, h, k}, find $\mathbf{B} \cap \mathbf{A} \cap \mathbf{C}$.

Solution:

$\mathbf{B} \cap \mathbf{A} \cap \mathbf{C}$ = {b, c}. The elements b, c are the only elements that are in the three sets. The element a cannot be in the intersection of all three sets since a is only in the sets $\mathbf{A}$ and $\mathbf{B}$.

5.2 - Example 3: If $\mathbf{E}$ = {a, c, e, f} and $\mathbf{A}$ = {a, f}, find $\mathbf{E} \cap \mathbf{A}$.

Solution:

$\mathbf{E} \cap \mathbf{A}$ = {a, f} = $\mathbf{A}$. All the elements of $\mathbf{A}$ are also elements of $\mathbf{E}$.

5.2 - Example 4: If $\mathbf{P}$ = {7, 0, 5}, $\mathbf{E}$ = {1, 2, 3}, find $\mathbf{P} \cap \mathbf{E}$.

Solution:

P∩E = ϕ. Since the sets **E** and **P** have no elements in common the intersection is empty.

5.2 - Example 5: If **R** = {1, 2, 3, 4, 5}, **I** = {1, 2, 3, 4}, **M** = {1, 3} find **R∩M∩I**.

Solution:

R∩M∩I = {1, 3}. Note that all the elements of **M** are in **R** and **I**. Also, all the elements of **I** are elements of **R**. Therefore, **R∩M∩I** = **M**.

5.2 - Example 6: If **A** = {(b,r), (r,b), (r,r), (b,b)}, **B** = {(r,b), (b,b)}, Find **A∩B**.

Solution:

A∩B = {(r,b), (b,b)}. All the elements of **B** are also elements of **A**. Therefore, **A∩B** = **B**.

Solved Problems

Solved Problem 1: If **A** = {M, J, H, B, K} and **B** = {J, B, F, H}, find **A∩B**.

Solution:

A ∩B = {J,B,H}. The intersection of sets is the set of all elements that are in common to all sets.

Solved Problem 2: If **A** = {M, J, H, B, K} and **B** = {J, B, F, H}, find **B∩A∩B**.

Solution:

B∩A∩B = {B, H, J}. Even though the set **B** is repeated twice, each element of **B** can only be represented once.

Solved Problem 3: If **E** = {1, 2, 3, 4, 5} and **D** = {4,6,8,10}, find **E∩D**.

Solution:

E∩D = {4}. The element 4 is the only element in both sets **E** and **D**.

Solved Problem 4: If **P** = {(h,h), (t,t)}, **E** = {h, t}, find **P∩E**.

Solution:

P∩E = ϕ. There are no elements in common to both sets. (h,h) is a different element then h.

Solved Problem 5: If **R** = {a, b, c, d, e, f}, **I** = {a, b, c, d}, **M** = {c, d}, find **R∩M∩I**.

Solution:

R∩M∩I = {c, d} = **M**. All elements of **I**, **M** are also elements of **R**.

Solved Problem 6: If **T** = {(r,r,b), (r,b,r), (b,r,r)}, **V** = {(r,b,r), (r,r,r)}, **R** = {(r,b,r)} find **T∩V∩R**.

Solution:

T∩V∩R = {(r,b,r)}. The element (r,b,r) is the only element that is in all three sets.

Unsolved Problems with Answers

5.2 - Problem 1: If **A** = {New York, Washington D.C., Miami, Berkeley} and **B** = {New York, Miami, Las Vegas}, find **A∩B**.

Answer:

A∩B = {Miami, New York}.

⇑ *Refer back to* **5.2 - Example 1 & 5.2 - Solved Problem 1.**

5.2 - Problem 2: If **A** = {(1,2,3), (1,2,6), (1,3,5)} and **B** = {(1,2,3), (1,1,1), (1,3,5)}, find **B∩A∩B**.

Answer:

B∩A∩B = {(1,2,3), (1,3,5)}.

⇑ *Refer back to* **5.2 - Example 2 & 5.2 - Solved Problem 2.**

5.2 - Problem 3: If **E** = {{1, 2, 3}, {2, 4, 5}} and **D** = {{3, 4, 5}, {1, 2, 3}, {1}}, find **E∩D**.

Answer:

E∩D = {{1, 2, 3}}.

⇑ *Refer back to* **5.2 - Example 3 & 5.2 - Solved Problem 3**.

5.2 - Problem 4: If **P** = {Mary Smith, Howard Jones}, **E**={Mary, Smith, Howard, Jones}, find **P∩E**.

Answer:

P∩E = ϕ

⇑ *Refer back to* **5.2 - Example 4 & 5.2 - Solved Problem 4**.

5.2 - Problem 5: If **R** = {1, 2, 3, 4, 5,...}, **I** = {2, 3, 4,...}, **M** = {4, 5, 6,...}, find **R∩M∩I**.

Answer:

R∩M∩I = {4, 5, 6,...}.

⇑ *Refer back to* **5.2 - Example 5 & 5.1 - Solved Problem 5**.

5.2 - Problem 6: If **T** = {(1,1), (1,2), (1,3), (1,4), (1,5), (1,6)},

S = {(1,1), (2,1), (3,1), (4,1), (5,1), (6,1)},
find S∩T.

Answer:

S∩T = {(1, 1)}

⇑ *Refer back to* **5.2 - Example 6 & 5.1 - Solved Problem 6**.

Before, explaining the complement operator, we need to define and explain when a set is a subset or a proper subset of another set. Also, we need to define the idea of a universal set.

5.3 - What is a Subset?

A set **C** is a subset of a set **D** if all the elements of **C** are also members of **D**. The symbol for subset is ⊆.
If **C** is a subset of **D** we write **C** ⊆ **D**.

5.3 - Example 1: If **F** = {1, 4, 6} and **B** = {8, 11, 4, 6, 1} is **F** ⊆ **B**?

Solution:

Yes. All the elements of **F** are also elements of **B**.

5.3 - Example 2: Write out all subsets of **H** = {1, 4}.

Solution:

{1, 4} ⊆ {1, 4}

{1} ⊆ {1, 4}

{4} ⊆ {1, 4}

φ ⊆ {1, 4}

Therefore, all subsets are {1, 4}, {1}, {4}, φ.

Note: It can be shown that the empty set is a subset of all sets (See supplementary problem 13).

5.3 - Example 3: Write out all the subsets of **H** = {J, K, L}.

Solution:

$\{J\} \subseteq \{J, K, L\}$

$\{K\} \subseteq \{J, K, L\}$

$\{L\} \subseteq \{J, K, L\}$

$\{J, K\} \subseteq \{J, K, L\}$

$\{J, L\} \subseteq \{J, K, L\}$

$\{K, L\} \subseteq \{J, K, L\}$

$\{J, K, L\} \subseteq \{J, K, L\}$

$\phi \subseteq \{J, K, L\}$

Therefore, all subsets are $\{J\}, \{K\}, \{L\}, \{J, K\}, \{J, L\}, \{K, L\}, \{J, K, L\}, \phi$.

Note: We always assume that the empty set is a subset of all sets.

5.3 - Example 4: Write out all subsets of **W** = $\{(r,r), (b,b)\}$.

Solution:

$\{(r,r)\} \subseteq \{(r,r), (b,b)\}$

$\{(b,b)\} \subseteq \{(r,r), (b,b)\}$

$\{(r,r),(b,b)\} \subseteq \{(r,r), (b,b)\}$

$\phi \subseteq \{(r,r), (b,b)\}$

Therefore, all subsets are $\{(r,r)\}, \{(b,b)\}, \{(r,r),(b,b)\}, \phi$.

Solved Problems

5.3 - Solved Problem 1: Is **F** = $\{a, b, c, d,\} \subseteq$ **B** = $\{0, 1, a, b, 3, c, d\}$?

Solution: Yes. All elements of **F** are also elements of **B**.

5.3 - Solved Problem 2: Find all subsets of **U** = $\{a, b\}$.

Solution:

$\phi \subseteq \{a, b\}$

{a} ⊆ {a, b}

{b} ⊆ {a, b}

{a, b} ⊆ {a ,b}

Therefore, φ,{a},{b},{a, b} are all subsets of {a, b}.

5.3 - Solved Problem 3: Find all subsets of **U** = {a, b, c}.

Solution:

φ ⊆ {a, b, c}
{a} ⊆ {a, b, c}
{b} ⊆ {a, b, c}
{c} ⊆ {a, b, c}
{a, b} ⊆ {a, b, c}
{a, c} ⊆ {a, b, c}
{b, c} ⊆ {a, b, c}
{a, b, c} ⊆ {a, b, c}

Therefore, φ, {a}, {b}, {c}, {a, b}, {a, c}, {b, c}, {a, b, c} are the subsets of {a, b, c}.

5.3 - Solved Problem 4: Write out all subsets of **E** = {(r,r,b), (r,b,r), (b,r,r)}.

Solution:

{(r,r,b)} ⊆ {(r,r,b), (r,b,r), (b,r,r)}

{(r,b,r)} ⊆ {(r,r,b), (r,b,r), (b,r,r)}

{(b,r,r)} ⊆ {(r,r,b), (r,b,r), (b,r,r)}

{(b,r,r), (r,b,r)} ⊆ {(r,r,b), (r,b,r), (b,r,r)}

{(b,r,r), (r,r,b)} ⊆ {(r,r,b), (r,b,r), (b,r,r)}

{(r,r,b), (r,b,r)} ⊆ {(r,r,b), (r,b,r), (b,r,r)}

{(r,r,b), (r,b,r), (b,r,r)} ⊆ {(r,r,b), (r,b,r), (b,r,r)}

φ ⊆ {(r,r,b), (r,b,r), (b,r,r)}

Therefore, the following are subsets of **E**:

φ, {(r,r,b)}, {(r,b,r)}, {(b,r,r)}, {(b,r,r),(r,b,r)}, {(b,r,r),(r,r,b)},{(r,r,b), (r,b,r)}, {(r,r,b), (r,b,r), (b,r,r)}.

Unsolved problems with answers

5.3 - Problem 1: Is F = {a, b, c, d, e} $\subseteq$ B = {a, b, c, d}?

Answer:

No.

⇑ *Refer back to* **5.3 - Example 1& 5.3 - Solved Problem 1**.

5.3 - Problem 2: Find all subsets of U = {{a, b}, {a, c}}

Answer:

ϕ, {{a,b}}, {{a,c}}, {{a,b},{a,c}}

⇑ *Refer back to* **5.3 - Example 2 & 5.3 - Solved Problem 2**.

5.3 - Problem 3: Find all subsets of U = {1, 2, 3}.

Answer:

ϕ, {1}, {2}, {3}, {1,2}, {1,3}, {2,3}, {1,2,3}

⇑ *Refer back to* **5.3 - Example 3 & 5.3 - Solved Problem 3**.

5.3 - Problem 4: Write out all subsets of E = {(r,r,b), (r,b,r), (b,r,r), (r,r,r)}.

Answer:

ϕ,

{(r,r,b)}, {(r,b,r)}, {(b,r,r)}, {(r,r,r)},

{(b,r,r), (r,b,r)}, {(b,r,r), (r,r,b)}, {(r,r,b), (r,b,r)},

{(r,r,r), (r,b,r)}, {(r,r,r), (r,r,b)}, {(r,r,r), (b,r,r)},

{(b,r,r), (r,b,r), (r,r,r)}, {(b,r,r), (r,r,b), (r,r,r)},

{(r,r,b), (r,b,r), (r,r,r)}, {(r,r,b), (r,b,r), (b,r,r)}, {(r,r,b), (r,b,r), (b,r,r), (r,r,r)}

⇑ *Refer back to* **5.3 - Example 4 & 5.3 - Solved Problem 4**.

5.4 - What is a Proper Subset?

A set **C** is a proper subset of a set **D** if all the elements of **C** are also members of **D** and there is at least one element of **D** that is not in **C**. The symbol for the proper subset is $\subset$. If **C** is a proper subset of **D** then we indicate this by **C** $\subset$ **D**.

Note: If **A** is a proper subset of B then **A** is also a subset of **B**.

5.4 - Example 1: If **F** = {1, 4, 6} and **B** = {8, 11, 4, 6, 1} is **F** $\subset$ **B**?

Solution:

Step 1: The elements of **F** are 1,4,6 which are also elements of **B**. Therefore, **F** is a subset of **B**.

Step 2: The set **B** has elements 8,11 that are not in **F**.

Therefore, **F** is a proper subset of **B**: **F** $\subset$ **B**.

5.4 - Example 2: Find all proper subsets of **W** = {1, 3, 5}.

Solution:

$\{1\} \subset \{1,3,5\}$
$\{3\} \subset \{1, 3, 5\}$
$\{5\} \subset \{1, 3, 5\}$
$\{1, 3\} \subset \{1, 3, 5\}$
$\{1, 5\} \subset \{1,3,5\}$
$\{3, 5\} \subset \{1, 3, 5\}$
$\phi \subset \{1, 3, 5\}$

Therefore, {1}, {3}, {5}, {1, 3}, {1, 5}, {3, 5}, ϕ are proper subsets of **W**.

5.4 - Example 3: Find all proper subsets of **W** = {(h,h), (t,t)}.

Solution:

$\{(h,h)\} \subset \{(h,h), (t,t)\}$

$\{(t,t)\} \subset \{(h,h), (t,t)\}$

$\phi \subset \{(h,h), (t,t)\}$

Therefore, {(h,h)}, {(t,t)}, ϕ are the proper subsets of **W**.

Solved Problems

5.4 - Solved Problem 1: If **F** = {1, 2, 3} and **B** = {0, 1, 2, 3,4, 5} is **F** ⊂ **B** ?

Solution:

Step 1: The elements of **F** are 1,2,3 which are also elements of **B**. Therefore, **F** is a subset of **B**.

Step 2: The set **B** has elements 0,4,5 that are not in **F**.

Therefore, **F** is a proper subset of **B**: **F** ⊂ **B**.

5.4 - Solved Problem 2: Find all proper subsets of **W** = {a, b, z}.

Solution:

ϕ ⊂ {a, b, z}
{a} ⊂ {a ,b, z}
{b} ⊂ {a ,b, z}
{z} ⊂ {a, b, z}
{a, b} ⊂ {a, b, z}
{a, z} ⊂ {a, b, z}
{b, z} ⊂ {a, b, z}

Therefore, the proper subsets are ϕ, {a}, {b}, {z}, {a, b}, {a, z}, {b, z}.

Each of these subsets lack at least one element of **W**.

5.4 - Solved Problem 3: Find all proper subsets of **W** = {(h,h,h), (t,t,t)}.

Solution:

ϕ ⊂ {(h,h,h), (t,t,t)}
{(h,h,h)} ⊂ {(h,h,h), (t,t,t)}
{(t,t,t)} ⊂ {(h,h,h), (t,t,t)}

Therefore, all proper subsets of **W** are ϕ, {(h,h,h)}, {(t,t,t)}.

Unsolved Problems with Answers.

5.4 - Problem 1: If **F** = {8, 1, 2, 3} and **B** = {0, 1, 2, 3, 4, 5} is **F** ⊂ **B** ?

Answer:

no.

⇑ *Refer back to* **5.4 - Example 1 & 5.4 - Solved Problem 1**.

5.4 - Problem 2: Find all proper subsets of **W** = {Billy, Sally, Jill}.

Answer:

φ, {Billy}, {Sally}, {Jill}, {Billy,Sally}, {Billy,Jill}, {Sally,Jill}

⇑ *Refer back to* **5.4 - Example 2 & 5.3 - Solved Problem 2**.

Problem 3: Find all proper subsets of **W** = {(1,1,1), (2,2,2), (1,2,1)}.

Answer:

φ,{(1,1,1)}, {(2,2,2)}, {(1,2,1)}, {(1,1,1), (2,2,2)}, {(1,1,1), (1,2,1)}, {(2,2,2), (1,2,1)}

⇑ *Refer back to* **5.4 - Example 3 & 5.4 - Solved Problem 3** .

5.5 - What is a Universal Set?

Assume **A,B,C.** etc are sets that are subsets of the same set indicated by **U** .

The set **U** is called the universal set.

5.5 - Example 1: Assume **A** = {1, 3, 5, 7, 9}, **B** = {1, 6, 8, 2}. Is **C** = {1, 2, 3, 4, 5, 6, 7, 8, 9, 10} a universal set?

Solution:

Since {1, 3, 5, 7, 9} ⊂ {1, 2, 3, 4, 5, 6, 7, 8, 9, 10} and {1, 6, 8, 2} ⊂ {1, 2, 3, 4, 5, 6, 7, 8, 9, 10} then

{1, 2, 3, 4, 5, 6, 7, 8, 9, 10} = **U**   is a universal set.

5.5 - Example 2: Assume **A** = {b, c, d}, **F** = {a, b, d, e, g}. Is {a, b, c, d, e, f} a universal set?

Solution: Since **F** contains the element g but g is not a member of {a, b, c, d, e, f} then {a, b, c, d, e, f} is not a universal set.

5.5 - Example 3: Find the smallest universal set of **Q** = {1, 2, 3, 4, 5}, and **P** = {0, 1, 2, 3}.

Solution:

The smallest universal set containing the sets **Q** and **P** can be formed by taking the union of these two sets:

Q∪**P** ={1, 2, 3, 4, 5}∪{0, 1, 2, 3} = {0, 1, 2, 3, 4, 5}.

Therefore, **U** = **Q**∪**P** = {0, 1, 2, 3, 4, 5} is the smallest universal set that only contains the elements in the sets **Q** and **P**.

5.5 - Example 4: Find the smallest universal set of

Q = {(w,w,l), (w,l,w), (l,w,w,)},

P = {(l,l,w), (l,w,l), (w,l,l)},

E = {(w,w,w), (l,l,l)}.

Solution:

The smallest universal set containing the sets **Q** and **P** can be formed by taking the union of these three sets:

Q∪**P**∪**E** = {(w,w,l), (w,l,w), (l,w,w,)}∪{(l,l,w), (l,w,l), (w,l,l)}∪{(w,w,w), (l,l,l)}

Therefore, **U** = **Q**∪**P**∪**E** = {(w,w,l), (w,l,w), (l,w,w,), (l,l,w), (l,w,l), (w,l,l), (w,w,w), (l,l,l)}.

Solved Problems.

5.5 - Solved Problem 1: Assume **A** = {e, f, g,...,z}, **B** = {w, x, y, z}. Is {a, b, c, d,..., z} a universal set?

Solution:

Since {e, f, g,..., z} ⊂ {a, b, c, d,..., z} and {w, x, y, z}⊂ {a, b, c, d,..., z}, the set **U** = {a, b, c, d,..., z} is a universal set of these two sets.

5.5 - Solved Problem 2: Assume **A** = {7, 8, 9}, **F** = {0, 1, 2, 3}. Is {1, 2, 3, 4, 5, 6, 7, 8, 10} a universal set?
Solution:

Since 0 is an element of **F** but not a member of {1, 2, 3, 4, 5, 6, 7, 8, 10}, the set {1, 2, 3, 4, 5, 6, 7, 8, 10} is not a universal set of these two sets.

5.5 - Solved Problem 3: Find the smallest universal set of **Q** = {a, b, c}, and **P** = {1, 2}.

Solution:

The union of **P** and **Q** is the smallest universal set:

$$Q \cup P = \{a, b, c\} \cup \{1, 2\} = \{a, b, c, 1,2\} = \mathbf{U} .$$

5.5 - Solved Problem 4: Find the smallest universal set of

Q = {(r,r,g), (g,r,r), (r,g,r)},

P = {(g,g,r), (g,r,g), (r,g,g)},

E = {(b,b,r), (b,r,b), (r,b,b)}.

Solution:

The union of **Q,P,E** is the smallest universal set:
$\mathbf{U} = Q \cup P \cup E =$ {(r,r,g), (g,r,r), (r,g,r)}∪{(g,g,r), (g,r,g), (r,g,g)}∪ {(b,b,r), (b,r,b), (r,b,b)} =
{(r,r,g), (g,r,r), (r,g,r), (g,g,r), (g,r,g), (r,g,g), (b,b,r), (b,r,b), (r,b,b)}.

Unsolved problems with Answers

5.5 - Problem 1: Assume **A** = {horses}, **B** = {cattle, pigs}. Is {horses, cattle, pigs} a universal set?

Answer:

Yes.

⇑ *Refer back to* **5.5 - Example 1 & 5.5 - Solved Problem 1**.

5.5 - Problem 2: Assume **A** = {horses}, **B** = {cats, cattle, pigs}. Is {horses, cattle, pigs} a universal set?

Answer:

no.

⇑ *Refer back to* **5.5 - Example 2 & 5.5 - Solved Problem 2**.

5.5 - Problem 3: Find the smallest universal set of **Q** = {(h,h), (h,t), (t,h), (t,t)}, and **P** = {h, t}.

Answer:

$\mathbf{U}$ = {(h,h), (h,t), (t,h), (t,t), h, t}

⇑ *Refer back to* **5.5 - Example 3 & 5.5 - Solved Problem 3**.

5.5 - Problem 4: Find the smallest universal set of

Q = {(h,h), (h,t), (t,h), (t,t)},

P = {(h,h), (t,t)},

E = {(t,h), (h,t)}.

Answer:

U = {(h,h), (h,t), (t,h), (t,t)}

⇑ *Refer back to* **5.5 - Example 4 & 5.5 - Solved Problem 4**.

5.6 - What is the Complement of a Set?

Assume **D** is a subset of the universal set **U** (**D** ⊆ **U**). The complement of **D** is the set consisting of all elements in **U** not contained in **D**. The complement of **D** is noted by **D′**.

5.6 - Example 1: Assume **U** = {a, b, c, d, e, f}, and **A** = {b, c, d}. Find **A′**.

Solution:

Step 1: **A** = {b, c, d} ⊂ {a, b, c, d, e, f}

Step 2: **A′** are the elements of {a, b, c, d, e, f} that are not in **A:** {a,e,f}.

Step 3: Therefore, **A′** = {a, e, f}.

5.6 - Example 2: Assume **U** = {(h,h), (t,t,), (h,t), (t,h)} and **C** = {(h,t), (t,h)}. Find **C′**.

Solution:

Step 1: **C** = {(h,t), (t,h)} ⊂ {(h,h), (t,t,), (h,t), (t,h)}

Step 2: **C′** are the elements of {(h,h), (t,t,), (h,t), (t,h)} that are not in **C:** {(h,h), (t,t)}

Step 3: Therefore, **C′** = {(h,h), (t,t)}.

Solved Problems

5.6 - Solved Problem 1: Assume **U** = {Harry, Mary, Jane, Bill, Frankie, Howard, Seth} and **K** = {Harry, Mary, Jane, Bill, Frankie}. Find **K′**.

Solution:

Step 1: K = {Harry, Mary, Jane, Bill, Frankie} ⊂ {Harry, Mary, Jane, Bill, Frankie, Howard, Seth}

Step 2: K′ are the elements of {Harry, Mary, Jane, Bill, Frankie, Howard, Seth} that are not in **K**:
{Howard, Seth}.

Step 3: Therefore, **K′** = {Howard, Seth}.

5.6 - Solved Problem 2: If **U** = {(h,h,h), (t,t,t), (h,h,t), (h,t,h), (t,h,h), (t,t,h), (t,h,t), (h,t,t)} and **A** = {(h,h,h), (t,t,t)}, find **A′**.

Solution:

Step 1: A = {(h,h,h), (t,t,t)} ⊆ {(h,h,h), (t,t,t), (h,h,t), (h,t,h), (t,h,h), (t,t,h), (t,h,t), (h,t,t)}

Step 2: A′ are the elements of {(h,h,h), (t,t,t), (h,h,t), (h,t,h), (t,h,h), (t,t,h), (t,h,t), (h,t,t)}

that are not in **A**: {(t,t,h), (t,h,t), (h,t,t), (h,h,t), (h,t,h), (t,h,h)}.

Step 3: Therefore, **A′** = {(t,t,h), (t,h,t), (h,t,t), (h,h,t), (h,t,h), (t,h,h)}.

Unsolved Problems with Answers

5.6 - Problem 1: If **U** = {a, b, c, d} and **D** = {a, d} Find **D′**.

Answer:

D′ = {b, c}

⇑ *Refer back to* **5.6 - Example 1 & 5.6 - Solved Problem 1**.

5.6 - Problem 2: If **U** = {(h,h,h), (t,t,t), (h,h,t), (h,t,h), (t,h,h), (t,t,h), (t,h,t), (h,t,t)} and **A** = {(h,h,t), (h,t,h), (t,h,h)} find **A′**.

Answer:

A′ = {(h,h,h), (t,t,t), (t,t,h), (t,h,t), (h,t,t)}

⇑ *Refer back to* **5.6 - Example 2 & 5.6 - Solved Problem 2** .

5.7 - Combining operators.

The three set operators can be combined using parentheses ( ) to determine the order of operations. The rule is to evaluate the set operations inside the parentheses first.

For the following examples, assume **U** = {1, 2, 3, 4, 5, 6, 7, 8, 9, 10}, A = {1, 3, 5, 7, 9}, **B** = {7, 8, 9, 10}, and **D** = {2, 4, 6, 8, 10}.

5.7 - Example 1: Find (A∪B)′.

Solution:

Step 1: First find the union:

A∪B = {1, 3, 5, 7, 9}∪{7, 8, 9, 10} = {1, 3, 5, 7, 8, 9, 10}.

Step 2: Next take the complement of the event in step 1:

(A∪B)′ = {1, 3, 5, 7, 8, 9, 10}′ = {2, 4, 6}.

5.7 - Example 2: Find (A∩ B)′.

Solution:

Step 1: First find the intersection:

A∩B = {1, 3, 5, 7, 9}∩{7, 8, 9, 10} = {7, 9}.

Step 2: Next find the complement of the event in step 1:

(A∩B)′ = {7, 9}′ = {1, 2, 3, 4, 5, 6, 8, 10}.

5.7 - Example 3: Find (A∪B)∩D.

Solution:

Step 1: First find the union of **A** and **B**: A∪B = {1, 3, 5, 7, 9}∪{7, 8, 9, 10} = {1, 3, 5, 7, 9, 8, 10}

Step 2: Next find the intersection of **D** with the event in step 1:

(A∪B)∩D = {1, 3, 5, 7, 9, 8, 10}∩{2, 4, 6, 8, 10} = {8, 10}

5.7 - Example 4: Find $(A \cap B) \cup D$.

Solution:

Step 1: First find the intersection of **A** and **B**:

$A \cap B = \{1, 3, 5, 7, 9\} \cap \{7, 8, 9, 10\} = \{7, 9\}$.

Step 2: Next find the union of **D** with the event in step 1:

$(A \cap B) \cup D = \{7, 9\} \cup \{2, 4, 6, 8, 10\} = \{2, 4, 6, 8, 10, 7, 9\}$.

5.7 - Example 5: Find $(A \cup B)' \cap D$.

Solution:

Step 1: First find the union of **A** and **B**:

$A \cup B = \{1, 3, 5, 7, 9\} \cup \{7, 8, 9, 10\} = \{1, 3, 5, 7, 8, 9, 10\}$.

Step 2: Next find the complement of the event in step 1:

$(A \cup B)' = \{1, 3, 5, 7, 8, 9, 10\}' = \{2, 4, 6\}$

Step 3: Find the intersect of **D** with the event above:

$(A \cup B)' \cap D = \{2, 4, 6\} \cap \{2, 4, 6, 8, 10\} = \{2, 4, 6\}$

5.7 - Example 6: Find $(A \cup B) \cap D'$.

Solution:

Step 1: First find the union of **A** and **B**:

$A \cup B = \{1, 3, 5, 7, 9\} \cup \{7, 8, 9, 10\} = \{1, 3, 5, 7, 8, 9, 10\}$

Step 2: Find the complement of **D**:

$D' = \{1, 3, 5, 7, 9\}$

Step 3: Find the intersection of the event in step 2 with the event in step 1:

$(A \cup B) \cap D' = \{1, 3, 5, 7, 8, 9, 10\} \cap \{1, 3, 5, 7, 9\} = \{1, 3, 5, 7, 9\}$

Solved Problems

For the following examples, assume $\mathbf{U} = \{a, e, i, o, u\}$, $A = \{a, e, i\}$, $B = \{a, o\}$, and $D = \{i, u\}$.

5.7 - Solved Problem 1: Find $(A \cup B)'$.

Solution:

Step 1: $A \cup B$ = {a, e, i}$\cup${a, o} = {a, e, i, o}

Step 2: $(A \cup B)'$ = {a, e, i, o}$'$ = {u}

Step 3: Therefore, $(A \cup B)'$ = {u}.

5.7 - Solved Problem 2: Find $(A \cap B)'$.

Solution:

Step 1: $A \cap B$ = {a, e, i}$\cap${a, o} = {a}

Step 2: $(A \cap B)'$= {a}$'$ = {e, i, o, u}

Step 3: Therefore, $(A \cap B)'$= {e, i, o, u}.

5.7 - Solved Problem 3: Find $(A \cup B) \cap D$.

Solution:

Step 1: $(A \cup B)$ = {a, e, i}$\cup${a, o} = {a, e, i, o}

Step 2: $(A \cup B) \cap D$ ={a, e, i, o}$\cap${i, u} = {i}

Step 3: Therefore, $(A \cup B) \cap D$ = {i}

5.7 - Solved Problem 4: Find $(A \cap B) \cup D$.

Solution:

Step 1: $(A \cap B)$ = {a, e, i}$\cap${a, o} = {a}.

Step 2: $(A \cap B) \cup D$ = {a}$\cup${i, u} = {a, i, u}

Step 3: Therefore, $(A \cap B) \cup D$ = {a, i, u}.

5.7 - Solved Problem 5: Find $(A \cup B)' \cap D$.

Solution:

$(A \cup B)' \cap D$ = {a, e, i, o}$'\cap${i, u} = {u}$\cap${i, u} = {u}

Step 1: $A \cup B$ = {a, e, i} $\cup$ {a, o} = {a, e, i, o}

Step 2: $(A \cup B)'$ = {a, e, i, o}$'$ = {u}

Step 3: $(A \cup B)' \cap D$ = {u} $\cap$ {i, u} = {u}

Step 4: Therefore, $(A \cup B)' \cap D$ = {u}

5.7 - Solved Problem 6: Find $(A \cup B) \cap D'$.

Solution:

$(A \cup B) \cap D'$ = {a, e, i, o} $\cap$ {a, e, o} = {a, e, o}

Step 1: $A \cup B$ = {a, e, i} $\cup$ {a, o} = {a, e, i, o}

Step 2: D' = {i, u}$'$ = {a, e, o}

Step 3: $(A \cup B) \cap D'$ = {a, e, i, o} $\cap$ {a, e, o} = {a, e, o}

Step 4: Therefore, $(A \cup B) \cap D'$ = {a, e, o}.

Unsolved Problems with Answers

For the following examples, assume **U** = {1, 2, 3, 4, 5,..., 100}, **A** = {2, 4, 6, 8, 10,..., 100}, **B** = {1, 3, 5, 7, 9, **...** , 99}, and **D** = {1, 2, 3, 4, 5}.

5.7 - Problems 1: Find $(D \cup B)'$.

Answer:

{6, 8, 10, 12,..., 100}

⇑ *Refer back to* **5.7 - Example 1 & 5.7 - Solved Problem 1**.

5.7 - Problem 2: Find $(D \cap B)'$.

Answer:

{2, 4, 6, 7, 8, 9,..., 100}

⇑ *Refer back to* **5.7 - Example 2 & 5.7 - Solved Problem 2**.

5.7 - Problem 3: Find $(A \cup B) \cap D$.

Answer:

{1, 2, 3, 4, 5}

⇑ *Refer back to* **5.7 - Example 3 & 5.7 - Solved Problem 3**.

5.7 - Problem 4: Find (A∩B)∪D.

Answer:

{1, 2, 3, 4, 5}

⇑ *Refer back to* **5.7 - Example 4 & 5.7 - Solved Problem 4**.

5.7 - Problem 5: Find (A∪D)′∩B.

Answer:

{7, 9, 11,.., 99}

⇑ *Refer back to* **5.7 - Example 5 & 5.7 - Solved Problem 5**.

5.7 - Problem 6: Find (A∪B)∩D′.

Answer:

{6, 7, 8, 9,..., 100}

⇑ *Refer back to* **5.7 - Example 6 & 5.7 - Solved Problem 6**.

5.8 - DeMorgan Laws

The following laws of DeMorgan are very useful when combining sets:

$$(A \cup B)' = A' \cap B'$$

$$(A \cap B)' = A' \cup B'$$

5.8 - Example 1: Assume **U** = {1, 2, 3, 4, 5,..., 10}, **A** = {4, 5, 6, 7, 8}, and **B** = {1, 2, 4, 5, 9}. Show

(a). $(A \cup B)' = A' \cap B'$.

(b). $(A \cap B)' = A' \cup B'$.

Solutions:

➤ **(a).**
Left Side:

Step 1: $A \cup B = \{4, 5, 6, 7, 8\} \cup \{1, 2, 4, 5, 9\} = \{1, 2, 4, 5, 6, 7, 8, 9\}$

Step 2: $(A \cup B)' = \{1, 2, 4, 5, 6, 7, 8, 9\}' = \{3, 10\}$

Right Side:

Step 1: $A' = \{4, 5, 6, 7, 8\}' = \{1, 2, 3, 9, 10\}$

Step 2: $B' = \{1, 2, 4, 5, 9\}' = \{3, 6, 7, 8, 10\}$

Step 3: $A' \cap B' = \{3, 10\}$

Step 4: Therefore, $(A \cup B)' = A' \cap B'$.

➤ **(b).**
Left Side:

Step 1: $A \cap B = \{4, 5, 6, 7, 8\} \cap \{1, 2, 4, 5, 9\} = \{4, 5\}$

Step 2: $(A \cap B)' = \{4, 5\}' = \{1, 2, 3, 6, 7, 8, 9, 10\}$

Right Side:

Step 1: $A' = \{4, 5, 6, 7, 8\}' = \{1, 2, 3, 9, 10\}$

Step 2: $B' = \{1, 2, 4, 5, 9\}' = \{3, 6, 7, 8, 10\}$

Step 3: $A' \cup B' = \{1, 2, 3, 9, 10\} \cup \{3, 6, 7, 8, 10\} = \{1, 2, 3, 6, 7, 8, 9, 10\}$

Step 4: Therefore, $(A \cap B)' = A' \cup B'$.

Solved Problems

5.8 - Solved Problem 1: Assume $\mathbf{U} = \{a, e, i, o, u\}$, $\mathbf{A} = \{a, e, i\}$, and $\mathbf{B} = \{a, o\}$. Show

(a). $(A \cup B)' = A' \cap B'$.

(b). $(A \cap B)' = A' \cup B'$.

Solutions:

➤ **(a).**
Left Side:

Step 1: $A \cup B = \{a, e, i\} \cup \{a, o\} = \{a, e, i, o\}$

Step 2: $(A \cup B)' = \{a, e, i, o\}' = \{u\}$

Right Side:

Step 1: $A' = \{a, e, i\}' = \{o, u\}$

Step 2: $B' = \{a, o\}' = \{e, i, u\}$

Step 3: $A' \cap B' = \{u\}$

Step 4: Therefore, $(A \cup B)' = A' \cap B'$.

➤ **(b).**
Left Side:

Step 1: $(A \cap B) = [\{a, e, i\} \cap \{a, o\}] = \{a\}$

Step 2: $(A \cap B)' = [\{a, e, i\} \cap \{a, o\}]' = \{a\}' = \{e, i, o, u\}$

Right Side:

Step 1: $A' = \{a, e, i\}' = \{o, \ u\}$

Step 2: $B' = \{a, o\}' = \{e, i, \ u\}$

Step 3: $A' \cup B' = \{o, \ u\} \cup \{e, i, u\} = \{e, i, o, \ u\}$

Step 4: Therefore, $(A \cap B)' = A' \cup B'$.

Unsolved Problems with Answers

5.8 - Problem 1: Assume $\mathbf{U} = \{(h,h), (h,t), (t,h), (t,t)\}$, $\mathbf{A} = \{(h,h), (h,t), (t,h)\}$, and $\mathbf{B} = \{(h,t), (t,h)\}$. Show

(a). $(A \cup B)' = A' \cap B'$.

(b). $(A \cap B)' = A' \cup B'$.

Answers:

➤ **(a).**

Left Side:

$(A \cup B)' = \{(t,t)\}$

Right Side:

$A' \cap B' = \{(t,t)\}$

➤ **(b).**
Left Side:

$(A \cap B)' = \{(h,h), (t,t)\}$

Right Side:

$A' \cup B' = \{(h,h), (t,t)\}$

⇑ *Refer back to* **5.8 - Example 1 & 5.8 - Solved Problem 1**.

Supplementary Problems

For the following questions, assume arbitrary sets.

1. If $A \subseteq B$ and $B \subseteq A$, show that $A = B$.

For questions 2 - 11, perform the indicated operations and simplify.

2. $D \cap D$ 3. $F \cap F'$ 4. $F \cup F'$ 5. ϕ' 6. $\phi \cap G$ 7. $\phi \cup G$ 8. $U \cup G$ 9. $U \cap G$ 10. U' 11. F''

12. If $A_k = \{-100 + k, ..., 0, 1,..., 100 - k\}$, $k = 0, 1, 2, 3, ..., 100$, then simplify $A_0 \cap A_1 \cap ... \cap A_{100}$.

13. Show that the null set ϕ is a subset of all sets.

14. Show that

a. $(A \cup B)' \cap B = \phi$.

b. $(A \cap B)' \cup B = U$.

15. Let $U = \{0,1,2,3,...,10\}$, $A = \{2,3,4,8,9\}$, $B = \{1,3,5,7,9\}$, $C = \{1,3,6,10\}$.

Simplify:

$E =$
$(A \cap B \cap C) \cup (A \cap B' \cap C) \cup (A \cap B \cap C') \cup (A \cap B' \cap C') \cup (A' \cap B \cap C) \cup (A' \cap B' \cap C) \cup (A' \cap B \cap C') \cup (A' \cap B'$

∩**C**′)

16. Show the following are true:

a. **A**′′ = **A**

DeMorgan's laws:

b. (**A**∪**B**)′ = **A**′∩**B**′

c. (**A**∩**B**)′ = **A**′∪**B**′

distributive laws:

d. **A**∩(**B**∪**C**) = (**A**∩**B**)∪(**A**∩**C**)

e. **A**∪(**B**∩**C**) = (**A**∪**B**)∩(**A**∪**C**)

f. (**A**∪**B**)∩(**C**∪**D**) = (**A**∩**C**)∪(**A**∩**D**)∪(**B**∩**C**)∪(**B**∩**D**)

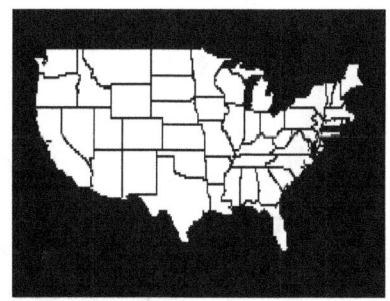
6.1 - What is a Venn diagram?

Venn diagrams allow us to see a graphical picture of a universal set and related subset using circles and rectangle. Each set will be represented by a circle and the universal set will be represented by a rectangle.

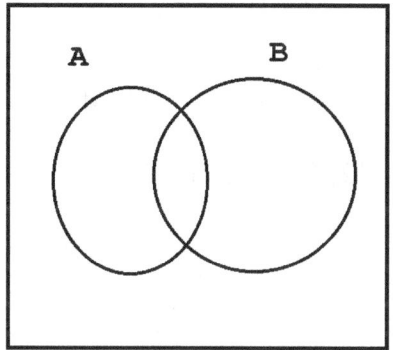

6.1 - Example 1: The following Venn diagram represents a universal set containing the counting numbers
1 through 15. Underline the numbers that represent the following sets:

(a). **A**

(b). **B**

(c). **A∪B**

(d). **(A∪B)′**

(e). **A′∩B**

(f). **A∩B′**

(g). **(A′∩B)∪(A∩B′)**

(h). **A∩B**

(i). **(A∩B)′**

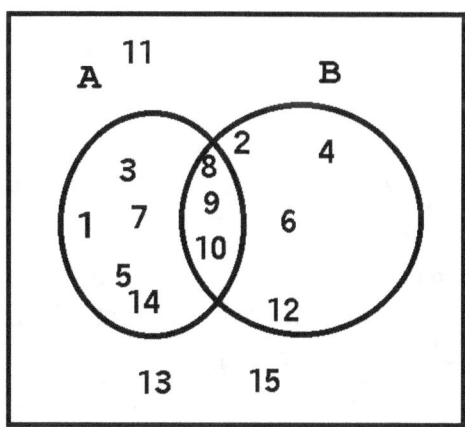

Solutions:

(a).

➤ **(a).**
The bold underlined numbers, totally make up the elements of the set **A.**
None of these numbers lie outside of **A.** Therefore,
A = {1, 3, 5, 7, 8, 9, 10,14}.

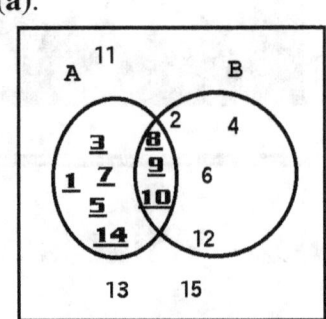

➤ **(b).**
The bold underlined numbers totally make-up the elements of the set **B.** None of these number lie outside of B. Therefore,
B = {2, 4, 6, 8, 9, 10, 12}.

(b).

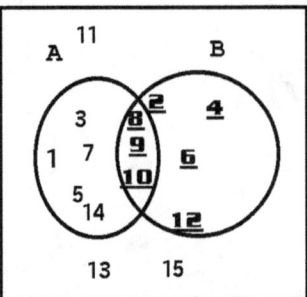

➤ **(c).**
The bold underlined numbers totally make-up the elements of both **A** and **B.** Therefore, these numbers make up the set **A∪B.** Therefore,
A∪B = {1, 2, 3, 4, 5, 6, 7, 8, 9, 10, 12, 14}.

(c).

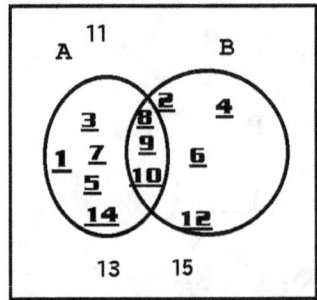

➤ **(d).**
The bold underlined numbers entirely make-up the numbers outside of the union **A∪B.** Therefore,
(**A∪B**)′ = {11, 13, 15}.

(d).

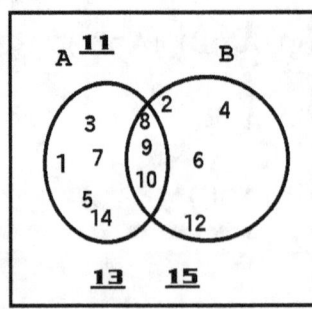

➤ **(e).**

The bold underlined numbers make- up all the numbers in B that
do not lie in **A**. This is the set **A′∩B** = {2, 4, 6, 12}.

(e).

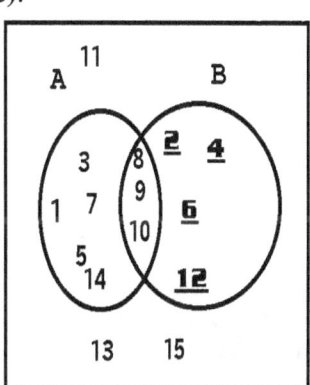

➤ **(f).**

The bold underlined numbers make- up all the numbers in **A** that
do not lie in **B**. This is the set **A∩B′** = {1, 3, 5, 7, 14}.

(f).

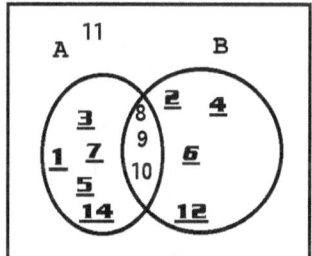

➤ **(g).**

The bold underlined numbers lie in either **A′∩B** or **A∩B′**.
Therefore, we take the union of these two sets: **(A′∩B)∪(A∩B′)**
= {1, 3, 5, 7, 14 ,2, 4, 6, 12}.

(g).

➤ **(h).**

The bold underlined numbers make-up those numbers that lie both in set **A** and set **B.** Therefore,
the underlined numbers is the intersection of these two sets: **A∩B** = {8,9,10}.

(h).

(i).

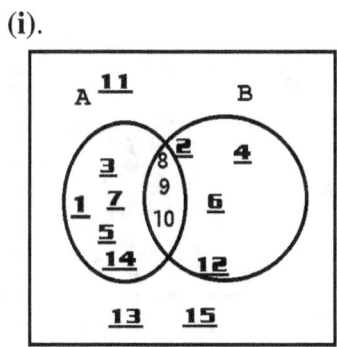

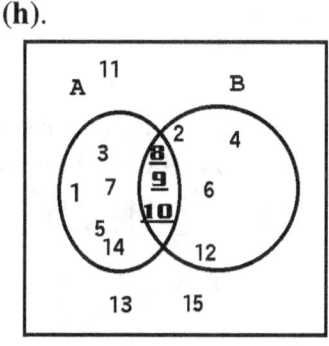

➤ **(i).**

The bold underlined numbers are all numbers that lie outside A∩B. This set of numbers can be presented as the complement of A∩B. Therefore,
(A∩B)′ = {1, 2, 3, 4, 5, 7, 11, 12, 13, 14, 15}.

6.1 - Example 2: The following Venn diagram represents a universal set containing the alphabetic letters
a through z. Underline the letters that represent the following sets:

(a). **A**

(b). **A∩B**

(c). **(A∪B)′∩C**

(d). **A∩C∩B′.**

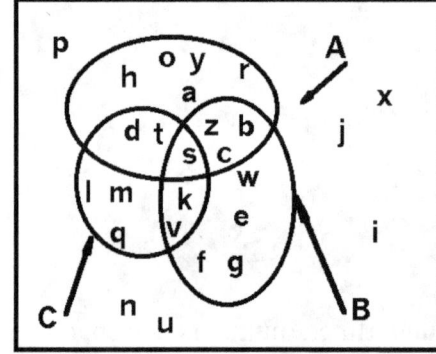

Solutions:

➤ **(a).**

From the Venn diagram, we see that the top circle is the set **A**. Set **A** is made up of all letter contained in this top circle. Therefore, **A** = {h ,o, y, a, r, z, s, c, b, d, t}.

   (a).

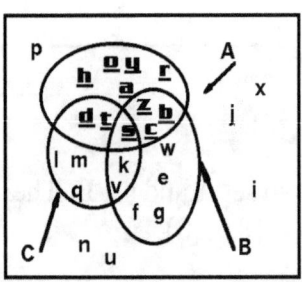

                                              (b).

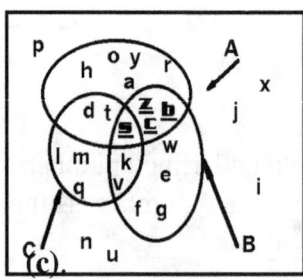

(c).

➤ **(b).**

From the Venn diagram, we see the intersection of **A** and **B** include the letters that lie in both the circles representing A and B. Therefore, A∩B = {s, z, c, b}.

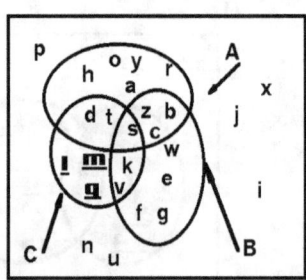

➤ (c).

Step 1: Take the union of **A** and **B**.

Step 2: Take the complement of the union of **A** and **B** and intersect this set with **C**.

Step 3: (A∪B)′∩C = {l, m, q}

➤ (d).

Step 1: Take the complement of **B**.

(d).

Step 2: Intersect the complement of **B** with **A** intersection **C**. Therefore, A∩C∩B′ = {d, t}.

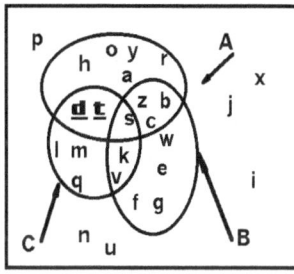

Solved Problems

6.1 - Solved Problem 1: The following Venn diagram represents a universal set containing the counting numbers 1 to 15. Underline the numbers that represent the following sets:

(a). **A**

(b). **B**

(c). **A∪B**

(d). **(A∪B)′**

(e). **A′∩B**

(f). **A∩B′**

(g).**(A′∩B)∪(A∩B′)**

(h). **A∩B**

(i). **(A∩B)′**

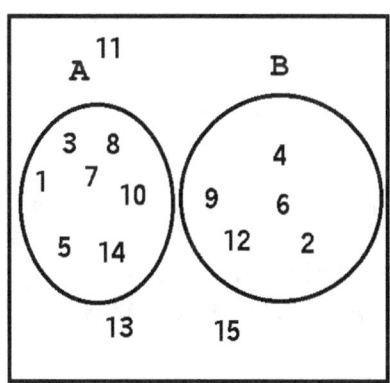

(a).

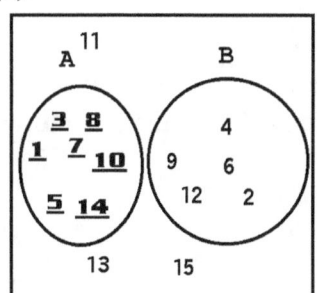

Solutions:

➤ **(a).**
The underlined numbers, make-up the elements of the set **A**. Therefore,
A = {1, 3, 5, 7, 8, 10, 14}.

(b).

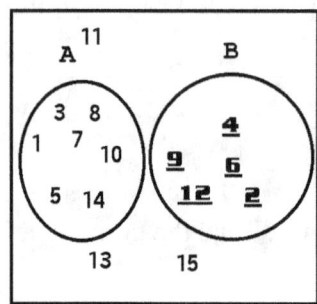

➤ **(b).**
The underlined numbers, make-up the elements of the set **B.** Therefore,
 B = {2, 4, 6, 9, 12}.

(c).

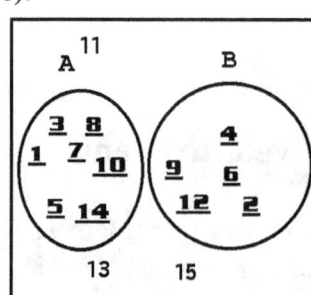

➤ **(c).**

The underlined numbers make-up the elements of both **A** and **B**.
Therefore,
A∪**B** = {1,3,5,7,8,10,14,2,4,6,9,12}.

(d).

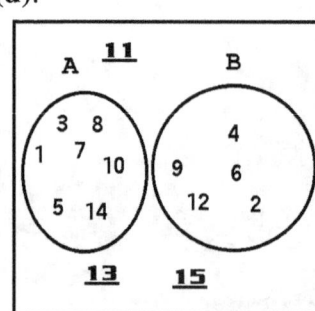

➤ **(d).**

The underlined numbers make-up the numbers outside of the
union **A**∪**B**. Therefore, (**A**∪**B**)′ = {11,13,15}.

(e).

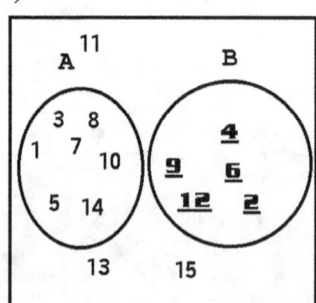

➤ **(e).**
Since **A**∩**B** = ϕ then **A**′∩**B** = **B** = {2, 4, 6, 9, 12}.

➤ **(f).**
Since **A∩B = φ**, then **A∩B′ = A** = {1, 3, 5, 7, 8, 10, 14}.

(f).

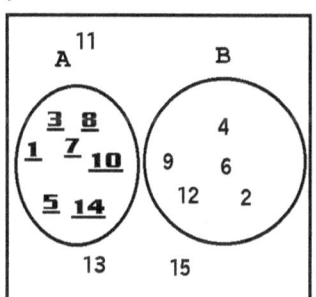

➤ **(g).**
From (e) and (f), **(A∩B′)∪(A′∩B) = A∪B**.

➤ **(h).**
Since **A∩B = φ**, no numbers are underlined.

(g).

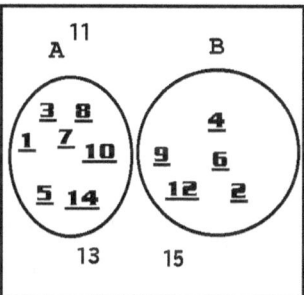

➤ **(i).**
Since **A∩B = φ**, **(A∩B)′ = U**. Therefore, all the numbers are underlined.

(**i**).

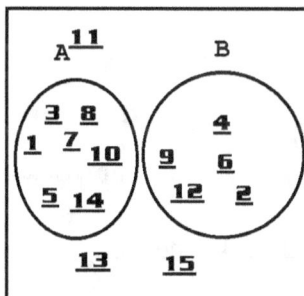

6.1 - Solved Problem 2: The following Venn diagram represents a universal set containing the alphabetic letters a through z. Underline the letters that represent the following sets:

(a). **B**

(b). **B∩C**

(c). **(A∪C)′∩B**

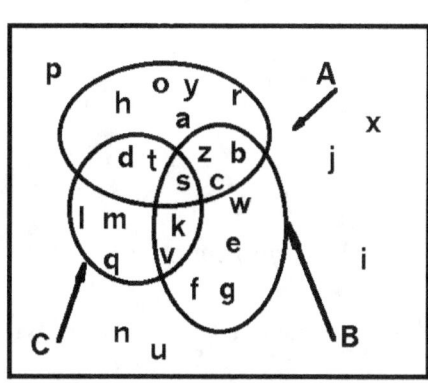

(d). **A∩C′∩B**

(a).

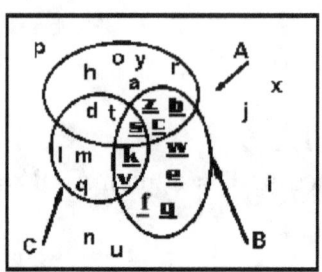

Solutions:

➤ (a).
From the Venn diagram, we see that the bottom right circle is the set **B**. Therefore,
B = {z, c, b, s, w, k, v, e, g, f}.

(b).

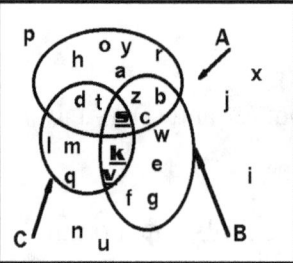

➤ (b).
From the Venn diagram, we see the intersection of **B** and **C** include the letters that lie in both these circles that represent **B** and **C**. Therefore,
   B∩C = {v, k, s}.

(c).

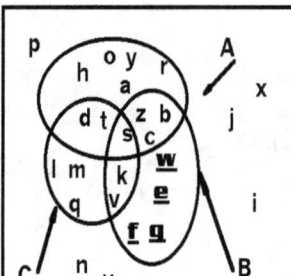

➤ (c).
Step 1: Take the union of sets **A** and **C**.

Step 2: Take the complement of the union of **A** and **C** and intersect it with **B**.

Step 3: (A∪C)′∩B = {f, e, g, w}

➤ (d).

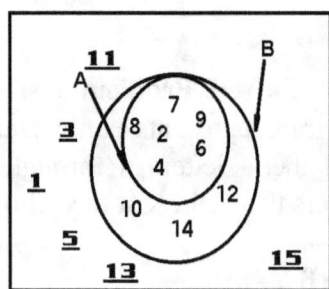

➤ (d).
Step 1: Take the complement of **C**.

Step 2: Intersect the complement of **C** with the intersection of **A** and **B**.

Step 3: $C' \cap A \cap B = \{b,c,z\}$

Unsolved Problems with Answers

6.1 - Problem 1: The following Venn diagram represents a universal set containing the counting numbers 1 through 15. Underline the numbers that represent the following sets:

(a). **A**

(b). **B**

(c). **A∪B**

(d).**(A∪B)′**

(e). **A′∩B**

(f). **A∩B′**

(g). **(A′∩B)∪(A∩B′)**

(h). **A∩B**

(i). **A′**

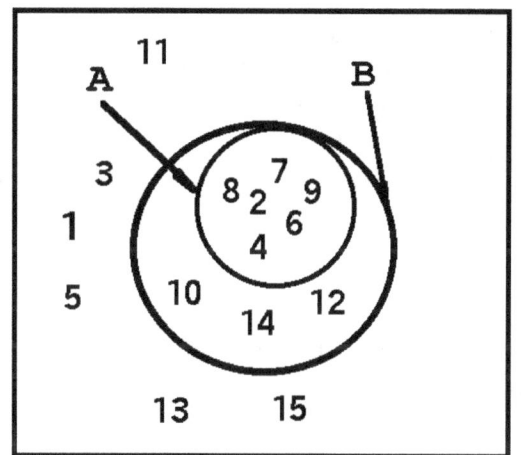

Answers: ➤ **(a).**

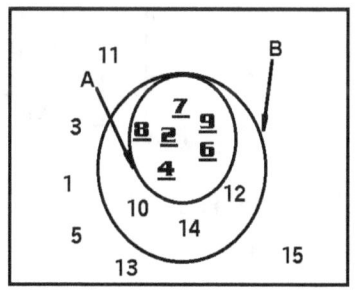

➤ **(b).**

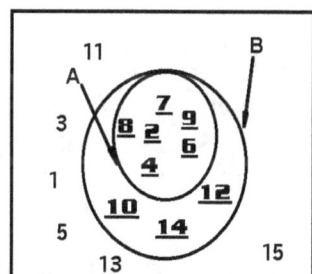

➤ **(c).**

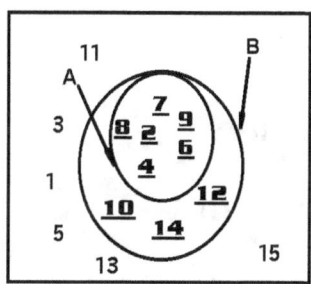

➤ **(e).**

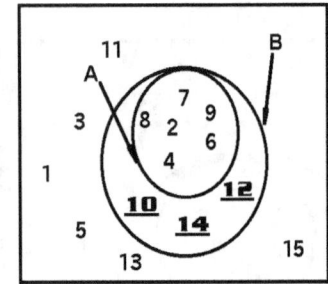

➤ **(g).**

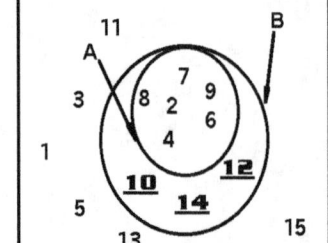

➤ **(f).** This is an empty set.

➤ **(h).**

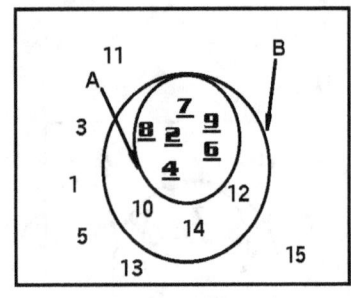

➤ **(i).**

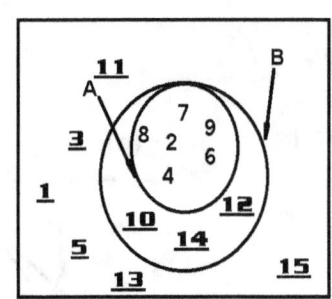

⇑ *Refer back to* **6.1 - Example 1 & 6.1 - Solved Problem 1.**

6.1 - Problem 2: The following Venn diagram represents a universal set containing the alphabetic letters a through z. Underline the letters that represent the following sets:

(a). **C**

(b). **A∩B∩C**

(c). **(A∪B∪C)′**

(d). **A′∩B′∩C**

Answers:

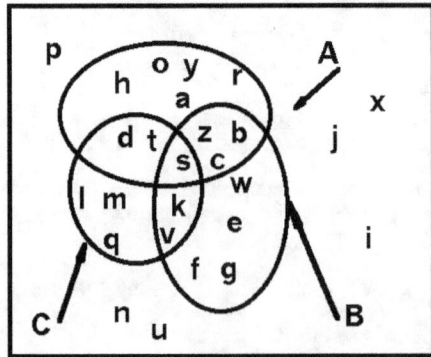

➤ **(a).**

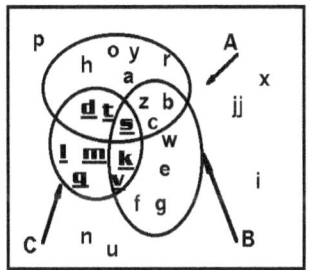

➤ **(c).**

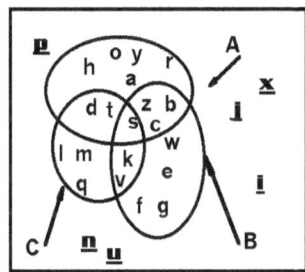

⇑ *Refer back to* **6.1 - Example 2 & 6.1 - Solved Problem 2.**

➤ **(b).**

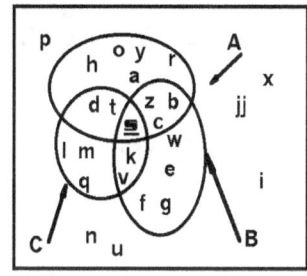

➤ **(d).**

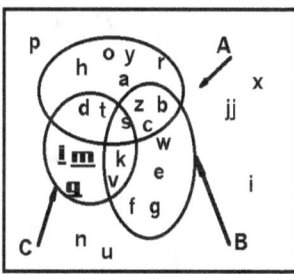

Supplementary Problems

1. For the Venn diagram, using combinations of **A,B**,∩,∪,′ find all possible non-empty subsets.

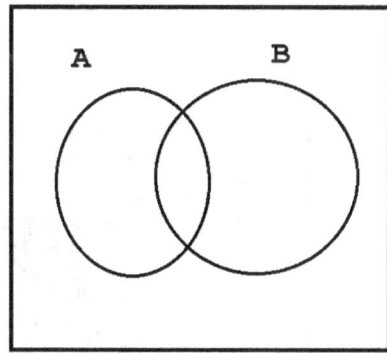

2. For the Venn diagram below, using combinations of **A,B,C**∩,∪,′ find all possible non-empty subsets.

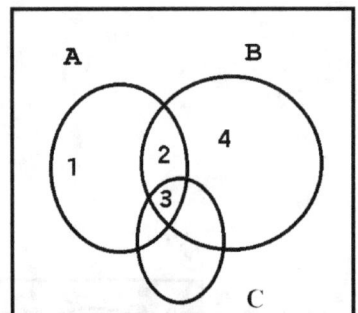

In the following Venn diagrams, rectangles represent sets **A,B,C**. Using **A,B,C**, unions, intersections, and complements, represent the shaded areas.

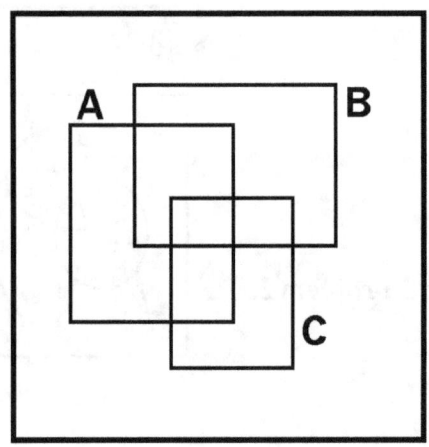

3.

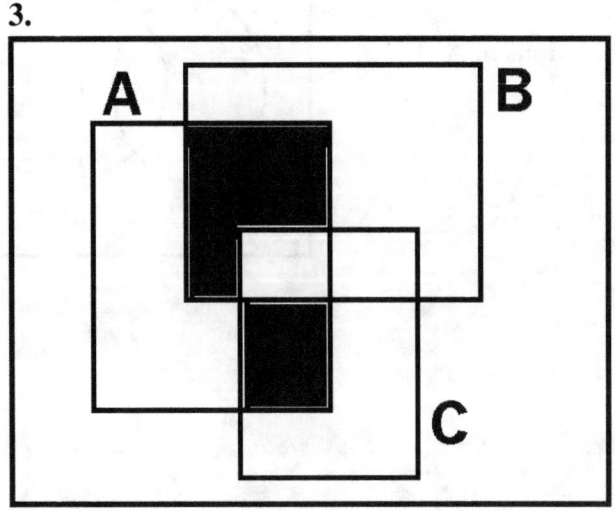

4

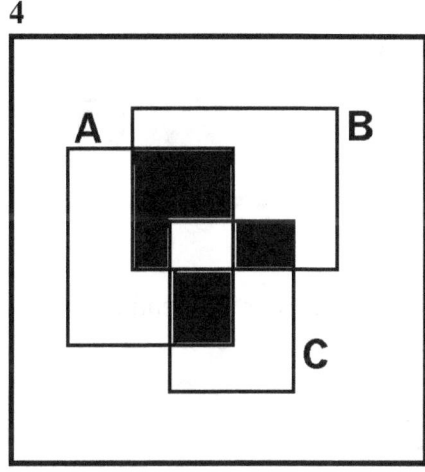

5.

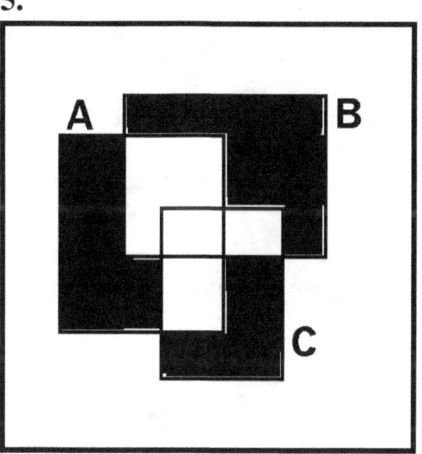

Using the following Venn diagram show

6. (A∩B)∩C = A∩(B∩C)

7. (A∪B)∪C = A∪(B∪C)

8. A∩(B∪C) = (A∩B)∪(A∩C)

9. A∪(B∩C) = (A∪B)∩(A∪C)

10. (A∪B)′ = A′∩B′

11. (A∩B)′ = A′∪B′

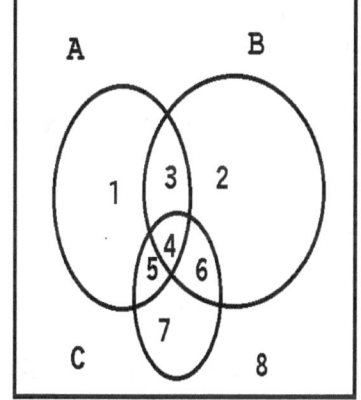

12. A subdivision of A∪B = (A∩B′)∪(A′∩B)∪(A∩B). Find a subdivision for A∪B∪C.

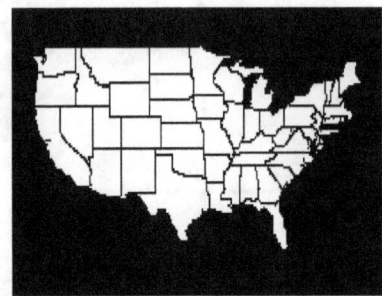
In this lesson, we discuss in some detail the cardinality of sets, subsets and their applications.[1]

7.1 - What is the Cardinality of a Set?

> The cardinality of a set is the number of elements in a set. The cardinality of the set **A** is indicated by #**A**.

7.1 - Example 1: If **A** = {Mary, Jane, Harry, bill}, find #**A**.

Solution:

By simply counting, we see there are four elements of this set. Therefore, #**A = 4.**

7.1 - Example 2: If **F** = {(b,b.), (g,g), (g,b), (b,g) }, find #**F**.

Solution:

Each member of this set is a pair of letters. But, each pair counts as only one member. Therefore, there are 4 pairs and #**F** = 4.

7.1 - Example 3: Find #ϕ.

Solution:

The empty set has no elements. Therefore, #ϕ = 0.

Solved Problems

7.1 - Solved Problem 1: If **A** = {2, 4, 6, 8,.., 20}, find #**A**.

---

[1] We assume that the sets in this chapter have only a finite number of elements.

Solution:

#**A** = 10. This set consists of all even numbers from 2 to 20. Since 20 is even, the number of even numbers is half of 20.

7.1 - Solved Problem 2: If **F** = {(h,h,h), (t,t,t), (t,h,h), (h,t,h), (h,h,t), (h,t,t), (t,h,t), (t,t,h)}, find #**F**.

Solution:

#**F** = 8. Each element is a triple but only counted as one element.

7.1 - Solved Problem 3: If **K** = {1, 2, 3, 4, 5}, find #(**K**∩**K**′).

Solution:

Since **K**∩**K**′ = ϕ then #**K**∩**K**′ = #ϕ = 0.

Unsolved Problems with answers

7.1 - Problem 1: Assume **A** = {a, b, c, d,...}. Find #**A**.

Answer:

#**A** = 26.

⇑ *Refer back to* **7.1 - Example 1 & 7.1 - Solved Problem 1**.

7.1 - Problem 2: If **F** = {(h,h,h,h}, (h,h,h,t}, (h,h,t,h}, (h,t,h,h), (h,h,t,t), (h,t,h,t), (h,t,t,h), (h,t,t,t)},find #**F**.

Answer:

#**F** = 8.

⇑ *Refer back to* **7.1 - Example 2 & 7.1 - Solved Problem 2**.

7.1 - Problem 3: If **K** = {1, 2, 3, 4, 5}, **W** = {6, 7, 8, 9, 10}, **T** = {1, 2, 3, 4, 5, 6, 7, 8, 9, 10}, and **U** = {1, 2, 3, 4, 5, 6, 7, 8, 9, 10}, find #[(**K**∪**W**)′∩**T**].

Answer:

0.

⇑ *Refer back to* **7.1 - Example 3 & 7.1 - Solved Problem 3**.

7.2 - Rules on Cardinality of Sets

Rule 1: $\#(A \cup B) = \#A + \#B - \#(A \cap B)$.

7.2 - Example 1: If $\#A = 20$, $\#B = 10$, and $\#(A \cap B) = 4$, find $\#(A \cup B)$.

Solution:

The Venn diagram shows that by counting the number of elements in A and the number of elements in B causes the elements in $A \cap B$ to be counted twice. Therefore we need to subtract off $A \cap B$.

$\#(A \cup B) = \#A + \#B - \#(A \cap B) = 20 + 10 - 4 = 26$

Rule 2: $\#(A' \cap B) = \#B - \#(A \cap B)$.

7.2 - Example 2: If $\#B = 10$, $\#(A \cap B) = 4$, and $\#A = 4$, find $\#(A' \cap B)$.

Solution:

$\#(A' \cap B) = \#B - \#(A \cap B) = 10 - 4 = 6$

Rule 3: $\#A' = \#\mathbf{U} - \#A$.

7.2 - Example 3: If $\#\mathbf{U} = 100$, and $\#A = 24$, find $\#A'$.

Solution:

$\#A' = \#\mathbf{U} - \#A = 100 - 24 = 76$

Rule 4: $\#(A \cup B) = \#(A' \cap B) + \#(A \cap B') + \#(A \cap B)$.

7.2 - Example 4: If $\#(A \cap B) = 10$, $\#(A \cap B') = 14$ and $\#(A' \cap B) = 7$,

find $\#(A \cup B)$.

Solution:

1.

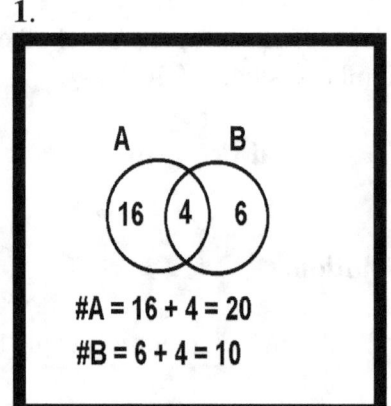

#A = 16 + 4 = 20
#B = 6 + 4 = 10

2.

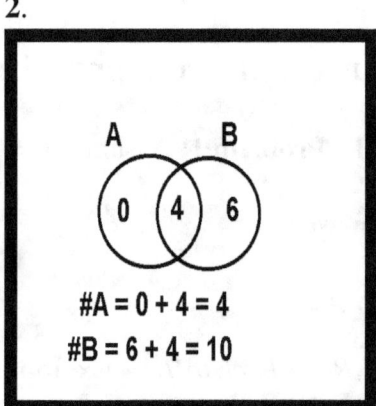

#A = 0 + 4 = 4
#B = 6 + 4 = 10

3.

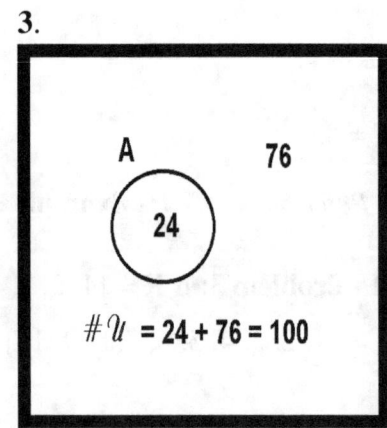

$\#\mathcal{U} = 24 + 76 = 100$

$\#(A \cup B) = \#(A' \cap B) + \#(A \cap B') + \#(A \cap B) = 7 + 14 + 10 = 31.$

Rule 5: $\#[(A \cup B)'] = \#(A' \cap B').$

7.2 - Example 5: If $\#(A' \cap B') = 10$, find $\#[(A \cup B)']$.

Solution:

$\#[(A \cup B)'] = \#(A' \cap B') = 10$

Rule 6: $\#[(A \cap B)'] = \#(A' \cup B').$

The following is a summary of the six rules:

> **Rule 1:** $\#(A \cup B) = \#A + \#B - \#(A \cap B).$
>
> **Rule 2:** $\#(A' \cap B) = \#B - \#(A \cap B).$
>
> **Rule 3:** $\#A' = \#\mathbf{U} - \#A.$
>
> **Rule 4:** $\#(A \cup B) = \#(A' \cap B) + \#(A \cap B') + \#(A \cap B).$
>
> **Rule 5:** $\#[(A \cup B)'] = \#(A' \cap B').$
>
> **Rule 6:** $\#[(A \cap B)'] = \#(A' \cup B').$

4.

$\#A = 14 + 10 = 24$
$\#B = 7 + 10 = 17$

7.2 - Example 6: Assume $\#\mathbf{U} = 15$, $\#A = 5$, $\#B = 8$ and $\#(A \cap B) = 2$. Find:

(a). $\#(A' \cap B)$, (b). $\#(A \cap B')$, (c). $\#(A \cup B)$, (d). $\#(A \cap B)'$, (e). $\#A'$

Solutions:

➤ **(a).**
Use Rule 2, $\#(A' \cap B) = \#B - \#(A \cap B) = 8 - 2 = 6.$

➤ **(b).**
Use Rule 2. $\#(A \cap B') = \#A - \#(A \cap B) = 5 - 2 = 3.$

➤ **(c).**

Use Rule 1. #(A∪B) = #A + #B - #(A∩B) = 8 + 5 - 2 = 11.

➤ **(d).**

Use Rule 3. #(A∩B)′ = #**U** - #(A∩B)= 15 - 2 = 13.

➤ **(e).**

Use Rule 3. #A′ = #**U** - #A= 15 - 5 = 10.

7.2 - Example 7: From the following Venn Diagram, find:

(a). #(A′∩B) (b). #(A∩B′) (c). #A (d). #B

(e). #(A′∩B′) (f).#(A∩B)′ (g). #**U**

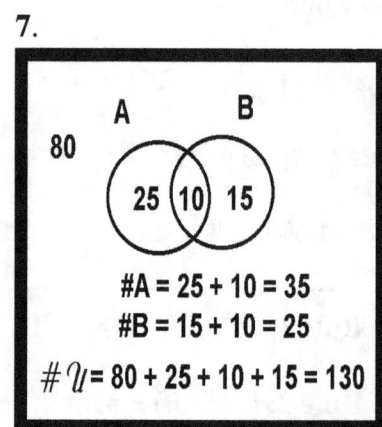

7.

#A = 25 + 10 = 35
#B = 15 + 10 = 25
#𝒰 = 80 + 25 + 10 + 15 = 130

Solutions:

➤ **(a).**
#(A′∩B) = 15. **B** is divided into two parts where

#B = #(A′∩B) + #(A∩B) = 15 + 10 = 25.

➤ **(b).**
#(A∩B′) = 25. **A** is divided into two parts where

#A = #(A∩B′) + #(A∩B).

➤ **(c).**
#A = 35. #A = #(A∩B′) + #(A∩B) = 25 + 10 = 35

➤ **(d).**
#B = 25. #B = #(A′∩B) + #(A∩B) = 15 + 10 = 25

➤ **(e).**
#(A′∩B′) = 80. From Rule 5, #(A′∩B′)= #[(A∪B)′] = 80.

➤ **(f).**
#(A∩B)′ = 120. From the Venn Diagram,

#(A∩B)′ = 25 + 15 + 80 = 120.

➤ **(g).**
#**U** = 130. From the Venn Diagram,

#**U** = 80 + 25 + 10 + 15 = 130.

Solved Problems

7.2 - Solved Problem 1: If #**A** = 40, #**B** = 10, and #(**A**∩**B**) = 1, find #(**A**∪**B**).

Solution:

From Rule 1, #(**A**∪**B**) = 49.

#(**A**∪**B**) = #**A** + #**B** - #(**A**∩**B**) = 40 + 10 - 1 = 49.

7.2 - Solved Problem 2: If #**B** = 25, and #(**A**∩**B**) = 15, find #(**A**′∩**B**).

Solution:

From Rule 2, #(**A**′∩**B**) = 10.

#(**A**′∩**B**) = #**B** - #(**A**∩**B**) = 25 - 15 = 10

7.2 - Solved Problem 3: If #**U** = 50, and #**A** = 30, find #**A**′.

Solution:

#**A**′= 20. From Rule 3:

#(**A**′) = #**U** - #**A** = 50 - 30 = 20.

7.2 - Solved Problem 4: If #(**A**∩**B**) = 30, #(**A**∩**B**′) = 4 and #(**A**′∩**B**) = 10, find #(**A**∪**B**).

Solution:

#(**A**∪**B**) = 44. From Rule 4:

#(**A**∪**B**) = #(**A**′∩**B**) + #(**A**∩**B**′) +#(**A**∩**B**) = 30 + 4 + 10 = 44

7.2 - Solved Problem 5: Assume #(**A**′∩**B**′) = 25. Find #[(**A**∪**B**)′].

Solution:

#[(**A**∪**B**)′] = 25. From Rule 5: #[(**A**∪**B**)′] = #(**A**′∩**B**′) = 25.

7.2 - Solved Problem 6: Assume #**U** = 35, #**A** = 5, #**B** = 5

3.

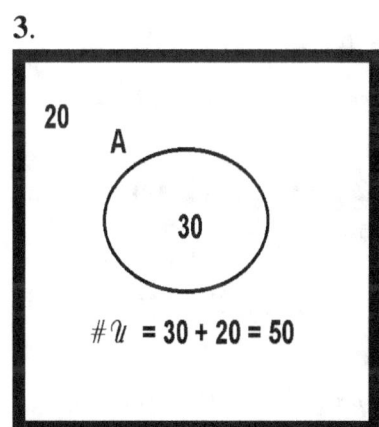

4.

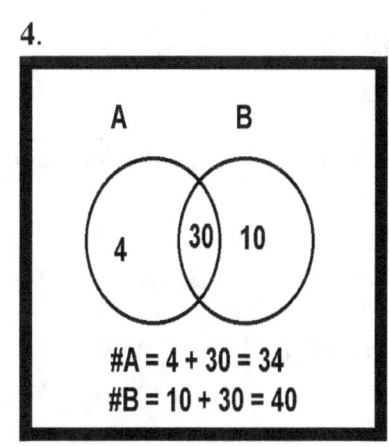

and #(A∩B) = 1.

Find:

(a). #(A′∩B), (b). #(A∩B′), (c). #(A∪B), (d). #(A∩B)′, (e). #A′

Solutions:

➤ **(a).**
Use Rule 2. #(A′∩B) = #B - #(A∩B) = 5 - 1 = 4.

➤ **(b).**
Use Rule 2. #(A∩B′) = #A - #(A∩B) = 5 - 1= 4.

➤ **(c).**
Use Rule 4. #(A∪B) = #A + #B - #(A∩B) = 5 + 5 - 1 = 9.

➤ **(d).**
Use Rule 3. #(A∩B)′ = #**U** - #(A∩B)′= 35 - 1 = 34.

➤ **(e).**
Use Rule 3. #A′ = #**U** - #A= 35- 5 = 30.

7.2 Solved Problem 7: From the following Venn Diagram, find

(a). #(A′∩B) (b). #(A∩B′) (c). #A (d). #B

(e). #(A′∩B′) (f). #(A∩B)′ (g). #**U**

7.

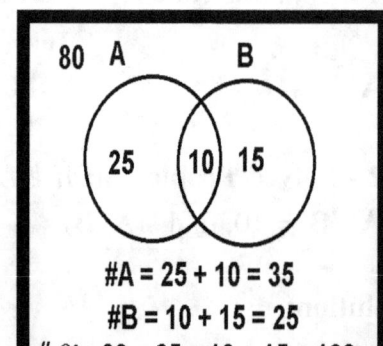

#A = 25 + 10 = 35
#B = 10 + 15 = 25
𝒰 = 80 + 25 + 10 + 15 = 130

Solutions:

➤ **(a).**
 #(A′∩B) = 15. B is divided into two parts where

#B = #(A′∩B) + #(A∩B)

➤ **(b).**
#(A∩B′) = 25. A is divided into two parts where #A = #(A∩B′) + #(A∩B).

➤ **(c).**
 #A = 35

 #A = #(A∩B′) + #(A∩B) = 25 + 10 = 35

➤ **(d)**.
#**B** = 25.

#**B** = #(**A**′∩**B**) + #(**A**∩**B**) = 15 + 10 = 25

➤ **(e)**.
#(**A**′∩**B**′)= #[(**A**∪**B**)′] = #**U** - #(**A**∪**B**) = 130 - 50 = 80

➤ **(f)**.
#(**A**∩**B**)′= 120. From the Venn Diagram we see

#(**A**∩**B**)′ = 80+ 25+ 15 = 120.

➤ **(g)**.
#**U** = 130. From the Venn Diagram, we see #**U** = 80 + 25+ 10 + 15.

Unsolved Problems with answers

7.2 - Problem 1: Assume #**A** = 120, #**B** = 50, and #(**A**∩**B**) = 0. Find #(**A**∪**B**).

Answer:

#(**A**∪**B**) = 170.

⇑ *Refer back to* **7.2 - Example 1 & 7.2 - Solved Problem 1**.

7.2 - Problem 2: Assume #**B** = 55, and #(**A**∩**B**) = 0. Find #(**A**′∩**B**).

Answer:

#(**A**′∩**B**) = 55.

⇑ *Refer back to* **7.2 - Example 2 & 7.2 - Solved Problem 2.**

7.2 - Problem 3: Assume #**U** = 150, and #**A** = 10. Find #**A**′.

Answer:

#**A**′= 140.

⇑ *Refer back to* **7.2 - Example 3 & 7.2 - Solved Problem 3**.

7.2 - Problem 4: Assume #(**A**∩**B**) = 10, #(**A**∩**B**′) = 15 and #(**A**′∩**B**) = 20. Find #(**A**∪**B**).

Answer:

#(A∪B) = 45.

⇑ *Refer back to* **7.2 - Example 4 & 7.2 - Solved Problem 4.**

7.2 - Problem 5: Assume #(A′∩B′) = 1. Find #[(A∪B)′].

Answer:

#[(A∪B)′] = 1.

⇑ *Refer back to* **7.2 - Example 5 & 7.2 - Solved Problem 5.**

7.2 - Problem 6: If #**U** = 55, **#A** = 15, **#B** = 5 and #(A∩B) = 0, find

(a). #(A′∩B), (b). #(A∩B′), (c). #(A∪B), (d). #(A∩B)′, (e). #A′

Answers:

➤ **(a).** #(A′∩B) = 5.

➤ **(b).** #(A∩B′) = 15.

➤ **(c)..** #(A∪B) = 20.

➤ **(d)..** #(A∩B)′ = 55.

➤ **(e).** #A′ = 40. ⇑ *Refer back to* **7.2 - Example 6 & 7.2**

7.2 - Problem 7: From the following Venn Diagram, find

(a). #**U** (b).**#A** (c). **#B** (d). #(A′∩B) (e). #(A∩B′)

(f). #(A∩B)′(g). #(A′∩B′)

Answers:

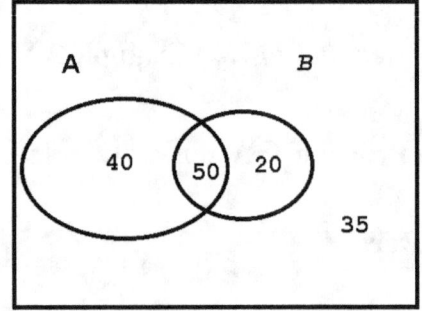

➤ **(a).** #**U** = 145.

➤ **(b).** #A = 90.

➤ **(c).** #B = 70.

➤ **(d).** #(A′∩B) = 20.

➤ (e). #(A∩B′) = 40.

➤ (f). #(A∩B)′ = 95.

➤ (g). #(A′∩B′) = 35.

⇑ *Refer back to* **7.2 - Example 7 & 7.2 - Solved Problem 7.**

7.3 - Counting Applications.

Ms. Apple teaches third grade. Her class of 35 students was surveyed as to the type of ice cream flavors they like. She discovered that 20 students like chocolate, 15 like strawberry and 5 liked both. Assume the set **C** represents the students that like chocolate ice cream, the set **S** represents the students that like strawberry ice cream and the universal set **U** represents all students in her class. The following Venn diagram is a graphic representation of this survey:

7.3 - Example 1: From the Venn diagram, we can find the answer to the following questions:

From these 2 flavors,

(a). How many students only like chocolate ice cream?

(b). How many students only like strawberry ice cream?

(c). How many students like at least one of these flavors?

(d). How many students like neither flavor?

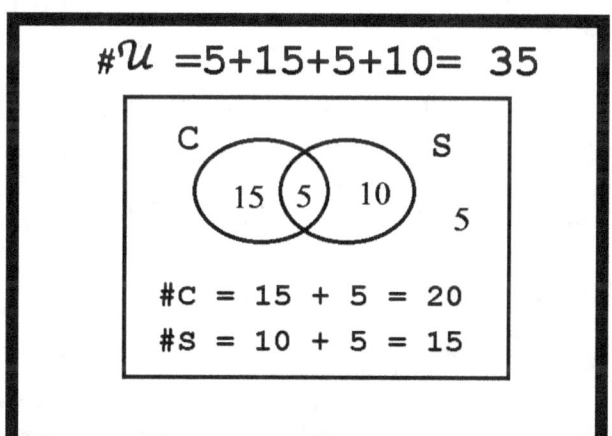

$$\#\mathcal{U} = 5+15+5+10 = 35$$

#C = 15 + 5 = 20
#S = 10 + 5 = 15

Solutions:

➤ (a).
The subset representing students that only like chocolate ice cream is **C∩S′**. Thus the number of students
that only like chocolate ice cream is #(**C∩S′**) = #**C** - #(**C∩S**) = 20 - 5 = 15.

➤ (b).
The subset representing students that only like strawberry ice cream is **S∩C′** Since #(**S∩C′**) = #**S** - #(**C∩S**) = 15 - 5 = 10, the number of students that only like strawberry ice cream is 10.

➤ (c).

The set that represents all students that like at least one of these flavors is C∪S. Since #(C∪S) = #C + #S - #(S∩C) = 20 + 15 - 5 = 30, the number of students that like at least one of these flavors is 30.

➤ (d).

The set that represents all students that like neither of these two flavors is (C∪S)'.

Since #(C∪S)' = #U - #(C∪S) = 35 - 30 = 5,. the number of students that like neither of these two flavors is 5.

7.3 - Example 2: A recent survey of 150 men were taken to find out their participation in the following sports: football, baseball, and ice hockey. From the survey the following information was given:

13 men participated in all three sports.

20 men participated in football and baseball.

30 men participated in football and ice hockey.

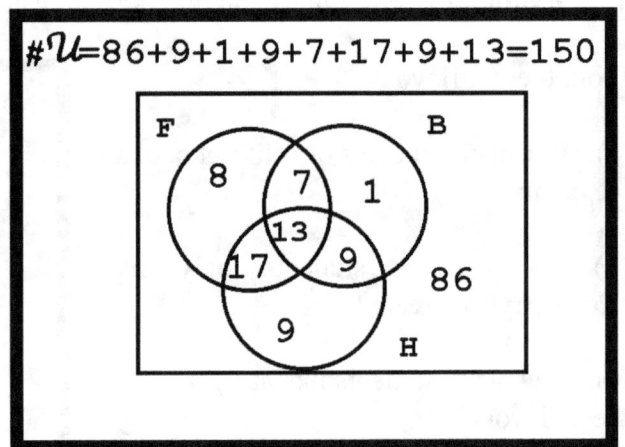

22 men participated in ice hockey and baseball.

45 men participated in football.

48 men participated in ice hockey.

30 men participated in baseball.

The following Venn diagram is a graphic representation of this survey.

From the Venn diagram, we can find the answers to the following questions:

How many men participated:

(a). in only football and ice hockey?

(b). in only football and baseball?

(c). in only baseball and ice hockey?

(d). in only baseball?

(e). in only football?

(f). in only hockey?

(g). in none of these sports?

To assist in solving these problems, we assume that **H** represents the set of all men that participated in ice hockey, **F** the set of all men that participated in football and **B** the set of all men that participated in baseball.

Solutions:

➤ (a). T
The subset representing men that participated in only football and ice hockey is $(H \cap F) \cap B'$. Since $\#[(H \cap F) \cap B'] = \#(H \cap F) - \#[(H \cap F) \cap B] = 30 - 13 = 17$, the number of men that participated in only football and ice hockey is 17.

➤ (b).
The subset representing men that participated in only football and baseball is $(F \cap B) \cap H'$. Since $\#[(F \cap B) \cap F'] = \#(F \cap B) - \#[(F \cap B) \cap H] = 20 - 13 = 7$ the number of men that participated in only football and
baseball is 7.

➤ (c).
The subset representing men that participated in only baseball and ice hockey is $(H \cap B) \cap F'$. Since $\#[(H \cap B) \cap F'] = \#(H \cap B) - \#[(H \cap B) \cap F] = 22 - 13 = 9$, the number of men that participated in only baseball and ice hockey is 9.

➤ (d).
The subset representing men that participated in only baseball is $B \cap (H' \cap F')$. From the Venn diagram, $\#[B \cap (H' \cap F')] = 30 - 7 - 9 - 13 = 1$.

➤ (e).
The subset representing men that participated in only football is $F \cap (H' \cap B')$. From the Venn diagram, $\#[F \cap (H' \cap B')] = 45 - 17 - 7 - 13 = 8$.

➤ (f).
The subset representing men that participated in only hockey is $H \cap (B' \cap F')$. From the Venn Diagram, $\#[H \cap (B' \cap F')] = 48 - 17 - 9 - 13 = 9$.

➤ (g).
The subset representing men that participated in non of these sports is $(F \cup B \cup H)'$. From the Venn diagram,
$\#(F \cup B \cup H)' = 150 - 8 - 7 - 13 - 17 - 1 - 9 - 9 = 86$.

Solved Problems

7.3 - Solved Problem 1: A recent survey of 200 students in an English class showed that 25 like biographies, 45 liked novels, and 10 liked both. Find the number of students that:

a. only like novels.

b. only like biographies.

c. at least one of these.

d. like neither.

Solutions.

Constructing a Venn diagram for this survey gives:

Let **N** = "likes novels".

Let **B** = "likes biographies".

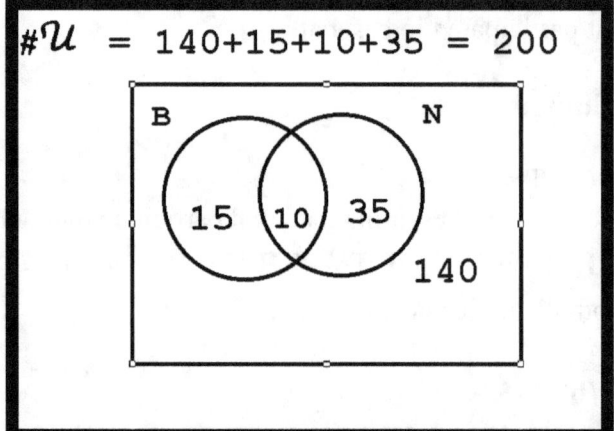

➤ **(a).**
E = "only like novels" = **N**∩**B'**.

From the Venn diagram, #(**N**∩**B'**) = 35.

➤ **(b).**
E = "only like biographies" = **N'**∩**B**.

From the Venn diagram, = #(**N'**∩**B**) = 25 - 10 = 15.

➤ **(c).**
E = "at least one of these" = **N**∪**B**.

From the Venn diagram, #(**N**∪**B**)= 35 + 10 + 15 = 60.

➤ **(d).**
E = "likes neither" =(**N**∪**B**)'

From the Venn diagram, #(**N**∪**B**)' = 140.

7.3 - Solved Problem 2: A survey at a local high school of 200 graduating seniors produced the following results:

25 will be going to college
30 will be searching for work.
35 will go on summer vacation.

15 will be going to college and searching for work.

10 will be going to college and will also go on summer vacation.

5 will be searching for work and will also go on summer vacation

1 will be doing all three.

Find the number of students that will:

(a). be going to college but will not search for work.

(b). be searching for work and going to college but will not be going on vacation.

(c). do none of these.

(d). all or none of these.

(e). do only one of these.

(f). do only two of these.

(g). do at least one of these.

Solutions.

Constructing a Venn Diagram for this survey gives:

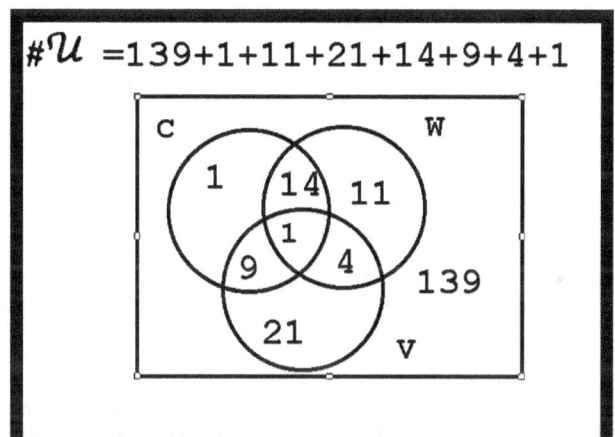

$$\#\mathcal{U} = 139+1+11+21+14+9+4+1$$

C: "Students going to college"

V: "students going on vacation"

W: "students searching for work".

➤ **(a).**
Step 1:E: "Students that will be going to college but will not search for work."

$$E = C \cap W'$$

Step 2: From the Venn diagram, $\#E = \#(C \cap W') = 1 + 9 = 10$.

➤ **(b).**
Step 1: E: "Students that will be searching for work and going to college but will not be going on vacation."

$$E = W \cap C \cap V'.$$

Step 2: From the Venn diagram, $\#E = \#(W \cap C \cap V') = 14$.

➤ **(c).**
Step 1: E: "students that will do none of these."

$E = (W \cup C \cup V)'$

Step 2: From the Venn diagram, #E = #($W \cup C \cup V$)' = 139.

➤ **(d).**
Step 1: E: "all or none of these."

$E = (W \cap C \cap V) \cup (W \cup C \cup V)'$

Step 2: From the Venn diagram,

#E =#[$(W \cap C \cap V) \cup (W \cup C \cup V)'$] = # $(W \cap C \cap V)$ + # $(W \cup C \cup V)'$] = 1 + 139 = 140.

➤ **(e).**
Step 1: E: "do only one of these."

$E = [W \cap (C \cup V)'] \cup [C \cap (W \cup V)'] \cup [V \cap (C \cup W)']$

Step 2: From the Venn diagram,

#E =#{[$W \cap (C \cup V)'] \cup [C \cap (W \cup V)'] \cup [V \cap (C \cup W)']$]} = 11 + 1 + 21 = 33.

➤ **(f).**
Step 1: E: "do only two of these."

$E = [W \cap C \cap V'] \cup [W \cap C' \cap V] \cup [W' \cap C \cap V]$

Step 2: From the Venn diagram,

#E =#{[$W \cap C \cap V'] \cup [W \cap C' \cap V] \cup [W' \cap C \cap V]$]} =#[$W \cap C \cap V'$] + #[$W \cap C' \cap V$] + #[$W' \cap C \cap V$]} = 14 + 4 +9 = 27.

➤ **(g).**
Step 1: E: "do at least one of these."

$E = W \cup C \cup V$

Step 2: #E = #[$W \cup C \cup V$] = 1 + 9 + 1 + 14 + 21 + 4 + 11 = 61.

Unsolved Problems with Answers

7.3 - Problem 1: A recent survey of 100 customers at a local fast food restaurant showed that 35 order tacos, 55 order hamburgers, and 15 order both. Find the number of customers that

a. order only tacos.

b. order only hamburgers.

c. order at least one of these food items.

d. order neither tacos nor hamburgers.

Answers:

➤ **(a).** 20

➤ **(b).** 40

➤ **(c).** 75

➤ **(d).** 25

⇑ *Refer back to* **7.3 - Example 1 & 7.3 - Solved Problem 1.**

7.3 - Problem 2: A survey of farmers produced the following results:

120 raised chickens
104 raised hogs
94 raised sheep
66 raised both chickens and hogs
57 raised both chickens and sheep
47 raised both hogs and sheep
10 raised all three.

 Find the number of farmers that

(a). raised only one type of these animals.

(b). raised at least two type of these animals.

(c). raised no sheep.

(d). raised sheep or hogs

(e). raised sheep or hogs but not chickens.

(f). raised none of these animals.

(g). raised at least one of these animals.

Answers:

➤ **(a).** 8

➤ **(a).** 150

➤ **(c).** 64

➤ **(d).** 151

➤ **(e).** 38

➤ **(f).** can not be determined.

➤ **(g).** 158

⇑ *Refer back to* **7.3 - Example 2 & 7.3 - Solved Problem 2.**

Supplementary Problems.

In a survey of 40 students, the following data was collected:

20 were over six foot
11 were blonde and over six foot.
16 were blonde
7 were over six foot and male.
19 were male.
8 were blonde and male.
3were over six foot, male and blonde.

1. How many students are not over six foot?
2. How many students were over six foot and male, but not blonde?
3. How many students are female?
4. How many female students are blonde and not over six foot?
5. How many students are blonde or over 6 feet?
6. How many females in the survey are not blonde?

Marshall interviewed customers in a shopping center to find out some of their cooking habits. He obtained the following results:

58 use microwave ovens.
63 use electric ranges.
58 use gas ranges.
19 use microwave ovens and electric ranges.
4 use both gas and electric ranges.
20 use gas and microwave ovens.
1 uses all three.
2 cook only with solar energy.

How many

7. only use Microwave ovens?

8. use exactly one type?

9. uses microwave and electric but not gas?

10. people were interviewed?

Assume $A \subset B \subset C \subset U$ and $\#U = 100$, $\#B = 40$, $\#A = 20$, and $\#C' = 10$. For problems 11-13, find

11. $\#(A' \cap B)$

12. $\#C$

13. $\#(B' \cap C)$

14. If $\#A \le \#B$ show $\#B' \le \#A'$.

15. For any sets A, B, C:

a. Show $A \cap C' \subseteq (A \cap B') \cup (B \cap C')$

b. Show $\#(A \cap C') \le \#(A \cap B') + \#(B \cap C')$

16. Find the cardinality of the following sets:

a. $\{\phi\}$

b. $\{\phi, \{\phi, \{\phi, 1\}\}\}$

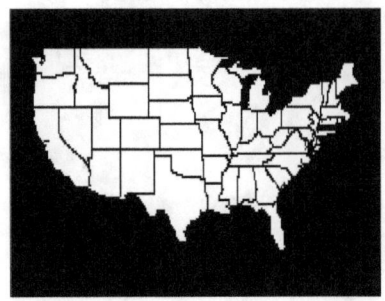

8.1 - What is an Experiment?

> An experiment is a well defined, step by step, process that can produce a finite or an infinite number of single outcomes. The actual experiment is normally not performed.

8.1 - Example 1: Assume a coin is to be tossed once. Find all possible single outcomes.

Solution:

There are two possible single outcomes: heads or tails. Therefore this experiment would result in the following single outcomes: h, t, where h is heads appears and t is tails appears.

8.1 - Example 2: Assume a die (a cube numerically marked on each side) is to be tossed once. Find all possible single outcomes.

Solution:

There are six possible single outcomes: 1, 2, 3, 4, 5, 6. Therefore, this experiment would result in the following single outcomes: 1, 2, 3, 4, 5, 6.

8.1 - Example 3: Assume Ms. Jones receives at least two phone calls a day. Find all possible single outcomes.
Solution:

The following are all possible single outcomes: 2, 3, 4, 5, 6,....

8.1 - Example 4: An urn contains 4 red marbles and 6 white marbles. Two marbles are selected at random. Find all possible single outcomes.

Solution:

There are four possible single outcomes: (r,r), (r,w), (w,r), (w,w) where w = white marble and r = red marble.

Solved Problems

8.1 - Solved Problem 1: A bag contains one red apple and one green apple. An apple is selected at random. Find all possible single outcomes.

Solution:

All possible single outcomes are r, g where r is a red apple selected and g is a green apple selected.

8.1 - Solved Problem 2: An urn contains a red ball (r), blue ball (b), white ball (w),and a green ball (g). One ball is selected at random. Find all possible single outcomes.

Solution:

All possible single outcomes are: r, b, w, g.

8.1 - Solved Problem 3: Mr. Smith makes at most 10 phone calls a day. Find all possible single outcomes.

Solution:

All possible single outcomes are: 0, 1, 2, 3, 4, 5, 6, 7, 8, 9, 10. Here, at most means that the maximum number of phone calls he will make each day is 10. Naturally, he may not make any calls. Thus all possible number of calls is between 0 and 10 inclusive.

8.1 - Solved Problem 4: A coin is tossed twice. Find all possible single outcomes.

Solution:

All possible single outcomes are: (h,h), (h,t), (t,h), (t,t) where h is heads and t is tails.

Unsolved Problems with Answers

8.1 - Problem 1: A used car lot contains Fords, Hondas, and Datsuns. A customer purchases a car from this lot. List all possible single outcomes.

Answer:

Ford, Honda, Datsun.

⇑ *Refer back to* **8.1 - Example 1 & 8.1 - Solved Problem 1.**

8.1 - Problem 2: Mary, Jane, Bill, Henry, Gayle, and Frank belong to a stamp club. A member is selected to attend the national conference. Find all possible single outcomes.

Answer:

Mary, Jane, Bill, Henry, Gayle, Frank.

⇑ *Refer back to* **8.1 - Example 2 & 8.1 - Solved Problem 2.**

8.1 - Problem 3: Mr. Smith makes between 20 and 30 phone calls a day. Find all possible single outcomes.

Answer:

20, 21,..., 30

⇑ *Refer back to* **8.1 - Example 3 & 8.1 - Solved Problem 3.**

8.1 - Problem 4: A coin is tossed three times. Find all possible single outcomes.

Answer:

(h,h,h), (h,h,t), (h,t,h), (t,h,h), (t,t,h), (t,h,t), (h,t,t), (t,t,t).

⇑ *Refer back to* **8.1 - Example 4 & 8.1 - Solved Problem 4.**

8.2 - What is a Sample Space?

A sample space is a universal set where elements are created from some well defined experiment. The elements of the sample space are all possible single outcomes that result from the experiment. The symbol for the sample space is **S**.

8.2 - Example 1: Assume a coin is to be tossed once. Write out the sample space.

Solution:

There are two possible single outcomes: heads or tails. Therefore, this experiment would result in the following sample space: **S** = {H, T} where H is heads appears and T is tails appears.

8.2 - Example 2: Assume a die is to be tossed once. Write out the sample space.

Solution:

There are six possible single outcomes: 1, 2, 3, 4, 5, 6. Therefore, this experiment would result in the following sample space:

S = {1, 2, 3, 4, 5, 6}.

8.2 - Example 3: Assume Ms. Jones receives at least two phone calls a day. Write out the sample space.

Solution:

The following are all possible outcomes: 2, 3, 4, 5, 6,... . Therefore, the sample space is **S** = {2, 3, 4, 5, 6,.....}.

8.2 - Example 4: An urn contains 4 red marbles and 6 white marbles. Two marbles are selected at random. Write out the sample space.

Solution:

The sample space is **S** = {(r,r), (r,w), (w,r), (w,w)} where w = white marble and r = red marble.

Solved Problems

8.2 - Solved Problem 1: A bag contains one red apple and one green apple. An apple is selected at random. Write out the sample space.

Solution:

All possible single outcomes are r, g where r is a red apple selected and g is a green apple selected. Therefore, the sample space is **S** = {r, g}.

8.2 - Solved Problem 2: An urn contains a red ball (r), blue ball (b), white ball (w),and a green ball (g). One ball is selected at random. Write out the sample space. .

Solution:

All possible single outcomes are: r,b,w,g. Therefore, **S** = {r, b, w, g}.

8.2 - Solved Problem 3: Mr. Smith makes at most 10 phone calls a day. Write out the sample space.

Solution:

All possible single outcomes are: 0, 1, 2, 3, 4, 5, 6, 7, 8, 9, 10. Here, at most means that the maximum number of phone calls he will make each day is 10. Naturally, he may not make any calls. Thus all possible number of calls is between 0 and 10 inclusive.

S = {0, 1, 2, 3, 4, 5, 6, 7, 8, 9, 10}.

8.2 - Solved Problem 4: A coin is tossed twice. Write out the sample space.

Solution: All possible single outcomes are:(h,h), (h,t), (t,h), (t,t) where h = heads and t = tails. Therefore, **S** = {(h,h), (h,t), (t,h), (t,t)}.

Unsolved Problems with Answers

8.2 - Problem 1: A used car lot contains Fords, Hondas, and Datsuns. A customer purchases a car from this lot. Write out the sample space.

Answer:

S = {Ford, Honda, Datsun}.

⇑ *Refer back to* **8.2 - Example 1 & 8.2 - Solved Problem 1.**

8.2 - Problem 2: Mary, Jane, Bill, Henry, Gayle, and Frank belong to a stamp club. A member is selected to attend the national conference. Write out the sample space.

Answer:

S = {Mary, Jane, Bill, Henry, Gayle, Frank}.

⇑ *Refer back to* **8.2 - Example 2 & 8.2 - Solved Problem 2.**

8.2 - Problem 3: Mr. Smith makes between 20 and 30 phone calls a day. Write out the sample space.

Answer:

S = {20, 21,..., 30}

⇑ *Refer back to* **8.2 - Example 3 & 8.2 - Solved Problem 3.**

8.2 - Problem 4: A coins is tossed three times. Write out the sample space. .

Answer:

S = {(h,h,h), (h,h,t), (h,t,h), (t,h,h), (t,t,h), (t,h,t), (h,t,t), (t,t,t)}.

⇑ *Refer back to* **8.2 - Example 4 & 8.2 - Solved Problem 4.**

8.3 - What is an Event?

An event is a subset of a sample space **S**. Since an event is a subset, we denote it with a capital letter i.e.
   E, F, etc.

.**8.3 - Example 1**: Assume a die is to be tossed once. Find the event that a even number will appear.

Solution:

The sample space created by this experiment is **S** = {1, 2, 3, 4, 5, 6}. The event "that a even number will appear" is the subset of the sample space **S.** Using **E** to represent this subset, **E** = {2, 4, 6} and **E⊂S**.

8.3 - Example 2: Assume a coin is to be tossed twice. Find the event that both tosses resulted in the same face.

Solution:

The sample space created by this experiment is **S** = {(H,H), (H,T), (T,H), (T,T)}. The event "that both tosses resulted" in the same face is the event **F** = {(H,H), (T,T)} and **F⊂S**.

8.3 - Example 3: Assume Ms. Jones receives at least two phone calls a day. Find the event that:

(a). she receives at least 10 calls a day.

(b). she receives between 4 and 8 calls a day.

(c). she receives less than 10 calls a day.

Solutions:

The sample space created by this experiment is **S** = {2, 3, 4, 5, 6,...}.

➤ **(a).**
At least 10 means 10 or more: **A** = {10, 11, 12, 13,...} and **A⊂S**.

➤ **(b).**
Between 4 and 8 calls means 4, 5, 6, 7, or 8 calls: B = {4, 5, 6, 7, 8} and **B⊂S**.

➤ **(c).**
Less than 10 calls means 9 or less:

C = {2, 3, 4, 5, 6, 7, 8, 9} and **C⊂S**.

8.3 - Example 4: An urn contains 4 red marbles and 6 white marbles. Two marbles are selected at random. Find the event **E** that

(a). the colors are the same.

(b). the colors are different

(c). a red marble appears on the first drawing.

(d). only white appears.

(e). white appears at least once.

Solutions:

The sample space is $S = \{(r,r), (r,w), (w,r), (w,w)\}$ where w is a white marble and r is a red marble.

➤ **(a)**.
Since, $E \subset S$, $E = \{(r,r), (w,w)\}$

➤ **(b)**.
Since, $E \subset S$, $E = \{(r,w), (w,r)\}$

➤ **(c)**.
Since, $E \subset S$, $E = \{(r,r), (r,w)\}$

➤ **(d)**.
Since, $E \subset S$, $E = \{(w,w)\}$

➤ **(e)**.
Since, $E \subset S$, $E = \{(r,w), (w,r), (w,w)\}$

Solved Problems

8.3 - Solved Problem 1: Assume a die is to be tossed once. Find the event that an odd number will appear.

Solution:

$S = \{1, 2, 3, 4, 5, 6\}$, $E = \{1, 3, 5\}$ and $E \subset S$.

8.3 - Solved Problem 2: Assume a coin is to be tossed twice. Find the event that both tosses resulted in different faces.

Solution:

S = {(H,H), (H,T), (T,H), (T,T)}, **F** = {(H,T), (T,H)} and **F** ⊂ **S.**

8.3 - Solved Problem 3: Assume Ms. Jones receives at most five phone calls a day.

Find the event that:

(a). she receives at least 1 call a day.

(b). she receives at least 4 calls a day.

(c). she receives less than 3 calls a day.

The sample space created by this experiment is **S** = {0, 1, 2, 3, 4, 5}.

Solutions:

➤ **(a).**
The event "at least 1 call a day" means one or more calls a day and **A** ⊂ **S**. Therefore, **A** = {1, 2, 3, 4, 5}.

➤ **(b).**
The event "at least 4 calls a day" means from 4 to 5. Therefore, **B** = {4, 5}.

➤ **(c).**
The event "less than 3 calls a day" means 2 or less calls a day and **C** ⊂ **S**. It does not include 3 as well as
4 and 5. Therefore, **C** = {0, 1, 2}.

8.3 - Solved Problem 4: A pair of dice is tossed once. Find the event **E** that

(a). a 2 and 4 appear.

(b). both numbers are odd and different.

(c). The number 1 appears twice.

(d). only numbers greater than 4 occur.

(e). the sum of the two number is 3.

Solution:

Assume one die is colored red and the other blue. For each pair of numbers assume that the first number is from the die colored red and the second number is from the die colored blue. For example (5,6) means that the number 5 is from the die colored red and the number 6 is from the die colored blue.

➤ **(a)**.
$E = \{(2,4), (4,2)\}$

➤ **(b)**.
$E = \{(1,3), (1,5), (3,1), (3,5), (5,1), (5,3)\}$

➤ **(c)**.
$E = \{(1,1)\}$

➤ **(d)**.
$E = \{(5,5), (5,6), (6,5), (6,6)\}$

➤ **(e)**.
$E \{(1,2), (2,1)\}$

Unsolved Problems with Answers

8.3 - Problem 1: Assume a die is to be tossed once. Find the event that a number greater than 3 appears.

Answer:

$E = \{4, 5, 6\}$.

⇑ *Refer back to* **8.3 - Example 1 & 8.3 - Solved Problem 1.**

8.3 - Problem 2: Assume a coin is to be tossed twice. Find the event that the first toss results in heads.

Answer:

$F = \{(h,t), (h,h)\}$

⇑ *Refer back to* **8.3 - Example 2 & 8.3 - Solved Problem 2.**

8.3 - Problem 3: Assume Ms. Jones receives at most five phone calls a day.
Find the event that:

(a). she receives more than 1 call a day.

(b). she receives at least 4 calls a day.

(c). she receives no calls a day.

Answers:

➤ **(a). A** = {2, 3, 4, 5}

➤ **(b). B** = {4, 5}

(➤ **(c). C** = {0}

⇑ *Refer back to* **8.3 - Example 3 & 8.3 - Solved Problem 3.**

8.3 - Problem 4: Three coins are tossed once. Find the event **E** that

(a). all the faces are the same.

(b). only one head occurs.

(c). only two heads occurs.

(d). at least two tails occur.

(e). at most two tails occur.

Answers:

➤ **(a). E** = {(h,h,h), (t,t,t)}

➤ **(b). E** = {(h,t,t), (t,h,t), (t,t,h)}

➤ **(c). E** = {(h,h,t), (h,t,h), (t,h,h)}

➤ **(d). E** = {(h,t,t), (t,h,t), (t,t,h), (t,t,t)}

➤ **(e). E** = {(t,t,h), (t,h,t), (h,t,t), (h,h,t), (h,t,h), (t,h,h), (h,h,h)}

⇑ *Refer back to* **8.3 - Example 4 & 8.3 - Solved Problem 4.**

Supplementary Problems.

A fair coin is tossed until tails appears. Write out

1. the sample space.[1]

2. the event that it took at most four tosses.

──────────────

    [1]

1. In this experiment assume that tails will eventually happen.

3. the event that it took at least three tosses.

4. the event that it took between two and four tosses (inclusive).

A fair coin is tossed until two heads occur or tossed four times, whichever occurs first. Write out

5. the sample space.

6. the event that less than four tosses were made.

7. the event that more than three tosses were made.

8. the event that heads occur on the first toss.

 Mr. Jones tosses a pair of dice once. Write out

9. the event that he tosses a sum greater than 7.

10. the event that he tosses a sum less than 4.

11. the event that he tosses a sum between 7 and 9 (inclusive).

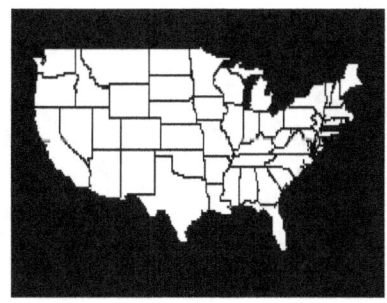

Set Theory
Lesson 9
Boolean Algebra of Sets

9.1 - What is Boolean algebra of Sets?

Boolean algebra is a powerful method for solving many problems in probability theory. Since we are studying sample spaces, we will define a Boolean algebra of sets in the following way: Given a family of sets in a sample space **A**, **B**, **C**, **D**, **E**..., we define two operations union $\cup$ and intersection $\cap$. Unless otherwise indicated with parentheses, intersections have priority over unions. For these operations, the following laws hold:

1. **Closure Laws:**

$A \cup B = E$

$A \cap B = F$,where **E** and **F** are unique.

2. **Commutative Laws:**

$A \cup B = B \cup A$

$A \cap B = B \cap A$

3. **Associative Laws:**

$(A \cup B) \cup C = A \cup (B \cup C)$

$(A \cap B) \cap C = A \cap (B \cap C)$

4. **Distributive Laws:**

$A \cup (B \cap C) = (A \cup B) \cap (A \cup C)$

$A \cap (B \cup C) = (A \cap B) \cup (A \cap C)$

5. **Identity Laws:**

For all sets **A** there exists sets denoted by ϕ and **S** where

$A \cup \phi = A$

$A \cap S = A$

6. Complement Laws:

For each set **A** there exists a set **A**$'$ where

$A \cup A' = S.$

$A \cap A' = \phi.$

For this algebra of sets, the set **S** is called the sample space and ϕ the null set.

From these laws, the following properties can be proven:

Uniqueness of the complement:

1. If $B \cap A = \phi$ and $B \cup A = S$ then $B = A'$.

DeMorgan Laws:

2. $(A \cup B)' = A' \cap B'$

and

$(A \cap B)' = A' \cup B'$

3. $A \cup A = A$

$A \cap A = A$

4. $A \cup S = S$

$A \cap \phi = \phi$

5. $A'' = A$

6. $S' = \phi$ and $\phi' = S$

Definition of subsets

A is said to be a subset of **B**, denoted by $A \subseteq B$, if $A \cap B' = \phi$.

A is said to be a proper subset of **B** if **A** is a subset of **B** but **B** is a not a subset of **A**.

Definition of equality of sets: $A = B$ if **A** is subset of **B** and **B** is a subset of **A**.

From the definition of subsets, we have the following properties which can be proven:

1. **Reflexive Law**: $A \subseteq A$ for all sets **A**.

2. **Anti-symmetric Law**: If $A \subseteq B$ and $B \subseteq A$ then $A = B$.

3. **Transitive Law**: If $A \subseteq B$ and $B \subseteq C$ then $A \subseteq C$.

4. **Complement Law**: If $A \subseteq B$ then $B' \subseteq A'$

5. **Universal Set S**: All the sets are subsets of **S**.

6. **Uniqueness law**:

If $A \cap B = \phi$ and $A \cup B = S$ then $B = A'$.

7. **Subdivision Law:**

$A \cup B = (A' \cap B) \cup (A \cap B') \cup (A \cap B)$

8. If $A \subseteq B$ then $B = B \cup A$

9. If $A \subseteq B$ then $A = B \cap A$

Applications

The sets in our boolean algebra are events in sample spaces that are generated by well defined experiments. To translate the events into boolean expressions, we use the following: the words **and**, **or** and **not** are used to express events in conjunction with intersections, union and complements. The following table shows the relationship of these expressions to the events:

Set	Expression	Key Word
A'	The event **A** does not occur.	NOT
$A \cap B$	The events **A** <u>and</u> **B** both occur.	AND
$A \cup B$	The events **A** <u>or</u> **B** or both occur.	OR

9.1 - Example 1: Assume two cards are drawn from an ordinary deck of cards. Assume K_1 is the event that the first card drawn is a king and K_2 is the event that the second card drawn is a king. Write out the event **E** that

(a). both cards drawn are kings.

(b). only one card is a king.

(c). no card drawn is a king.

(d). at least one card drawn is a king.

(e). Simplify $K_2 \cap (K_1 \cup K_2')$.

(f). Simplify $K_2 \cup (K_1 \cap K_2')$.

a.

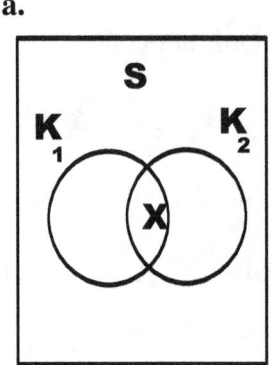

Solutions:

➤ **(a).**
The event **E** can be stated as follows: the first card and the second card is a king. Since the word and is associated with the intersection $\cap$, we can write $E = K_1 \cap K_2$.

➤ **(b).**
Since only one card is a king, we use the events K_1, K_2, K_1', K_2'. The event, "only one card is a king", means the following:

b.

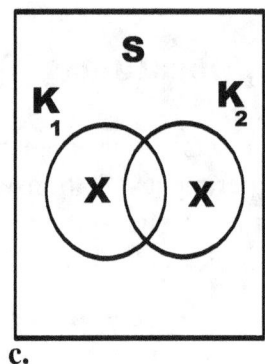

1. If a king occurs on the first drawing then a king cannot occur on the second drawing:$(K_1 \cap K_2')$.

or

2. If a king occurs on the second drawing then a king cannot occur on the first drawing:$(K_1' \cap K_2)$.

c.

3. We can write this event as $E = (K_1 \cap K_2') \cup (K_1' \cap K_2)$.

➤ **(c).**
The event , "no card drawn is a king", means that the first **and** the second card are not kings. Therefore, we can write this event as
$E = K_1' \cap K_2' = (K_1 \cup K_2)'$.

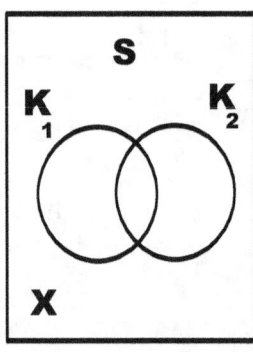

➤ **(d).**
The event, "at least one card drawn is a king, means that one **or** two cards drawn are kings. This can happen in the following ways:

1. The first card drawn is a king **and** second card is not a king: $K_1 \cap K_2'$.

or

2. The second card drawn is a king **and** first card is not a king: $K_1' \cap K_2$.

or   **d.**

3. Both cards drawn are kings: $K_1 \cap K_2$.

4. This can be written as $E = (K_1 \cap K_2') \cup (K_1' \cap K_2) \cup (K_1 \cap K_2) = K_1 \cup K_2$.

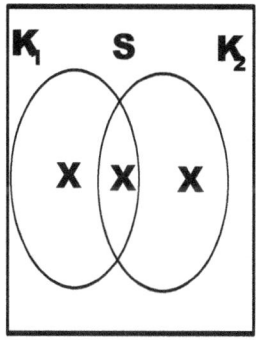

➤ **(e).**
From the distributive law , we can write

$$K_2 \cap (K_1 \cup K_2') = (K_2 \cap K_1) \cup (K_2 \cap K_2') = (K_1 \cap K_2) \cup \phi = K_1 \cap K_2.$$

➤ **(f).**
From the distributive law , we can write

$$K_2 \cup (K_1 \cap K_2') = (K_2 \cup K_1) \cap (K_2 \cup K_2') = (K_1 \cup K_2) \cap S = K_1 \cup K_2.$$

9.1 - Example 2: An urn contains several red, white and blue marbles. Three marbles are selected at random, one at a time. Let R_i, W_i, B_i (i = 1, 2, 3) represent the results of the i th color marble being selected. Find the events **E:**

(a). two marbles are blue and one is red.

(b). Exactly two marbles are blue.

(c). No red marble is selected.

(d). the first two marbles are red and the third is white.

(e). exactly one of each appears.

Solutions:

➤ **(a).**
The event **E** can happen in the following ways:

1. first is blue **and** second is blue **and** third is red: $B_1 \cap B_2 \cap R_3$.

or

2. first is blue **and** second is red **and** third is blue: $B_1 \cap R_2 \cap B_3$.

or

3. first is red **and** second is blue **and** third is blue: $R_1 \cap B_2 \cap B_3$.

4. Therefore, $E = (B_1 \cap B_2 \cap R_3) \cup (B_1 \cap R_2 \cap B_3) \cup (R_1 \cap B_2 \cap B_3)$.

➤ **(b).**
The event E can happen in the following ways:

1. The first and second marbles are blue and the third is not blue: $B_1 \cap B_2 \cap B_3'$.

or

2. The first and third marbles are blue and the second is not blue: $B_1 \cap B_2' \cap B_3$.

or

3. The Second and third marbles are blue and the first is not blue: $B_1' \cap B_2 \cap B_3$.

4. Therefore, $E = (B_1 \cap B_2 \cap B_3') \cup (B_1 \cap B_2' \cap B_3) \cup (B_1' \cap B_2 \cap B_3)$.

➤ **(c).**
This event can happen in the following way: no red marble on the first selection (R_1'), **and** no red marble on the second selection (R_2'), **and** no red marble on the third selection R_3'. Joining these events with intersections, we have $E = R_1' \cap R_2' \cap R_3'$.

➤ **(d).**
Since the first two selected are red and the last drawn is white, we can write the event as $E = R_1 \cap R_2 \cap W_3$.

➤ **(e).**
There are six ways this can happen:

1. The first selection is red, the second selection is blue and the third selection is white: $R_1 \cap B_2 \cap W_3$.

or

2. The first selection is red, the second selection is white and the third selection is blue: $R_1 \cap W_2 \cap B_3$.

or

3. The first selection is white, the second selection is blue and the third selection is red: $W_1 \cap B_2 \cap R_3$.

or

4. The first selection is white, the second selection is red and the third selection is blue: $W_1 \cap R_2 \cap B_3$.
or

5. The first selection is blue, the second selection is white and the third selection is red: $B_1 \cap W_2 \cap R_3$.

or

6. The first selection is blue, the second selection is red and the third selection is white: $B_1 \cap R_2 \cap W_3$.

7. Therefore,

$$E = (R_1 \cap B_2 \cap W_3) \cup (R_1 \cap W_2 \cap B_3) \cup (W_1 \cap B_2 \cap R_3) \cup (W_1 \cap R_2 \cap B_3) \cup (B_1 \cap W_2 \cap R_3) \cup (B_1 \cap R_2 \cap W_3).$$

9.1 - Example 3: Ms. Jones receives at least 10 calls a day.

Let **A** be the event she receives at most 15 calls a day;
Let **B** be the event she receives at least 12 calls a day;
Let **C** be the event she receives between 15 and 20 calls a day.

Using these events, find the following events **E**:

(a). She receives at least 16 calls a day.

(b). She receives 10 or 11 calls a day.

(c). She receives between 12 and 15 calls a day (inclusive).

(d). She receives exactly 15 calls a day.

Solutions:

➤ **(a).**
The sample space $S = \{10, 11, 12,...\}$.

Step 1: $A = \{10, 11, 12, 13, 14, 15\}$ and

Step 2: $A' = \{16, 17, 18,...\}$.

Step 3: Therefore, $E = A'$.

➤ **(b).**
Step 1: $B = \{12, 13, 14,...\}$

Step 2: $B' = \{10, 11\}$

Therefore, $E = B'$.

➤ **(c).**

Step 1: A = {10, 11, 12, 13, 14, 15}

Step 2: B = {12, 13, 14, 15, 16,...}

Step 3: A∩B = {12, 13, 14, 15}

Step 4: Therefore, **E = A∩B**.

➤ **(d)**.
Step 1: A = {10, 11, 12, 13, 14, 15}

Step 2: C = {15, 16, 17, 18, 19, 20}

Step 3: A∩C = {15}

Step 4: Therefore, **E = A∩C**.

9.1 - Example 4: Two urns each contain red and white marbles. An urn is selected at random and a red marble is selected. Let U_A represent the event that urn A was selected and **R** the event that a red marble is selected. Write **R** as an expression of **R**, U_A and U_A'.

Solution:

The event that a red marble is selected depends on which urn was selected.

1. We can write the event that a red marble is selected came from urn A as $R \cap U_A$.

or

2. We can write the event a red marble is selected from urn B as $R \cap U_A'$. Connecting these two event,
$R = (R \cap U_A) \cup (R \cap U_A')$.

9.1 - Example 5: Two cards are drawn from an ordinary deck of cards. Let H_1 be the event that a heart was drawn on the first drawing, H_2 be the event that a heart was drawn on the second drawing. Write H_2 as an expression of H_1 and H_2.

Solution:

1. A heart is drawn on the first **and** second drawing: $H_2 = H_1 \cap H_2$.

or

2. A heart is drawn on the second **but not** on the first drawing: $H_2 = H_1' \cap H_2$.

3. Connecting these two event gives $H_2 = (H_1 \cap H_2) \cup (H_1' \cap H_2)$.

Solved Problems

9.1 - Solved Problem 1: Assume three cards are drawn from an ordinary deck of cards. If K_i ($i = 1, 2, 3$) are the events that the i th card drawn is a king, write out the event **E** that

(a). all three cards drawn are kings.

(b). only one card is a king.

(c). no card drawn is a king.

(d). at least one card drawn is a king.

(e). Simplify $K_1 \cap [(K_1 \cap K_2) \cup (K_2 \cap K_1')]$.

(f). Simplify $K_1 \cup [(K_1 \cup K_2) \cap (K_2 \cup K_1')]$.

Solutions:

➤ **(a).**
The event **E** can be stated as follows: the first card **and** the second card **and** the third card is a king. Since the word **and** is associated with the intersection $\cap$, we can write $E = K_1 \cap K_2 \cap K_3$.

➤ **(b).**
The event, "only one card is a king", can occur in the following ways:

1. king is drawn first followed by two cards that are not kings: $K_1 \cap K_2' \cap K_3'$.

or
2. the first card drawn is not a king, the second card drawn is a king and the third card drawn is not a king:
$K_1' \cap K_2 \cap K_3'$.

or

3. the first card drawn is not a king, the second card drawn is not a king and the third card drawn is a king:
$K_1' \cap K_2' \cap K_3$.

Therefore, $E = (K_1 \cap K_2' \cap K'_3) \cup (K_1' \cap K_2 \cap K_3') \cup (K_1' \cap K_2' \cap K_3)$.

➤ **(c).**
The event, "no card drawn is a king", can occur in the following way: the first card drawn is not a king, the second card drawn is not a king and the third card drawn is not a king: $K_1' \cap K_2' \cap K_3'$.

➤ **(d).**
Step 1: E: the event that at least one card drawn is a king.

E′: the event that no card drawn is a king.

Step 2: From (c) we have $E' = K_1' \cap K_2' \cap K_3'$.

Step 3: $E = E'' = (K_1' \cap K_2' \cap K_3')' = K_1'' \cup K_2'' \cup K_3'' = K_1 \cup K_2 \cup K_3$.

➤ **(e).**
Using the distributive law,

$K_1 \cap [(K_1 \cap K_2) \cup (K_2 \cap K_1')] = [K_1 \cap (K_1 \cap K_2)] \cup [K_1 \cap (K_2 \cap K_1')]. = (K_1 \cap K_2) \cup (\phi \cap K_2) = K_1 \cap K_2$.

➤ **(f).**
Using the distributive

$K_1 \cup [(K_1 \cup K_2) \cap (K_2 \cup K_1')] = [K_1 \cup K_1 \cup K_2)] \cap [K_1 \cup (K_2 \cup K_1')] = (K_1 \cup K_2) \cap (S \cup K_2) = (K_1 \cup K_2) \cap S = K_1 \cup K_2$.

9.1 - Solved Problem 2: Mrs. Jones teaches third grade. In a recent survey from her students, she found that the only flavors of ice cream the students like are strawberry, chocolate, and raspberry. Three children are selected one at a time. Find the event **E**.

(a). two students like chocolate and one likes raspberry.

(b). Exactly two students like raspberry.

(c). No student likes strawberry.

(d). the first two students selected like strawberry and the third student likes chocolate.

(e). One student likes chocolate, one student likes strawberry and one student likes raspberry.

Solutions:

We use the following sets:

R_i: the i th child selected likes raspberry.

C_i: the i th child selected likes chocolate.

S_i: the i th child selected likes strawberry.

➤ **(a).**
The event **E** can happen in the following ways:

1. first student likes chocolate **and** second likes chocolate **and** third likes raspberry: $C_1 \cap C_2 \cap R_3$.

or

2. first student likes chocolate **and** second likes raspberry **and** third likes chocolate: $C_1 \cap R_2 \cap C_3$.

or

3. first student likes raspberry **and** second likes chocolate **and** third likes chocolate: $R_1 \cap C_2 \cap C_3$.

Therefore, $E = (C_1 \cap C_2 \cap R_3) \cup (C_1 \cap R_2 \cap C_3) \cup (R_1 \cap C_2 \cap C_3)$.

➤ **(b).**
The event E can happen in the following ways:

1. The first **and** second students like raspberry **and** the third student does not likes raspberry: $R_1 \cap R_2 \cap R_3'$.

or

2. The first **and** third students like raspberry **and** the second student does not like raspberry: $R_1 \cap R_2' \cap R_3$.

or

3. The second **and** third students like raspberry **and** the first student does not like raspberry: $R_1' \cap R_2 \cap R_3$.

Connecting these three expressions, we get $E = (R_1 \cap R_2 \cap R_3') \cup (R_1 \cap R_2' \cap R_3) \cup (R_1' \cap R_2 \cap R_3)$.

➤ **(c).**
This event can happen in the following way: the first student does not like strawberry (S_1'), **and** the second student does not like strawberry (S_2'), **and** the third student does not like strawberry S_3'.

Joining these events with intersections, we have $E = S_1' \cap S_2' \cap S_3'$.

➤ **(d).**
Since the first two students selected like strawberry and the last student likes chocolate, we can write the event as $E = S_1 \cap S_2 \cap C_3$.

➤ **(e).**
There are six ways this can happen:

1. The first likes chocolate, the second student likes strawberry and the third student likes raspberry: $C_1 \cap S_2 \cap R_3$, **or**

2. The first likes chocolate, the second student likes raspberry and the third student likes strawberry: $C_1 \cap R_2 \cap S_3$. **or**

3. The first likes strawberry, the second student likes raspberry and the third student likes chocolate: $S_1 \cap R_2 \cap C_3$. **or**

4. The first likes strawberry, the second student likes chocolate and the third student likes raspberry: $S_1 \cap C_2 \cap R_3$. **or**

5. The first likes raspberry, the second student likes chocolate and the third student likes strawberry: $R_1 \cap C_2 \cap S_3$. **or**

6. The first likes raspberry, the second student likes strawberry and the third student likes raspberry: $R_1 \cap S_2 \cap C_3$. .

7. Therefore, $E = (C_1 \cap S_2 \cap R_3) \cup (C_1 \cap R_2 \cap S_3) \cup (S_1 \cap R_2 \cap C_3) \cup (S_1 \cap C_2 \cap R_3) \cup (R_1 \cap C_2 \cap S_3) \cup (R_1 \cap S_2 \cap C_3)$.

9.1 - Solved Problem 3: Mr. Jones has at least three grandchildren.

let **A** be the event he has at most 10 grandchildren.

Let **B** be the event he has at least 5 grandchildren.

Let **C** be the event he has between 10 and 25 grandchildren.

Using these events, find the following events **E**:

(a). he has at least 11 grandchildren.

(b). he has 3 or 4 grandchildren.

(c). he has between 5 and 10 grandchildren.

(d). he has exactly 10 grandchildren.

Solutions:

➤ **(a).**
The sample space $S = \{3, 4, 5, 6, ...\}$.

Step 1: $A = \{3, 4, ..., 10\}$

Step 2: $A' = \{11, 12, ...\}$

Therefore, $E = A'$.

➤ **(b).**
Step 1: $B = \{5, 6, 7...\}$

Step 2: $B' = \{3, 4\}$

Therefore, $E = B'$.

➤ **(c).**
Step 1: $A = \{3, 4, 5,..., 10\}$

Step 2: $B = \{5, 6, 7,...\}$

Step 3: $A \cap B = \{5, 6, 7, 8, 9, 10\}$

Step 4: Therefore, $E = A \cap B$.

➤ **(d).**
Step 1: $A = \{3, 4, 5,..., 10\}$

Step 2: $C = \{10, 11,..., 25\}$

Step 3: $A \cap C = \{10\}$

Step 4: Therefore, $E = A \cap C$.

9.1 - Solved Problem 4: Three urns each contain red marbles, white marbles and black marbles. A marble is selected from one of these urns. Let U_A represent the event that urn A was selected, U_B represent the event that urn B was selected and, U_C represent the event that urn C was selected. If R is the event that a red marble is selected, write R as an expression of R and U_A, U_B, U_C.

Solution:

The event that a red marble is selected depends on which urn was selected.
1. We can write the event that a red marble is selected came from urn A as: $R \cap U_A$.

or

2. We can write the event a red marble is selected from urn B as: $R \cap U_B$.

or

3. We can write the event a red marble is selected from urn C as: $R \cap U_C$.

Connecting these three event, $R = (R \cap U_A) \cup (R \cap U_B) \cup (R \cap U_C)$.

9.1 - Solved Problem 5: Two computer chips are selected from a large shipment. Let D_1 be the event that the first chip drawn is defective and D_2 be the event that the second chip drawn is defective. Write D_2 as an expression of D_1 and D_2.

Solution:

1. The first and second computer chips drawn are both defective: $D_1 \cap D_2$.

or

2. The first computer chip drawn is not defective **but** the second computer chip drawn is defective: $D_1' \cap D_2$.

3. Connecting these two event gives $D_2 = (D_1 \cap D_2) \cup (D_1' \cap D_2)$.

Unsolved Problems with Answers

9.1 - Problems 1: Assume four computer chips are drawn from a box. If D_j is the event that the j th chip is defective, write out the event **E** that

(a). all of the chips are defective.

(b). only one chip is defective.

(c). none of the chips are defective.

(d). at least one chip is defective.

(e). Simplify $(D_1 \cap D_2 \cap D_3 \cap D_4) \cap [D_1' \cup D_2' \cup D_3' \cup D_4']$.

(f). Simplify $(D_1 \cup D_2 \cup D_3 \cup D_4) \cup [D_1' \cap D_2' \cap D_3' \cap D_4']$.

Answers:

➤ (a). $D_1 \cap D_2 \cap D_3 \cap D_4$

➤ (b). $(D_1 \cap D_2' \cap D_3' \cap D_4') \cup (D_1' \cap D_2 \cap D_3' \cap D_4') \cup (D_1' \cap D_2' \cap D_3 \cap D_4') \cup (D_1' \cap D_2' \cap D_3' \cap D_4)$

➤ (c). $D_1' \cap D_2' \cap D_3' \cap D_4'$

➤ (d). $D_1 \cup D_2 \cup D_3 \cup D_4$

➤ (e). ϕ

➤ (f). S

⇑ *Refer back to* **9.1 - Example 1 & 9.1 - Solved Problem 1.**

9.1 - Problem 2: A manufacturer of television sets purchases its television picture tubes from three different manufacturers, each located in different areas of the United States. Assume three picture tubes are selected at random, one at a time. Let E_i represent the event that the i th tube selected came from the East coast, W_i represent the event that the i th tube came from the West coast, and M_i represent the event that the i th tube came from the Midwest. Write out the expression for the event **F**:

(a). Two picture tubes selected came from the East coast and one from the West coast.

(b). Exactly two picture tubes selected came from the Midwest.

(c). No picture tubes selected came from the West coast.

(d). The first two picture tubes selected came from the west coast and the last selected came from the East coast.
(e). Each picture tube selected came from a different area.

Answers:

➤ **(a).** $(E_1 \cap E_2 \cap W_3) \cup (E_1 \cap W_2 \cap E_3) \cup (W_1 \cap E_2 \cap E_3)$

➤ **(b).** $(M_1 \cap M_2 \cap M_3') \cup (M_1 \cap M_2' \cap M_3) \cup (M_1' \cap M_2 \cap M_3)$

➤ **(c).** $W_1' \cap W_2' \cap W_3'$

➤ **(d).** $W_1 \cap W_2 \cap E_3$

➤ **(e).** $(W_1 \cap E_2 \cap M_3) \cup (W_1 \cap M_2 \cap E_3) \cup (E_1 \cap W_2 \cap M_3) \cup (E_1 \cap M_2 \cap W_3) \cup (M_1 \cap W_2 \cap E_3) \cup (M_1 \cap E_2 \cap W_3)$

⇑ *Refer back to* **9.1 - Example 2 & 9.1 - Solved Problem 2**.

9.1 - Problem 3: The minimum speed on the San Diego freeway is 45 mph. Assume a car is observed traveling on this freeway at the minimum speed or faster, $S = \{45, 46, 47, 48,...\}$.

let **A** be the event it is traveling less than 75 mph.

Let **B** be the event it is traveling at least 47 mph .

Let **C** be the event it is traveling between 74 and 80 mph (inclusive). sing these events, find the following events **E**:

(a). the car is traveling at least 75 mph.

(b). the car is traveling 45 or 46 mph.

(c). it is traveling between 47 and 74 mph.

(d). it is traveling at exactly 74 mph.

Answers:

➤ **(a).** A'

➤ **(b). B$'$**

➤ **(c). A∩B**

➤ **(d). A∩C**

⇑ *Refer back to* **9.1 - Example 3 & 9.1 - Solved Problem 3.**

9.1 - Problem 4: A television manufacturer receives its television tubes from three independent suppliers. From a shipment of tubes, a tube was tested for flaws. Assume **T** represents the event that a tube is tested and found flawed. Let **A**, **B**, and **C** represent the events that the tube came from one of these suppliers respectively. Write **T** as an expression of **T, A, B, C**.

Answer:

T = (T∩A)∪(T∩B)∪(T∩C)

⇑ *Refer back to* **9.1 - Example 4 & 9.1 - Solved Problem 4.**

9.1 - Problem 5: A die is tossed twice. Let S_1 represent the event that a six occurs on the first toss and S_2 represent the event that a six occurs on the second toss. Write S_2 in terms of S_1 and S_2.

Answer:

$S_2 = (S_1 \cap S_2) \cup (S_1' \cap S_2)$

⇑ *Refer back to* **9.1 - Example 5 & 9.1 - Solved Problem 5**.

Supplementary Problems

1. Restate the DeMorgan laws for three sets **A, B, C**.

2. Four cards are selected from an ordinary deck. If D_i is the event that the i th card drawn is a diamond, simplify the event $E = [(D_1 \cup D_2 \cup D_3 \cup D_4)] \cap [D_1 \cup (D_2' \cap D_3' \cap D_4')]$.

3. A die is tossed until two 6s' occur or four times, whichever occurs first. If A_j (j = 1, 2, 3, 4) are the events that a six occurs on the j th toss, express the event **E** that

a. three tosses occurred.

b. there was at most three tosses.

c. there was four tosses.

4. Write the expression $\{(A' \cap B') \cup [C \cup (D \cap E)]' \cup [F' \cap (G' \cup H')]\}'$ without complements.

For questions 5 & 6, apply the distributive laws:

5. $A \cap (B \cup C \cup D \cup E)$

6. $A \cup (B \cap C \cap D \cap E)$

For questions 7 & 8, prove the identity:

7. $(A \cup B) \cap (C \cup D) = (A \cap C) \cup (A \cap D) \cup (B \cap C) \cup (B \cap D)$

8. $(A \cap B) \cup (C \cap D) = (A \cup C) \cap (A \cup D) \cap (B \cup C) \cap (B \cup D)$

9. Assume $A \subseteq B \subseteq C \subseteq D \subseteq E$ then $(B \cap A') \cup (C \cap B') \cup (D \cap C') \cup (E \cap D')$ can be simplified to the intersection of two sets. Find this intersection.

10. An urn contains 6 red marbles and 4 blue marbles. Assume three marbles are drawn from the urn at random.

Let R_i be the event that the ith (i = 1, 2, 3) marble drawn is red. For problems a-d, express the events in terms of R_i, unions, intersection, and complements.

a. Exactly two of the marbles drawn are blue.

b. At most two of the marbles drawn are red.

c. Exactly two of the marbles are the same color.

d. A marble drawn on the second drawing is red.

11. Assume C is the event that a person will have a hamburger and D is the event that a person will have a frankfurter at a drive-in restaurant. Using C, D, unions, intersections and complements, express the following events:

a. a person will order either a hamburger or a frankfurter at this restaurant.

b. a person will order neither a hamburger nor a frankfurter at this restaurant.

c. a person will not order a hamburger at this restaurant.

d. a person will order a hamburger but not a frankfurter.

e. a person will order a frankfurter but not a hamburger.

f. a person will order one or the other but not both.

12. In a certain town, consider the events that a driver will receive one, two, three, four, or five or more tickets within one year. Assume $T1$, $T2$, $T3$, $T4$, $T5$ represent each of these events

respectively. Using these events as well as unions, intersections, and complements, express the event the driver, within one year will:

a. receive one or two traffic citations.

b. receive at most one traffic citation.

c. receive at least three traffic citations.

d. receive no traffic citations.

e. not get three traffic citations.

13. Ms. Jones loves to talk on the phone. Let Ck be the event that on a given day, she makes at least k phone calls (k=1, 2, 3, 4, 5,...). Express in your own words the meaning of the following events:

a. **C1**

b. **C5′**

c. **C7 ∩C8′**

d. **C10∪C1′**

e. **C10′∩C9**

f. **C10′∪C12**

g. **C1′**

h. **(C8′∪C15)′**

14. Three cards are drawn from a 52 card deck. Let K_i be the event that the ith card drawn is a king, let Q_i be the event that the ith card drawn is a queen; and let J_i be the event that the ith card drawn is a jack (i=1, 2, 3). Using unions, intersections and complements, express the following events:

a. all cards drawn are kings.

b. no cards drawn are kings.

c. at least two cards drawn are Queens.

d. exactly one Jack is drawn.

e. two cards drawn are Queens and one card is a Jack.

f. exactly two cards drawn are Queens.

g. exactly two cards drawn are Queens and the other is a Jack or a King.

h. the first two cards drawn are kings.

i. no face cards are drawn.

If **F** is the event that a student will get financial aid, **J** is the event that he will find a part-time job, and **G** is the event that he will graduate, express the following events:

15. A student who gets financial aid will also graduate or get a part-time job.

16. A student who gets financial aid will not graduate and will not get a part-time job.

17. A student will not graduate or will get a part-time job but will not get financial aid.

A machine is producing ball bearings. Each hour, three ball bearings are selected at random, one at a time. Let D_i (i = 1, 2, 3) represent the event that the i th ball bearing is defective. Write out the expression for the event **E**:

18. two ball bearings are defective.

19. at least one ball bearings is defective.

20. no ball bearings are defective.

21. the first two drawn are defective and the last drawn is not defective.

Assume three cards are drawn from an ordinary deck of cards. If K_i (i = 1, 2, 3) are the events that the ith card drawn is a king and Q_i (i = 1, 2, 3) are the events that the i th card drawn is a queen. Write out the event **E** that

22. no king or queen is drawn.

23. one card is a king and two are queens.

24. only one king and queen is selected.

25. only one queen and no kings are selected.

26. Two urns, containing several red and white marbles are sitting on a table. A marble is selected from the first urn and placed in the second urn. Then a marble is selected from the second urn. Let R_j
(j = 1, 2) represent the event that the marble selected from urn j is red. Write R_2 as an expression of **R_1 and R_2**
A recent survey of men was taken to find out their participation in the following sports: football,

baseball, and ice hockey. A member of this group is selected. Let **B** represent the event that he plays baseball, **F** represent the event that he plays football, and **I** represent the event that he plays ice hockey. Using these events along with unions, intersections and complements, find the following events 27 - 31 :

27. he only plays football and ice hockey.

28. he only plays baseball.

29. he only plays one of these sports.

30. he does not play any of these sports.

31. Express the event $(\mathbf{F} \cap \mathbf{I}') \cup (\mathbf{F}' \cap \mathbf{I})$.

32. Show $(\mathbf{A} \cup \mathbf{B}) \cap (\mathbf{C} \cup \mathbf{D}) = (\mathbf{A} \cap \mathbf{C}) \cup (\mathbf{A} \cap \mathbf{D}) \cup (\mathbf{B} \cap \mathbf{C}) \cup (\mathbf{B} \cap \mathbf{D})$.

33. Using the laws on sets, prove the following:

a. **Uniqueness of the complement:**

If $\mathbf{B} \cap \mathbf{A} = \phi$ and $\mathbf{B} \cup \mathbf{A} = \mathbf{S}$ then $\mathbf{B} = \mathbf{A}'$.

b. $\mathbf{A}'' = \mathbf{A}$

c. $\mathbf{A} \cap \mathbf{A} = \mathbf{A}$

d. $\mathbf{A} \cup \mathbf{A} = \mathbf{A}$

e. $\mathbf{S} \cup \mathbf{A} = \mathbf{S}$
f. $\mathbf{A} \cap \phi = \phi$

g. $\phi' = \mathbf{S}$

h. $\mathbf{S}' = \phi$

i. **DeMorgan Laws:**

$(\mathbf{A} \cup \mathbf{B})' = \mathbf{A}' \cap \mathbf{B}'$

$(\mathbf{A} \cap \mathbf{B})' = \mathbf{A}' \cup \mathbf{B}'$

j. **Anti-symmetric Law:** If $\mathbf{A} \subseteq \mathbf{B}$ and $\mathbf{B} \subseteq \mathbf{A}$ then $\mathbf{A} = \mathbf{B}$.

k. If $\mathbf{A} \subseteq \mathbf{B}$ then $\mathbf{B} = \mathbf{B} \cup \mathbf{A}$.

l. If $\mathbf{A} \subseteq \mathbf{B}$ then $\mathbf{A} = \mathbf{B} \cap \mathbf{A}$.

m. **Transitive Law:** If $\mathbf{A} \subseteq \mathbf{B}$ and $\mathbf{B} \subseteq \mathbf{C}$ then $\mathbf{A} \subseteq \mathbf{C}$.

n. **Complement Law:** If $\mathbf{A} \subseteq \mathbf{B}$ then $\mathbf{B}' \subseteq \mathbf{A}'$.

o. Assume **A** is an arbitrary set. Show that the null set $\phi \subseteq \mathbf{A}$.

34. A disjoint partition of $\mathbf{A} \cup \mathbf{B}$ is $\mathbf{A} \cup \mathbf{B} = (\mathbf{A} \cap \mathbf{B}') \cup (\mathbf{A}' \cap \mathbf{B}) \cup (\mathbf{A} \cap \mathbf{B})$.

a. Find a disjoint partition of $\mathbf{A} \cup \mathbf{B} \cup \mathbf{C}$, (Hint: Use a Venn diagram).

b. Assume

$A = \{1,2,3,4,\ a,g,m,d,j,\ \alpha,\gamma,\delta,\ \phi,\chi,\eta\}$,

$B = \{1,2,3,4,b,k,n,e,h,\ \beta,\varepsilon,\phi,\chi,\eta\}$,

$C = \{1,2,3,4,i,c,l,f,o,\ \alpha,\gamma,\delta,\beta,\varepsilon\}$.

Find the 7 subsets that make up the partition of $A \cup B \cup C$.

35. Using the laws of boolean algebra, show $(B \cap C) \cup (B \cap C') \cup (B' \cap C) \cup (B' \cap C') = S$.

36. Using the laws of boolean algebra, show

$(A \cap B \cap C) \cup (A \cap B \cap C') \cup (A \cap B' \cap C) \cup (A \cap (B' \cap C') = A$.

37. Show if $A \subseteq B$ then $A \subseteq B \cup C$.

38. If all proper subsets of a set A are subsets of a set B show A is a subset of B.

39. Assume 3 cards are drawn from an ordinary deck of cards. Let K_i ($i = 1,2,3$) the event that the i th card drawn is a king. Let

$E = (K_1 \cap K_2 \cap K_3) \cup (K_1' \cap K_2 \cap K_3) \cup (K_1 \cap K_2' \cap K_3) \cup (K_1 \cap K_2 \cap K_3') \cup (K_1' \cap K_2' \cap K_3) \cup (K_1' \cap K_2 \cap K_3') \cup$
$\cup (K_1 \cap K_2' \cap K_3') \cup (K_1' \cap K_2' \cap K_3')$

Show $E = S$.

40. Assume in a family of sets the following is true: if $A \subseteq B$ then $B \subseteq A$. Show this family of sets only contains the empty set ϕ.

10.1 - What is a Probability Sample Space?

A probability sample space is a sample space where each event **E** in the sample space has a number P(**E**) associated with it. The study of probability sample spaces allows for a systematic approach for solving probability problems. For the value P(**E**) we have the following laws:

1. $0 \leq P(\mathbf{E}) \leq 1$

2. $P(\mathbf{S}) = 1$

3. $P(\mathbf{E}') = 1 - P(\mathbf{E})$

4. $P(\mathbf{A} \cup \mathbf{B}) = P(\mathbf{A}) + P(\mathbf{B}) - P(\mathbf{A} \cap \mathbf{B})$

From these laws, one can show the following laws hold truth:

5. $P(\phi) = 0$

6. $P(\mathbf{E}) = 1 - P(\mathbf{E}')$

7. If $\mathbf{A} \cap \mathbf{B} = \phi$ then $P(\mathbf{A} \cup \mathbf{B}) = P(\mathbf{A}) + P(\mathbf{B})$

8. $P(\mathbf{A} \cap \mathbf{B}) = P(\mathbf{A}) + P(\mathbf{B}) - P(\mathbf{A} \cup \mathbf{B})$

To find the probability of an event P(**E**), the following systematic approach should be followed:

I. Express the event **E** as a Boolean expression of the events given.

II. Apply the above laws, to the Boolean expression.

The following examples and problems show how these laws are used to solve probability problems.

10.1 - Example 1: Ms. Jones receives at most 100 phone calls a day. Assume that the probability that she will receive at least 10 phone calls a day is 1/4.

(a). Find the probability that she will receive at most 9 calls a day.

(b). Interpret the meaning of the value P(**E**).

Solutions:

➤ **(a).**
I. First write out the sample space: **S** = {0, 1, 2, 3,..., 100}.

The event that she will receive at least 10 phone calls is **M** = {10, 11, 12, 13,..., 100}.

The event that she will receive at most 9 calls a day is **E** = {0, 1, 2,..., 9}.

We can write **E** in terms of **M** by seeing that **E** = **M**′ = {0, 1, 2, 3, 4, 5, 6, 7, 8, 9}.

II. To find the probability of P(**E**), we use the formula

$$P(\mathbf{E}) = P(\mathbf{M}') = 1 - P(\mathbf{M}) = 1 - \frac{1}{4} = \frac{3}{4}.$$

➤ **(b).**
On any day, there is a 75% chance that Ms. Jones will receive less than 10 calls a day.

10.1 - Example 2: Records of rain fall in Southern Iowa shows that the probability of rain fall on Monday is 0.65, on the following Tuesday is 0.70 and on both Monday and Tuesday, the chance of rain fall is 0.45.

(a). Find the probability that it will rain on Monday or the following Tuesday.

(b). Interpret the meaning of the value P(**E**).

Solutions:

➤ **(a).**
I. The sample space **S** consists of all possible weather conditions for Monday and the following Tuesday.

Let **M** represent the event that it will rain on Monday; Let **T** represent the event that it will rain on the following Tuesday; let **E** be the event that it will rain on Monday or the following Tuesday.

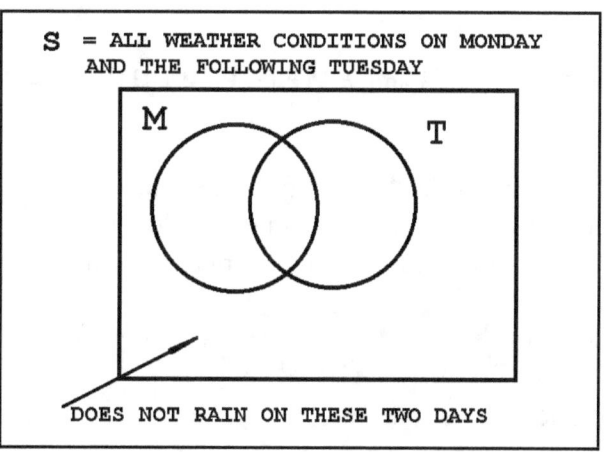

Therefore, $\mathbf{E} = \mathbf{M} \cup \mathbf{T}$.

II. From the example, we have $P(\mathbf{M}) = 0.65$, $P(\mathbf{T}) = 0.70$ and $P(\mathbf{M} \cap \mathbf{T}) = 0.45$.

From Step 2 and law 4,

$P(\mathbf{E}) = P(\mathbf{M} \cup \mathbf{T}) = P(\mathbf{M}) + P(\mathbf{T}) - P(\mathbf{M} \cap \mathbf{T}) = 0.65 + 0.70 - 0.45 = 0.90$.

➤ **(b).**
 For any week, there is a 90% chance that it will rain on Monday or Tuesday.

10.1 - Example 3: An urn contains red, white and blue marbles. A marble is selected at random. The chance of selecting a red marble is 1/3 and a white marble is 1/4. Find the probability that

(a). a red or white marble is selected.

(b). a blue marble is selected.

Solutions:

I. S = {red marble,white marble,blue marble}

➤ **(a).**
Let $\mathbf{R}$ stand for the event a red marble is selected; Let $\mathbf{W}$ stand for the event a white marble is selected. The event $\mathbf{E}$ that a red or white marble is selected can be written as $\mathbf{E} = \mathbf{R} \cup \mathbf{W}$.

Since both a red and white marble cannot be both selected, $\mathbf{R} \cap \mathbf{W} = \phi$.

II. Using law 7, we have $P(\mathbf{E}) = P(\mathbf{R} \cup \mathbf{W}) = P(\mathbf{R}) + P(\mathbf{W}) = \dfrac{1}{3} + \dfrac{1}{4} = \dfrac{7}{12}$.

➤ **(b).**
I. The event $\mathbf{E}$, a blue marble is selected can be written

$\mathbf{E} = \mathbf{W}' \cap \mathbf{R}' = (\mathbf{W} \cup \mathbf{R})'$, neither a white marble nor red marble was selected.

II. From laws 3 and 7, $P(\mathbf{E}) = P((\mathbf{W} \cup \mathbf{R})') = 1 - P(\mathbf{W} \cup \mathbf{R}) = 1 - \dfrac{7}{12} = \dfrac{5}{12}$.

10.1 - Example 4: A computer generates random whole numbers from 1 to 10. A study showed that the chance it will generate a random number from 1 to 5 is 0.60 and from 3 to 10 is 0.70. Find the probability that it will generate random numbers from 3 to 5 inclusive.

Solution:

I. The sample space is $\mathbf{S} = \{1, 2, 3, 4, 5, 6, 7, 8, 9, 10\}$.

Let **A** = {1, 2, 3, 4, 5}; **B** = {3, 4, 5, 6, 7, 8, 9, 10}.

The event **E** = {3, 4, 5} = A∩B.

II. Using law 8, P(**E**) = P(A∩B) = P(**A**) + P(**B**) - P(A∪B).

Since A∪B = {1, 2, 3, 4, 5}∪{3, 4, 5, 6, 7, 8, 9, 10} = **S**, we have P(A∪B) = P(**S**) = 1.

From the example, P(**A**) = 0.60, P(**B**) = 0.70 .

Therefore, P(**E**)= P(A∩B) = P(**A**) + P(**B**) - P(A∪B) = 0.60 + 0.70 - 1 = 0.30 .

10.1 - Example 5: Ms. Jones receives at least 10 calls a day.

Let **A** be the event she receives at most 15 calls a day;
Let **B** be the event she receives at least 12 calls a day;
Let **C** be the event she receives between 15 and 20 calls a day.

Assume: P(**A**) = 0.60, P(**B**) = 0.85, P(**C**) = 0.25, P(A∪C) = 0.40 .

Find the probability of the event **E** that

(a). she receives at least 16 calls a day.

(b). she receives 10 or 11 calls a day (inclusive).

(c). she receives between 12 and 15 calls a day (inclusive).

(d). she receives exactly 15 calls a day.

(e). she receives at least 16 calls a day or less than 12 calls a day.

Solutions:

➤ **(a).**
I. The sample space **S** = {10, 11, 12,...}.

A = {10, 11, 12, 13, 14, 15} and **A**′ = { 16, 17, 18,...}

E = {16, 17,...}

Therefore, **E** = **A**′.

II. P(**E**) = P(**A**′) = 1 - P(**A**) = 1 - 0.60 = 0.40 .

➤ **(b).**
I. B = {12, 13, 14,...}, **B**′ = {10, 11}

Therefore, $\mathbf{E} = \mathbf{B}'$.

II. $P(\mathbf{E}) = P(\mathbf{B}') = 1 - P(\mathbf{B}) = 1 - 0.85 = 0.15$.

➤ **(c).**
I. $\mathbf{A} = \{10, 11, 12, 13, 14, 15\}$, $\mathbf{B} = \{12, 13, 14, 15, 16,...\}$,

$\mathbf{E} = \{12, 13, 14, 15\}$

$\mathbf{A} \cap \mathbf{B} = \{12, 13, 14, 15\}$

Therefore, $\mathbf{E} = \mathbf{A} \cap \mathbf{B}$.

II. Using law 8, $P(\mathbf{E}) = P(\mathbf{A} \cap \mathbf{B}) = P(\mathbf{A}) + P(\mathbf{B}) - P(\mathbf{A} \cup \mathbf{B}) = 0.60 + 0.85 - P(\mathbf{A} \cup \mathbf{B})$.

Since $\mathbf{A} \cup \mathbf{B} = \mathbf{S}$,
$P(\mathbf{E}) = P(\mathbf{A} \cap \mathbf{B}) = P(\mathbf{A}) + P(\mathbf{B}) - P(\mathbf{A} \cup \mathbf{B}) = 0.60 + 0.85 - P(\mathbf{S}) = 0.60 + 0.85 - 1 = 0.45$.

➤ **(d).**
I. $\mathbf{A} = \{10, 11, 12, 13, 14, 15\}$, $\mathbf{C} = \{15, 16, 17, 18, 19, 20\}$

$\mathbf{E} = \mathbf{A} \cap \mathbf{C} = \{15\}$

II. Using law 8,

$P(\mathbf{E}) = P(\mathbf{A} \cap \mathbf{C}) = P(\mathbf{A}) + P(\mathbf{C}) - P(\mathbf{A} \cup \mathbf{C}) = 0.60 + 0.25 - 0.40 = 0.45$.

➤ **(e).**
I. $\mathbf{A}' = \{16, 17, ...\}$, $\mathbf{B}' = \{10,11\}$

$\mathbf{E} = \mathbf{A}' \cup \mathbf{B}'$

II. Since $\mathbf{A}' \cap \mathbf{B}' = \phi$, we use law 7:

$P(\mathbf{E}) = P(\mathbf{A}' \cup \mathbf{B}') = P(\mathbf{A}') + P(\mathbf{B}')$

By law 3,

$P(\mathbf{A}') = 1 - P(\mathbf{A}) = 1 - 0.60 = 0.4$

$P(\mathbf{B}') = 1 - P(\mathbf{B}) = 1 - 0.85 = 0.15$

Therefore, $P(\mathbf{E}) = P(\mathbf{A}') + P(\mathbf{B}') = 0.4 + 0.15 = 0.55$.

Solved Problems

10.1 - Solved Problem 1: Ms. Jones receives at most 100 phone calls a day. Assume that the probability that she will receive at least 1 phone call is 9/10.

(a). Find the probability that she will receive no calls.

(b). Interpret the meaning of the value P(**E**).

Solutions:

➤ (a).
I.: S = {0, 1, 2, 3,..., 100}

E = {0}

Let **M** = {1, 2, 3,..., 100}, **M**′ = {0}.

Therefore, **E** = **M**′.

II. $P(\mathbf{E}) = P(\mathbf{M}') = 1 - P(\mathbf{M}) = 1 - \dfrac{9}{10} = \dfrac{1}{10}$

➤ (b).
On any day, there is a 10% chance that Ms. Jones will receive no call a day.

10.1 - Solved Problem 2: A recent survey showed that 25% of all families in New York own dogs, 40% own cats and 10% own both dogs and cats.

(a). If a family from New York is selected at random, find the probability that the family owns dogs or cats.

(b). Assume a family from New York is randomly selected. Interpret the meaning of the value P(**E**).

Solutions:

➤ (a).
I. The sample space consists of all possible pets families in New York own.

Let **D** represent the event that the family owns dogs; Let **C** represent the event that the family owns cats ; let **E** be the event that the family owns dogs or cats.

Therefore, **E** = **D**∪**C**.

II. From the example, we have P(**D**) = 0.25, P(**C**) = 0.40 and P(**D**∩**C**) = 0.10.

From Step 2 and law 4,

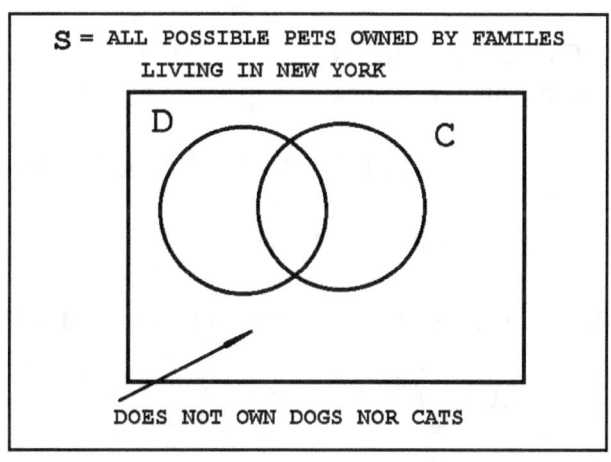

S = ALL POSSIBLE PETS OWNED BY FAMILES LIVING IN NEW YORK

DOES NOT OWN DOGS NOR CATS

$P(E) = P(D \cup C) = P(D) + P(C) - P(D \cap C) = 0.25 + 0.40 - 0.10 = 0.55.$

➤ **(b).** There is a 55% chance that the family will own a dog or a cat.

10.1 -Solved Problem 3: A statistics class has freshmen, junior, and senior students. 70% of the students are freshmen and 20% are seniors. A student is selected at random. Find the probability that

(a). a freshman or senior is selected.

(b). a junior is selected.

Solutions:

I. S = {freshmen, junior, senior}

➤ **(a).**
Let **F** stand for the event a freshman is selected; Let **S** stand for the event a senior is selected. The event E, a freshman or senior student is selected can be written as **E = F∪S**.

Since both a freshman and senior cannot be both selected, **F∩S = ϕ**.

II. Using law 7, we have $P(E) = P(F \cup S) = P(F) + P(S) = 0.70 + 0.20 = 0.90$.

➤ **(b).**
I. The event **J** a junior can be written **J = F′∩S′= (F∪S)′**, neither a freshman nor senior was selected.

II. From law 3 and law 7, $P(J) = P[(F \cup S)'] = 1 - P(F \cup S) = 1 - 0.90 = 0.10$.

10.1 - Solved Problem 4: A machine produces an important part to an automobile transmission. Each hour, it produces 10 such parts. A study showed that each hour, the chance it will produce at most 5 defective parts is 0.35 and more than 3 defective parts is 0.65. Find the probability that it will produce 4 or 5 defective parts.

Solution:

I. The sample space is **S** = {0, 1, 2, 3, 4, 5, 6, 7, 8, 9, 10}. This sample space represents the possible number of defective parts produced per hour.

Let **A** = {0, 1, 2, 3, 4, 5}, **B** = {4, 5, 6, 7, 8, 9, 10}.

The event **E** = {4, 5} = A∩B.

II. Using law 8, P(**E**)= P(A∩B) = P(A) + P(B) - P(A∪B).

Since A∪B = {0, 1, 2, 3, 4, 5}∪{4, 5, 6, 7, 8, 9, 10} = **S**, we have P(A∪B) = P(**S**) = 1 .

From the example, $P(A) = 0.35$, $P(B) = 0.65$.

Therefore, $P(E) = P(A \cap B) = P(A) + P(B) - P(A \cup B) = 0.35 + 0.65 - 1 = 0$.

10.1 - Solved Problem 5: The California Highway Patrol estimates that on a certain section of the San Diego freeway, highway patrolmen issue at least 100 speeding tickets a day.

let **A** be the event that at most 150 tickets are issued each day;
Let **B** be the event that at least 120 tickets are issued each day;
Let **C** be the event that between 150 and 200 tickets are issued each day.

Assume: $P(A) = 0.75$, $P(B) = 0.90$, $P(C) = 0.12$, $P(A \cup C) = 0.55$.

Find the probability of the event **E** that

(a). at least 151 tickets are issued a day.

(b). less than 120 tickets are issued a day.

(c). between 120 and 150 tickets are issued a day (inclusive).

(d). exactly 150 tickets are issued.

(e). more than 150 tickets are issued each day or less than 120 tickets are issued each day.

Solutions:

➤ **(a).**
I. The sample space $S = \{100, 101, 102,...\}$.

$A = \{100, 101,..., 150\}$, $A' = \{151, 152, 153,...\}$

$E = \{151, 152, 153,...\}$

Therefore, $E = A'$

II. $P(E) = P(A') = 1 - P(A) = 1 - 0.75 = 0.25$

➤ **(b).**
I. $B = \{120, 121, 122,...\}$, $B' = \{100, 101,..., 119\}$

$E = \{100, 101,..., 119\}$

Therefore, $E = B'$.

II. $P(E) = P(B') = 1 - P(B) = 1 - 0.90 = 0.10$

➤ **(c).**

I. B = {120, 121, 122,...}, **A** = {100, 101,..., 150}

E = {120, 121,..., 150}

A∩B = {120, 121,..., 150}

Therefore, **E** = **A∩B**.

II. Using law 8, P(**E**) = P(**A∩B**) = P(**A**) + P(**B**) - P(**A∪B**) = 0.75 + 0.90 - P(**A∪B**).

Since **A∪B** = **S** , P(**E**) = P(**A∩B**) = P(**A**) + P(**B**) - P(**A∪B**) = 0.75 + 0.90 - P(**S**) = 0.75 + 0.90 - 1 = 0.65 .

➤ **(d).**
I. A = {100, 101, 102,..., 150}, **C** = {150, 151,..., 200}

E = **A∩C** = {150}

II. Using law 8,

P(**E**) = P(**A∩C**) = P(**A**) + P(**C**) - P(**A∪C**) = 0.75 + 0.12 - 0.55 = 0.32 .

➤ **(e).**
I. More than 150 tickets are issued each day: **A′** = {151, 152, ...}.

Less than 120 tickets are issued each day: **B′** = {100,101, ..., 119}.

E = **A′∪B′**

II. Since **A′∩B′** = φ, we use law 7:

P(**E**) = P(**A′∪B′**) = P(**A′**) + P(**B′**).

By law 3,

P(**A′**) = 1 - P(**A**) = 1 - 0.75 = 0.25

P(**B′**) = 1 - P(**B**) = 1 - 0.90 = 0.10

Therefore, P(**E**) = P(**A′∪B′**) = P(**A′**) + P(**B′**) = 0.25 + 0.10 = 0.35 .

Unsolved Problems with Answers

10.1 - Problem 1: Ms. Romano's Latin class has 50 students. Assume that the probability that she will have at most 48 students at the end of the semester is 0.72.

(a). Find the probability that at the end of the semester, at least 49 students will finish the class.

(b). Interpret the meaning of the value P(**E**).

Answers:

➤ **(a).** 0.28

➤ **(b).** There is a 28% chance that at the end of the semester Ms. Romano's Latin class will have at least 49 students remaining.

⇑ *Refer back to* **10.1 - Example 1 & 10.1 - Solved Problem 1**.

10.1 - Problem 2: A computer manufacturer recently received a large shipment of computer chips. The chance that at least 10 chips are defective is 0.15; that between 5 and 20 chips are defective is 0.30 and between 10 and 20 chips are defective is 0.05.

(a). Find the probability that the box contains more than 4 defective chips.

(b). Interpret the value P(**E**).

Answers:

➤ **(a).** 0.40

➤ **(b).** There is a 40% chance that the box contains more than 4 defective chips.

⇑ *Refer back to* **10.1 - Example 2 & 10.1 - Solved Problem 2**.

10.1 - Problem 3: Jill is attending a local community college. To finish her degree she needs to take only one class from the following classes: American history, British history or Ancient Greek history. The chance that she will take American history is 0.53 and Ancient Greek history is 0.25. Find the probability that she will take
(a). British History.

(b). British History or Ancient Greek History.

Answers:

➤ **(a).** 0.22

➤ **(b).** 0.47

⇑ *Refer back to* **10.1 - Example 3 & 10.1 - Solved Problem 3**.

10.1 - Problem 4: A study of the World Historical Club estimates that a typical American history text book has many factual errors. The study further showed that the chance of at least 50 errors is 0.35 and the chance of less than 60 errors is 0.70. Find the probability that an American history book contains between 50 and 59 historical errors. (inclusive).

Answer: 0.05

⇑ *Refer back to* **10.1 - Example 4 & 10.1 - Solved Problem 4**.

10.1 - Problem 5: Johnny Roadster finished third in a local stock car race. The following events relate to his speed at the finish line where **S** = {100, 101, 102,..., 170}.

Let **A** be the event his speed was at most 160 mph.
Let **B** be the event his speed was at least 150 mph.
Let **C** be the event his speed was between 160 and 170 mph.

Assume: $P(\mathbf{A}) = 0.90$, $P(\mathbf{B}) = 0.85$, $P(\mathbf{C}) = 0.25$.

Find the probability of the event **E** that his speed was

(a). more than 160 mph.

(b). less than 150 mph.

(c). between 150 and 159 mph (inclusive).

(d). exactly 160 mph.

(e). less than 150 mph or between 160 and 170 mph.

Answers:

➤ **(a)**. 0.10

➤ **(b)**. 0.15

➤ **(c)**. 0.60

➤ **(d)**. 0.15

➤ **(e)**. 0.40

⇑ *Refer back to* **10.1 - Example 5 & 10.1 - Solved Problem 5**.

10.2 - Defining the Probability of Events when Elements of the Sample Space are equally likely to Occur.

Assume an experiment is to be defined where the sample space is finite and each element of the sample space has an equal chance to occur. The probability of an event **E** is defined as

$$P(E) \;=\; \frac{\#E}{\#S}.$$

Since each element of the event E has equal chance of being selected, this ratio gives the best numerical value to represent the probability that the event **E** will occur.

10.2 - Example 1: A fair coin is tossed once. Find the probability that a head occurs.

Solution:

I. The sample space is **S** = {h, t} where h stands for heads occurs and t for tails occurs.

Let **E** = {t}.

#E = 1 and **#S** = 2.

II. Since both elements t and h are equally likely to occur, P(E) = $\dfrac{\#E}{\#S}$ = $\dfrac{1}{2}$.

10.2 - Example 2: A fair coin is tossed twice. Find the following probability that:

(a). a head and tail occurs.

(b). at least one head occurs.

(c). two sides are the same.

Solutions:

I. S = {(h,h), (h,t), (t,h), (t,t)}

➤ **(a).**
E = {(h,t), (t,h)}

#E = 2 and **#S** = 4.

II. Since each element of the sample space has equal chance to occur,

$$P(E) \;=\; \frac{\#E}{\#S} \;=\; \frac{2}{4} \;=\; \frac{1}{2}.$$

➤ **(b).**
I. The event "at least one heads occurs" can be written **E** = {(h,t), (t,h), (h,h)}

II. Since each element of the sample space has equal chance to occur,

$$P(E) = \frac{\#E}{\#S} = \frac{3}{4}.$$

➤ **(c).**

I. The event "two sides are the same" can be written

$$E = \{(t,t), (h,h)\}$$

II. $P(E) = \dfrac{\#E}{\#S} = \dfrac{2}{4} = \dfrac{1}{2}.$

10.2 - Example 3: A local reading club has 25 members. A survey was taken as to their reading preferences. The following is the result of this survey:

Eleven members like historical novels.

Ten members like mysteries.

Ten members like romance novels.

Three members like history and romance novels.

Five members like mysteries and romance novels.

Six members like history novels and mysteries.

One member like all three.

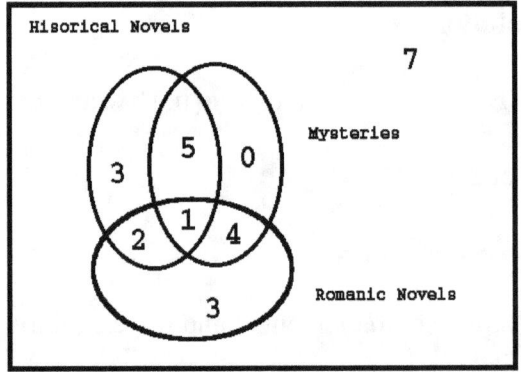

Assume a member is selected at random. Find the probability that the member:

(a). only likes history novels.

(b). likes mysteries and romance but does not like history novels.

(c). only likes one of these types of novels.

(d). likes history or romance novels but does not like mysteries.

(e). does not like any of these types of books.

Solutions:

We use the above Venn Diagram.

➤ **(a).**
Here we know that #S = 25. The event **E** "that the member only likes historical novels" has cardinality

#E = 3. Since each member of the club has equal chance of being selected,

$$P(E) = \frac{\#E}{\#S} = \frac{3}{25}.$$

➤ **(b).**
The event **E** "likes mysteries and romance but does not like history novels" has cardinality #**E** =

4 . Since each member of the club has equal chance of being selected,

$$P(E) = \frac{\#E}{\#S} = \frac{4}{25}.$$

➤ **(c).**
The event **E** "only likes one of these types of books" has cardinality #**E** = 3 + 3 + 0 = 6. Since each member of the club has equal chance of being selected,

$$P(E) = \frac{\#E}{\#S} = \frac{6}{25}.$$

(d).
The event **E**: "likes history or romance novels but does not like mysteries" has cardinality
#**E** = 3 + 2 + 3 = 8

Since each member of the club has equal chance of being selected,

$$P(E) = \frac{\#E}{\#S} = \frac{8}{25}.$$

(e).
The event **E**: "does not like any of these types of books" has cardinality #**E** = 7. Since each member of the club has equal chance of being selected,

$$P(E) = \frac{\#E}{\#S} = \frac{7}{25}.$$

10.2 - Example 4: One card is drawn from any ordinary deck of cards. Find the probability that it is a king or queen.

Solution:

I. K: The event a king is drawn.
Q: The event a queen is drawn.
E: The event a king or queen is drawn.

E = K∪Q

II. P(**K**) = 4/52

P(**Q**) = 4/52

Since $K \cap Q = \phi$, apply law 7: $P(E) = P(K \cup Q) = P(K) + P(Q) = 4/52 + 4/52 = 8/52$.

Solved Problems

10.2 - Solved Problem 1: An urn contains three marbles colored red, white and blue. A marble is selected at random. Find the probability that a red or white marble is selected.

Solution:

I. The sample space is $S = \{r, w, b\}$ where r stands for a red marble is being selected, w stands for a white marble being selected and b stands for a blue marble being selected.

Let $E = \{w, r\}$. Then $\#E = 2$ and $\#S = 3$.

II. Since each marble in the urn has equal chance of being selected,

$$P(E) = \frac{\#E}{\#S} = \frac{2}{3}.$$

10.2 - Solved Problem 2: A die is tossed twice. Find the following probability that:

(a). a one and six occurs.

(b). at least one three occurs.

(c). both numbers are the same.

Solutions:

I. S = {(1,1), (1,2), (1,3), (1,4), (1,5), (1,6),
        (2,1), (2,2), (2,3), (2,4), (2,5) ,(2,6),
        (3,1), (3,2), (3,3), (3,4), (3,5), (3,6),
        (4,1), (4,2), (4,3), (4,4), (4,5), (4,6),
        (5,1), (5,2), (5,3), (5,4), (5,5), (5,6),
        (6,1), (6,2), (6,3), (6,4), (6,5), (6,6)}

➤ **(a).**
The number 1 can occur on the first toss and 6 on the second toss **or** the number 6 can occur on the first toss and 1 on the second toss. Therefore, $E = \{(1,6), (6,1)\}$, $\#E = 2$ and $\#S = 36$.

II. Since each pair of numbers has equal chance of occurring,

$$P(E) = \frac{\#E}{\#S} = \frac{2}{36}.$$

➤ **(b).**
II. The event "at least one three occurs" can be written

$E = \{(3,1), (3,2), (3,3), (3,4), (3,5), (3,6), (1,3), (2,3), (4,3), (5,3), (6,3)\}$.

#E = 11.

II. Since each pair of numbers has equal chance of occurring,

$$P(E) = \frac{\#E}{\#S} = \frac{11}{36}.$$

➤ **(c).**

I. The event "both numbers are the same" can be written

E = {(1,1), (2,2), (3,3), (4,4), (5,5), (6,6)}.

#E = 6

II. Since each pair of numbers has equal chance of occurring,

$$P(E) = \frac{\#E}{\#S} = \frac{6}{36}.$$

10.2 - Solved Problem 3: A recent survey of 150 men was taken to find out their participation in the following sports: football, baseball, and ice hockey. From the survey, the following information was given:

13 men participated in all three sports.

20 men participated in football and baseball.

30 men participated in football and ice hockey.

22 men participated in ice hockey and baseball.

45 men participated in football.

48 men participated in ice hockey.

30 men participated in baseball.

Assume a man is selected at random. Find the probability that he:

(a). only likes football.

(b). likes ice hockey and baseball but does not like football.

(c). only likes one of these sports.

(d). likes football or baseball but does not like ice hockey.

(e). does not like any of these sports.

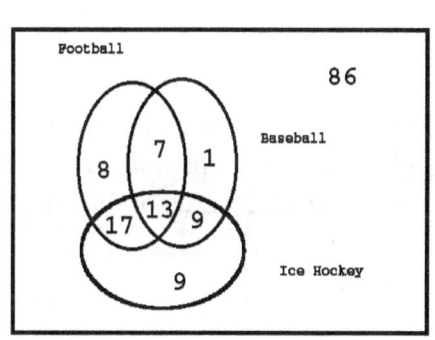

Solutions:

We use the Venn Diagram

➤ **(a).**

From the statement of the problem #**S** = 150. The event **E** "that he only like football" has cardinality #**E** = 8. Since each man has equal chance of being selected,

$$P(E) = \frac{\#E}{\#S} = \frac{8}{150}.$$

➤ **(b).**

The event **E** "he likes ice hockey and baseball but does not like football" has cardinality #**E** = 9. Since each man has equal chance of being selected,

$$P(E) = \frac{\#E}{\#S} = \frac{9}{150}.$$

➤ **(c).**

The event **E** "he likes only one of these sports" has cardinality #**E** = 8 + 1 + 9 = 18. Since each man has equal chance of being selected,

$$P(E) = \frac{\#E}{\#S} = \frac{18}{150}.$$

➤ **(d).**

The event **E** "likes football or baseball but does not like ice hockey has cardinality " #**E** = 8 +7 + 1 = 16 . Since each man has equal chance of being selected,

$$P(E) = \frac{\#E}{\#S} = \frac{16}{150}.$$

➤ **(e).**

The event **E** "does not like any of these sports" has cardinality #**E** = 86. Since each man has equal chance of being selected,

$$P(E) = \frac{\#E}{\#S} = \frac{86}{150}.$$

10.2 - Solved Problem 4: One card is drawn from an ordinary deck of cards. Find the probability that it is a king or diamond.

Solution:

I.
K: The event that a king is drawn.
D: The event that a diamond is drawn.
E: The event that a king or diamond is drawn.

E = K∪D

II. P(**K**) = 4/52

P(**D**) = 13/52

Since **K∩D** ≠ φ, apply law 4: P(**E**) = P(**K∪D**) = P(**K**) + P(**D**) - P(**K∩D**) = 4/52 + 13/52 - 1/52 = 16/52.

Unsolved Problems with Answers

10.2 - Problem 1: A card is randomly selected from an ordinary deck of cards. Find the probability that a diamond is selected.

Answer:

$$\frac{13}{52}$$

⇑ *Refer back to* **10.2 - Example 1 & 10.2 - Solved Problem 1.**

10.2 - Problem 2: A coin is tossed three times. Find the following probability that:

(a). two heads and a tail occurs.

(b). at least one tail occurs.

(c). all sides are the same.

Answers:

➤ (a). $\dfrac{3}{8}$

➤ (b). $\dfrac{7}{8}$

➤ (c). $\dfrac{2}{8}$

⇑ *Refer back to* **10.2 - Example 2 & 10.2 - Solved Problem 2.**

10.2 - Problem 3: A survey of 200 farmers produced the following results:

120 raised chickens.
104 raised hogs.
94 raised sheep.
66 raised both chickens and hogs.
57 raised both chickens and sheep.

47 raised both hogs and sheep.
10 raised all three.

A farmer is selected at random. Find the probability that the farmer

(a). only raises sheep.

(b). raises sheep and hogs but not chickens.

(c). only raises one of these animals.

(d). raises sheep or hogs but not chickens.

(e). does not raise any of these animals.

Answers:

➤ **(a).** 0

➤ **(b).** $\dfrac{37}{200}$

➤ **(c).** $\dfrac{8}{200}$

➤ **(d).** $\dfrac{38}{200}$

➤ **(e).** $\dfrac{42}{200}$

⇑ *Refer back to* **10.2 - Example 3 & 10.2 - Solved Problem 3.**

10.2 - Problem 4: One card is drawn from an ordinary deck of cards. Find the probability that it is a diamond or face card.

Answer:

22/52

⇑ *Refer back to* **10.2 - Example 4 & 10.2 - Solved Problem 4**.

Supplementary Problems

A roulette wheel has numbers 0,1,2,.., 36 and 00. Each time the wheel is spun, a ball falls on one of these numbers. Assume we spin the wheel three times. Find the following probabilities:

1. 00 appears all three times.

2. all the numbers are the same.

3. only two fives appear.

A survey of local teenagers produced the following results:

54 like Burger King.
28 like McDonald's.
36 like Wendy's.
13 like McDonald's and Wendy's.
18 like Wendy's and Burger King.
12 like McDonald's and Burger King.
5 like all three.
20 do not like any of the three.

Assume a teenager is selected at random. Find the probability

4. that the teenager likes McDonald's but does not like Burger King.

5. that the teenager likes McDonald's or does not like Burger King.

6. Given that the teenager likes Burger King and Wendy's, the teenager also likes McDonald.

7. Given that the teenager likes Burger King or Wendy's, the teenager does not like McDonald's

A die is tossed twice. Find the probability

8. that the sum of the numbers is either greater than 10 or equal to four.

9. the sum of the numbers is between 2 and 4 inclusive.

In September, 1988, the House of Representatives voted on an amendment requiring life imprisonment for drug-related murders. Results of the vote were reported as shown below:

	YEA	NAY	DID NOT VOTE	Total
Democrat	153	83	19	255
Republican	169	0	8	177
Total	322	83	27	432

10. What is the probability that a randomly selected representative voted for the amendment?

11. What is the probability that a randomly selected representative is both a Democrat and voted

in favor of the amendment?

12. Given that a representative voted for the amendment, what is the probability that he/she is a Democrat?

13. Show the formula

a. $P(A \cup B \cup C) = P(A) + P(B) + P(C) - P(B \cap C) - P(A \cap B) - P(A \cap C) + P(A \cap B \cap A \cap C)\}$.

b. A game of pool is being played. There are 10 balls on the table, each numbered 1 through 10 respectively. Assume a ball is randomly shot into a pocket. Use the above formula to find the probability that the ball is an odd number or a number between 5 to 9 (inclusive) or a number between 3 to 7 (inclusive).

14. Show that $P(A_1 \cup A_2 \cup A_3 \cup ... \cup A_n) \leq P(A_1) + P(A_2) + P(A_3) + ... + P(A_n)$.

15. Assume an experiment creates a sample space **S** and $P(A) = 0.60$, $P(B) = 0.5$ and $P(A \cup B) = 0.65$. Using combinations of **A**, **B**,$\cap$,$\cup$,$'$ find the probabilities of all possible non-empty events.

16. If $A \subseteq B$ show $P(A) \leq P(B)$.

17. If $P(A) \leq P(B)$ show $P(B') \leq P(A')$.

18. For events **A**,**B**,**C**, show $P(A \cap C') \leq P(A \cap B') + P(B \cap C')$.

19. For events **A**, **B**, **C**, show $P(A \cap B) \geq P(A \cap B \cap C)$.

20. For the two events **A** and **B**, show that $P(A) + P(B) - 2P(A \cap B)$ is the probability that only 1 of these events occurs.

21. Assume $P(A) = 1/3$ and $P(B) = 3/4$. Show $1/12 \leq P(A \cap B) \leq 1/3$.

22. Explain why the following statements are true or false.

a. For all **A**,**B**, if $P(A) = 0$, then $A = \varphi$.

b. For a sample space if $P(A) = 0.76$ and $P(B) = 0.92$ can $P(A \cup B) \leq 1$?

c. For a sample space, if $P(A \cap B) > 0$, can $P(A) = 0.76$ and $P(B) = 0.92$?

d. For all **A**,**B** , if $A \subseteq B$ and $P(A) = P(B)$ then $A = B$

e. If $A \cap B = \varphi$, can $P(A) = 0.76$ and $P(B) = 0.92$?

f. Can $P(A) = 0.26$ and $P(B) = 0.52$ and $P(A \cap B) = 0.80$?

23. Assume $P(A) = P(B) = P(C) = 0.50$, $P(A \cap B) = P(A \cap C) = P(B \cap C) = 0.25$, $P(A \cap B \cap C) = 0.125$.

a. Show $A \cap B = (A \cap B \cap C') \cup (A \cap B \cap C)$ (Hint: use Anti-symmetric law: if $M \subseteq N$ and $N \subseteq M$

then $M = N$).

b. Show $A = (A \cap B' \cap C') \cup (A \cap (B \cup C))$.

c. Find $P(A \cap B \cap C')$, $P(A \cap B' \cap C)$, $P(A' \cap B \cap C)$, $P(A \cap B' \cap C')$, $P(A' \cap B \cap C')$, $PA' \cap B' \cap C)$, $P(A' \cap B' \cap C')$.

24. Show $P(A_1 \cup A_2 \cup ... \cup A_n) = 1 - P(A_1' \cap A_2' \cap ... \cap A_n')$.

For the following problems assume a sample space **S** and various events **A, B, C** etc.

25. Assume $A \cap B = A \cap C = B \cap C = \varphi$. Show $P(A \cup B \cup C) = P(A) + P(B) + P(C)$.

26. Assume $A \cap (B \cup C) = \varphi$. Show $P(A \cup B \cup C) = P(A) + P(B) + P(C) - P(B \cap C)$.

27. Show $P(B \cap C) + P(B' \cap C) + P(B \cap C') + P(B' \cap C') = 1$.

28. Show $P(A \cap B \cap C) + P(A \cap B' \cap C) + P(A \cap B \cap C') + P(A \cap B' \cap C') = P(A)$.

29. Assume $A' \cap B' = \varphi$. Show $P(A) + P(B) = 1 + P(A \cap B)$.

30. Give an example of a sample space, and events **A, B** where $A' \cap B' = \varphi$.

11.1 - Counting Tasks

Assume two tasks have to be performed. If task 1 can be done in n ways and task 2 can be done in m ways then task1 and task2 can be done in mxn ways. This method of counting can be extended to N tasks:

> Assume r tasks have to be performed. If task 1 can be done in N_1 ways, task 2 can be done in N_2 ways, task 3 can be done in N_3 ways, etc., then all these tasks can be done in $N_1 x N_2 x...x N_r$ different ways.

11.1 - Example 1: Mr. and Ms. Smith are expecting a new born girl. In deciding the name of the child, Mr. Smith makes a list of ten first names that he likes while Ms. Smith makes a list of five middle names that she likes. Combining these lists together, find the number of possible names that can be given to the child.

Solution:

Step 1: Mr. Smith has the task of making up a list of ten first names: m = 10.

Step 2: Mrs. Smith has the task of making up a list of five middle names: n = 5.

Step 3: Combining these lists together give a possible number of names: mxn = (10)(5) = 50.

11.1 - Example 2: At a local college, Ms. Roman teaches three sections of Latin. The morning section has 25 students, the afternoon section has 32 students and the evening section has 76 students. She needs three students, one from each section, to help her on a special project. How large a list can she make up to select from?

Solution:

Step 1: From the morning section. she can make up a list of 25 students' names: $N_1 = 25$.

Step 2: From the afternoon section, she can make up a list of 32 students' names: $N_2 = 32$.

Step 3: From the evening section, she can make up a list of 76 students' names: $N_3 = 76$.

Step 4: Combining these lists, she can select from a list of $(25)(32)(76) = 60,800$.

Solved Problems

11.1 - Solved Problem 1: Five West coast football teams are to play 8 East coast football teams. Each team will play only once. Find the number of games that will be played.

Solution:

Step 1: Each West coast football team is to play each of the East coast team: $m = 5$.

Step 2: Each East coast football team is to play each of the West coast team: $n = 8$.

Step 3: Combining these teams together give $5 \times 8 = 40$ games.

11.1 - Solved Problem 2: Mr. Klien orders a pizza, salad, drink and appetizer from a fast food restaurant. The menu has 10 different kinds of pizzas, 6 different salads, 5 different drinks and 4 different appetizers. Find the total number of possible meals Mr. Klien can order.

Step 1: From the menu, he has 10 different possible pizzas to select from: $N_1 = 10$.

Step 2: From the menu. he has 6 different possible salads to select from: $N_2 = 6$.

Step 3: From the menu, he has 5 different possible drink to select from: $N_3 = 5$.

Step 4: From the menu, he has 4 different possible appetizers to select from: $N_4 = 4$.

Step 5: Combining these selections, there are $(10)(6)(5)(4) = 1,200$ meals.

Unsolved Problems with Answers

11.1 - Problem 1: Two urns sit on a table. One urn contains ten red marbles numbered 1 to 10 respectively and the other urn contains 15 blue marbles numbered 1 to 15. One marble is selected from each urn. Find the total number of possible selections.

Answer:

150

⇑ *Refer back to* **11.1 - Example 1** & **11.1 - Solved Problem 1**.

11.1 - Problem 2: This evening, Ms. Besch is going to attend a new production of Mozart's opera, The *Magic Flute*. From her wardrobe, she has the following selections: 12 hats, 15 pairs of shoes and 20 dresses. Find the total number of possible outfits she can select.

Answer:

3,600

⇑ *Refer back to* **11.1 - Example 2 & 11.1 - Solved Problem 2**.

11.2 - What is Sampling With Replacement From a Population?

A population is a collection of numeric or non-numeric data. Sampling with replacement is whenever an item is selected from a given population, we return the item each time to the population. A sample of size r is made up of r selections from the population. The formula for counting the number of possible samples of this type is

$$(n)(n)(n)...(n) = n^r$$

where n is the total number of items in the population and r is the number of selections. From such sampling, we can generate a sample space **S**, whose elements are the samples generated. The cardinality of **S** is #**S** = n^r.

11.2 - Example 1: An urn contains two marbles colored red and white respectively. Three marbles are sampled with replacement.

(a). List the original population.

(b). Using the above formula, find the number of possible samples and the cardinality of the sample space.

(c). List the sample space.

(d). Find the probability that all the marbles selected are red.

(e). Find the probability that all the marbles selected are the same color.

(f). Find the probability that two red and one white marbles are selected.

Solutions:

➤ (a).
The original population is red (r), white (w).

➤ (b).
Step 1: For the first marble selected, we have two possibilities: n = 2.

Step 2: For the second marble selected, we have two possibilities: n = 2.

Step 3: For the third marble selected, we have two possibilities: n = 2.

Therefore, the number of selections is r = 3 and the number of possible samples possible is 2^3 = 8. The cardinality of the sample space is also 8.

➤ **(c).**
Using the notation r for red and w for white, we have the sample space:

S = {(r,r,r), (w,r,r), (r,w,r), (r,r,w), (w,w,r), (w,r,w), (r,w,w), (w,w,w)}.(Note: **#S** = 2^3 = 8).

➤ **(d)..**
The event, "all the marbles are red" is **E** = {(r,r,r)}. Therefore,

$$P(E) = \frac{\#E}{\#S} = \frac{1}{8}.$$

➤ **(e).**
The event, "all the marbles are the same color" is **E** = {(r,r,r), (w,w,w)}. Therefore,

$$P(E) = \frac{\#E}{\#S} = \frac{2}{8}.$$

➤ **(f).**
The event, "the marbles selected are two red and one white" is **E** = {(r,r,w), (r,w,r), (w,r,r)}. Therefore,

$$P(E) = \frac{\#E}{\#S} = \frac{3}{8}.$$

11.2 - Example 2: Mr. Edsel recently received a four digit (0-9) secret code to access his brokerage account. Find the probability that his code consists of 3 fours and 1 two.

Solution:

The original population is 0, 1, 2, 3, 4, 5, 6, 7, 8, 9.

Step 1: For the first number in the code, we have ten possibilities: n = 10.

Step 2: For the second number in the code, we have ten possibilities: n = 10.

Step 3: For the third number in the code, we have ten possibilities: n = 10.

Step 4: For the fourth number in the code, we have ten possibilities: n = 10. Therefore, the number of selections is r = 4 and the number of samples in the sample space is **#S** = 10^4 = 10,000.

Step 5: The event " that his code consists of three fours and one two" is

$E = \{(4,4,4,2), (4,4,2,4), (4,2,4,4), (2,4,4,4)\}$.

Step 6: $P(E) = \dfrac{\#E}{\#S} = \dfrac{4}{10000} = \dfrac{1}{2500}$.

Solved Problems

11.2 - Solved Problem 1: Each day, Ms. Smith receives several phone calls from Germany, France, and England. On Christmas day, she received two calls.

(a). List the original population space.

(b). Using the above formula, find the number of possible samples and cardinality of the sample space.

(c). List the sample space.

(d). Find the probability that all the calls came from Germany.

(e). Find the probability that all the calls came from the same country.

(f). Find the probability that one call came from Germany and one from England.

Solutions:

Let e represent England, f represent France, and g represent Germany.

➤ **(a).**
The original population is e, f, g.

➤ **(b).**
Step 1: For the first call, we have three possibilities: n = 3.

Step 2: For the second call, we have three possibilities: n = 3.

Therefore, the number of calls is r = 2 and the number of samples possible is $3^2 = 9$ and the cardinality of this sample space is also 9.

➤ **(c).**
Using the above notation, we have the sample space:

$S = \{(e,e), (e,f), (e,g), (f,e), (f,f), (f,g), (g,e,), (g,f), (g,g)\}$.

.Note: $\#S = 3^2 = 9$.

➤ **(d).**
The event "all the calls come from Germany is" $E = \{(g,g)\}$ Therefore,

$$P(E) = \frac{\#E}{\#S} = \frac{1}{9}.$$

➤ **(e).**

The event, "all the calls came from the same country" is $E = \{(g,g), (f,f), (e,e)\}$. Therefore,

$$P(E) = \frac{\#E}{\#S} = \frac{3}{9}.$$

➤ **(f).**

The event, "one call came from Germany and one from England" is $E = \{(g,e), (e,g)\}$. Therefore,

$$P(E) = \frac{\#E}{\#S} = \frac{2}{9}.$$

11.2 - Solved Problem 2: In a recent election, Mr. Jones, Ms. Thorton and Mrs. Besch ran for Governor of the State. In a survey, 5 voters were asked which of the 3 candidates they voted for. Find the probability that four of them voted for Mrs. Besch and one for Mr. Jones.

Solution:

Let j stand for Mr. Jones, t for Ms. Thorton and b for Mrs. Besch.

The original population is j, t, b.

Step 1: For the first voter interviewed , we have 3 possibilities: n = 3.

Step 2: For the second interviewed, we have 3 possibilities: n = 3.

Step 3: For the third interviewed , we have 3 possibilities: n = 3.

Step 4: For the fourth interviewed, we have 3 possibilities: n = 3.

Step 5: For the fifth interviewed, we have 3 possibilities: n = 3.

Therefore, the number of interviews is r = 5 and the number of samples in the sample space is

$\#S = 3^5 = 243$.

Step 6: The event " four of them voted for Mrs. Besch and one for Mr. Jones " is

$E = \{(b,b,b,b,j), (b,b,b,j,b), (b,b,j,b,b), (b,j,b,b,b), (j,b,b,b,b)\}$.

Step 7: $P(E) = \dfrac{\#E}{\#S} = \dfrac{5}{243}$

Unsolved Problems with Answers

11.2 - Problem 1: A fair die is tossed 2 times.

(a). List the original population.

(b). Using the above formula, find the number of samples and cardinality of the sample space.

(c). List the sample space.

(d). Find the probability that all the tosses resulted in even numbers.

(e). Find the probability that all the tosses resulted in the same numbers.

(f). Find the probability that the sum of the two numbers equals 5.

Answers:

➤ **(a).** 1, 2, 3, 4, 5, 6

➤ **(b).** 36

➤ **(c).**
S = {(1,1), (1,2), (1,3), (1,4), (1,5), (1,6), (2,1), (2,2), (2,3), (2,4), (2,5), (2,6),
      (3,1), (3,2), (3,3), (3,4), (3,5), (3,6), (4,1), (4,2), (4,3), (4,4), (4,5), (4,6),
      (5,1), (5,2), (5,3), (5,4), (5,5), (5,6), (6,1), (6,2), (6,3), (6,4), (6,5), (6,6)}

➤ **(d).** 9/36

➤ **(e).** 6/36

➤ **(f).** 4/36

⇑ *Refer back to* **11.2 - Example 1 & 11.2 - Solved Problem 1**.

11.2 - Problem 2: At a certain downtown intersection, six drivers were stopped for speeding or driving under the influence. Find the probability that five drivers were stopped for speeding and one for driving under the influence.

Answer:

6/64

⇑ *Refer back to* **11.2 - Example 2 & 11.2 - Solved Problem 2**.

11.3 - What is Sampling Without Replacement?

Sampling without replacement is whenever an item is selected from an original population , we do not return the item each time to the population. The formula for counting the number of possible samples of this type is:

$$(n)(n - 1)(n - 2)...(n - r + 1),$$

where n is the total number of items in the original population and r is the number of selections. From such sampling, we can generate a sample space **S**, whose elements are the samples generated. The cardinality of **S** is #**S** = $(n)(n - 1)(n - 2)...(n - r + 1)$.

11.3 - Example 1: Three books are lying on a table. Three students took a book home.

(a). List the original population space.

(b). List the sample space.

(c). Find the number of samples and the cardinality of the sample space.

Solutions:

Let a, b, c represent the three books respectively.

➤ **(a).**
The original population is a, b, c.

➤ **(b).**
The sample space is **S** = {(a,b,c), (a,c,b), (b,a,c), (b,c,a), (c,a,b), (c,b,a)}

Note: (a, b, c) means that the first student took book a, the second student took book b and the third student took book c.

➤ **(c).**
Each triplet in the sample space represents a single element or sample. Using the above formula where n = 3 and r = 3, #**S** = (3)(2)(1) = 6, the number of samples as well as the cardinality of the sample space.

11.3 - Example 2: In Ms. Hayashi's Senior English class, three students were asked to read certain sections from Hamlet. Assuming her class has 15 students and the students were selected at random,

(a). Find the number of possible samples and the cardinality of the sample space.

(b). If there are ten boys in the class, find the probability that all three students selected were boys.

Solution:

➤ **(a).**
For this problem, n = 15 and r = 3.

Step 1: For the first student selected, we have 15 possibilities: n = 15.

Step 2: For the second student selected, we have 14 possibilities: n - 1 = 14.

Step 3: For the third student selected, we have 13 possibilities: n - 2 = 13.

Therefore, the total number of samples is (15)(14)(13) = 2,730 and **#S** = 2,730.

➤ **(b).**
For this problem, n = 10, r = 3.

Step 1: For the first boy selected, we have 10 possibilities: n = 10.

Step 2: For the second boy selected, we have 9 possibilities: n - 1 = 9.

Step 3: For the third boy selected, we have 8 possibilities: n - 2 = 8.

Therefore, the total number of samples is (10)(9)(8) = 720.

Step 4: The event "all three students selected were boys", **E** is made up of these 720 possible samples. Therefore,

$$\#E = 720 \text{ and } P(E) = \frac{720}{2730}.$$

11.3 - Example 3: Two cards are drawn from an ordinary deck of cards without replacement. Find the probability that the first card drawn is a king or the second card drawn is also a king.

Solution:

S: The sample space.
K$_1$: The event that the first card is a king.
K$_2$: The event that the second card is a king.

Solved Problems

11.3 - Solved Problem 1: At a magic show, Spingali handed two kings to two members of the audience.
(a). List the original population.

(b). List the sample space.

(c). Find the number of samples and cardinality of the sample space.

Solutions:

Let kh, kd, ks, kc represent the king of hearts, diamonds, spades, and clubs respectively.

➤ **(a).**
The original population is kh, kd, ks, kc.

➤ **(b).**
The sample space is

{(kh,kd), (kh,ks), (kh,kc), (kd,kh), (kd,ks), (kd,kc), (ks,kh), (ks,kd), (ks,kc), (kc,kh), (kc,kd), (kc,ks)}.

➤ **(c).**
Here, n = 4 kings and r = 2 kings selected. Using the above formula, **#S** = (4)(3) = 12.

11.3 - Solved Problem 2: Mr. Kahn was dealt three cards at random from an ordinary deck.

(a). find the number of possible samples and cardinality of the sample space.
(b). Find the probability that all three cards selected are kings.

Solutions:

➤ **(a).**
For this problem, n = 52 and r = 3.

Step 1: For the first card selected, we have 52 possibilities: n = 52.

Step 2: For the second card selected, we have 51 possibilities: n - 1 = 51.

Step 3: For the third card selected, we have 50 possibilities: n - 2 = 50.

Therefore, the total number of samples is (52)(51)(50) = 132600.

➤ **(b).**
For this problem, n = 4 kings, r = 3.

Step 1: For the first king selected, we have 4 possible kings: n = 4.

Step 2: For the second king selected, we have 3 possible kings: n - 1 = 3.

Step 3: For the third king selected, we have 2 possibilities: n -2 = 2.

Therefore, the total number of samples is (4)(3)(2) = 24.

Step 4: The event "all three cards selected are kings ", **E** is made up of these 24 possible samples. Therefore,

#**E** = 24 and P(**E**) = $\dfrac{24}{132600}$.

Unsolved Problems with Answers

11.3 Problem 1: An urn contains three marbles marked 1 to 3 respectively. Two marbles are selected at random, without replacement.

(a). List the original population.

(b). List the sample space.

(c). Find the sample size and cardinality of the sample space.

Answers:

➤ **(a).** 1, 2, 3

➤ **(b).** {(1,2), (1,3), (2,1), (2,3), (3,1), (3,2)}

➤ **(c).** #**S** = 6

⇑ *Refer back to* **11.3 - Example 1 & 11.3 - Solved Problem 1**.

11.3 - Problem 2: Five boys and 7 girls were the finalists in a spelling bee contest. At the end of the competition, three students received prizes.

(a). Find the number of possible samples.
(b). Find the probability that all three students receiving prizes were girls.

Answers:

➤ **(a).** 1320 samples

➤ **(b).** $\dfrac{210}{1320}$

⇑ *Refer back to* **11.3 - Example 2 & 11.3 - Solved Problem 2**.

Supplementary Problems

1. At a certain downtown intersection, six drivers were stopped for speeding or driving under the influence. Find the probability four drivers were stopped for speeding and two for driving under the influence.

2. A large Eastern state prints automobile license plates consisting of a mixture of ten digits and 26 letters. Each plate has five such characters. Mrs. Flower recently received a new license plate. Find the probability that the first two characters are digits and the last three are letters.

3. Five boys and seven girls were the finalists in a spelling bee contest. At the end of the competition, four students received different prizes. Find the probability that two boys and two girls win prizes.

4. A four card hand was dealt from an ordinary deck of cards. Find the probability that the hand contains

a. at least one king.

b. at most three jacks.

c. all the cards are face cards.

d. exactly three cards are queens.

5. N students are enrolled in a Spanish class. Find a formula that gives the probability that at least two students are born on the same day. Assume there are 365 days in the year and $N < 365$.

6. An urn contains three marbles colored red, white and blue respectively. Three marbles are sampled with replacement. Find the probability that all the marbles selected are all red or all white.

7. A computer selects three numbers, without replacement, from the numbers 1, 2, 3, 4, 5, 6, 7, 8, 9, 10 . Find the probability that:

a. the numbers 1, 2, 3 are selected.

b. only even numbers are selected.

c. exactly two even numbers are selected.

8. Four cards are drawn from an ordinary deck of cards. Each time a card is drawn, it is returned to the deck. Find the total number of possible sample hands.

9. In the state Ms. Jones lives in, the Department of Motor Vehicles issues license plates with 5 alpha-numeric characters. Find the probability that his license plate has 2 numbers and 3 letters where repetitions is allowed.

10. A coin is tossed until 2 heads occur or four times, whichever comes first. Write out the sample space **S**.

11. Five cards are drawn from an ordinary deck of cards without replacement. Find the probability that

a. all five cards are clubs.

b. four cards are clubs.

12. Three people recently joined a stamp club. Find the probability that

a. none of them were born in the same month.

b. at least two were born in the same month.

13. Cards are drawn, without replacement, until a diamond occurs or 5 cards are drawn, whichever occurs first. Find the cardinality of the sample space.

14. Twelve pairs of shoes are in a closet. Six shoes are selected at random. Find the probability that

a. at least 1 pair is selected.

b. Exactly 1 pair is selected.

15. A pair of dice is tossed N times where N < 36. Let **A** = {(1,1), (2,2), (3,3), (4,4), (5,5), (6,6)}.Find the probability

a. at least one of the pairs in **A** occurs.

b. all the pair of numbers are different (not in **A**).

16. Ms. Bullet, a sharpshooter for a local swat team, claims that she can hit a moving target, 1,000 yards away. To test her claim, she shot ten times at such a target and missed eight times. Find the number of ways this can happen.

17. Two cards are drawn, without replacement, from an ordinary deck of cards. Find the number of ways the event **E**, a king or diamond is selected on the first drawing and a king on the second drawing can happen.

18. The following game is played. A fair coin is tossed until 2 heads are tossed or 5 times, whichever occurs first. Find the cardinality of the sample space.

19. A fair coin is tossed n times.

a. Find the probability of each element of the sample space.

b. Find the probability that each toss alternates a different face. (Example: hthththt).

20. A fair die is rolled 3 times.

a. Find the probability of each element of the sample space.

b. Find the probability that the first number rolled is larger than the second number rolled and the second number rolled is larger than the third number rolled.

21. Three cards are randomly drawn, without replacement, from an ordinary deck of cards.

a. Find the probability of each element of the sample space.

b. Find the probability that only 2 kings are drawn.

22. A fair coin is tossed 6 times. Find the probability that exactly 2 heads appear and they appear in a row.

23. The third race at a local race track has 10 horses running. Assume each horse is numbered 1 to 10 respectively. Assume no ties for this race.

a. Find the cardinality of the sample space as to the order the horses cross the winning line.

b. A horse is said to be in the money if it is one of the first three crossing the winning line. Find the probability that horses number 5,8,2 are in the money.

Probability Theory
Lesson 12
Conditional Probability

Frequently the probability of an event can be changed by having an a priori or previous knowledge of other events. For example, the probability of a 4 appearing in the toss of a single die is 1/6. However, if it is known or assumed that an even number resulted from the experiment, then the probability of a 4 appearing changes to 1/3. since we know that an odd number could not occur. Another example is the probability of rain on Tuesday would be influenced if it is known that it rained the previous Monday.

12.1 - What is Conditional Probability?

The conditional probability of an event **E** is the probability of the event **E** given the a priori knowledge of the occurrence of one or more events. Assume we have a sample space where each single outcome has equal chance of occurring. In this sample space, we assume that the event **B** has occurred. The shaded area represents that portion of the event **E** that can happen. Knowing that **B** occurred precludes the non-shaded area from happening. Therefore, the conditional probability of the event **E** is

$$P(\mathbf{E}|\mathbf{B}) = \frac{\#(\mathbf{E} \cap \mathbf{B})}{\#\mathbf{B}}.$$

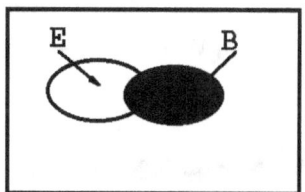

However, by dividing the numerator and the denominator by #**S**, we

get the fraction $\dfrac{\#(\mathbf{E} \cap \mathbf{B})}{\#\mathbf{B}} = \dfrac{\dfrac{\#(\mathbf{E} \cap \mathbf{B})}{\#\mathbf{S}}}{\dfrac{\#\mathbf{B}}{\#\mathbf{S}}} = \dfrac{P(\mathbf{E} \cap \mathbf{B})}{P(\mathbf{B})}.$

Using the above explanation We define, for any sample space, conditional probability of an event **E** given that the event **B** occurred as

$$P(E|B) = \frac{P(E \cap B)}{P(B)}.$$

Rule 1: $P(E'|B) = 1 - P(E|B)$.

Rule 2: $P(C \cup D|B) = P(C|B) + P(D|B) - P(C \cap D|B)$.

Throughout these lessons, sample spaces generated by drawing cards from an ordinary deck of cards provides a rich assortment of problems that can be explored for acquiring a deeper understand of the nature of the subject and developing methods for solving complex problems. The following presents the individual cards that make up a 52 deck of ordinary deck of cards:

<u>**Suites(4):**</u>

Diamonds (13): ace, 2, 3, 4, 5, 6, 7, 8, 9, 10, jack, queen, king
Hearts   (13): ace, 2, 3, 4, 5, 6, 7, 8, 9, 10, jack, queen, king
Clubs   (13): ace, 2, 3, 4, 5, 6, 7, 8, 9, 10, jack, queen, king
Spades   (13): ace, 2, 3, 4, 5, 6, 7, 8, 9, 10, jack, queen, king

Face cards (12):

Diamonds (3): jack, queen, king
Hearts   (3): jack, queen, king
Clubs   (3): jack, queen, king
Spades   (3): jack, queen, king

12.1 - Example 1: Let **M** be the event that it will rain on Monday and **T** the event that it will rain on Tuesday. Assume that the chance it will rain on Monday is 0.20 and that the chance it will rain on both Monday and Tuesday is 0.10. If it rains on Monday, find the probability that it will rain on Tuesday.

Solution:

Step 1: M: The event that it will rain on Monday.

T: The event that it will rain on Tuesday.

M∩T: The event that it will rain on both Monday and Tuesday.

Step 2: $P(M) = 0.20$

Step 3: $P(M \cap T) = 0.10$

Step 4: $P(T|M) = \dfrac{P(M \cap T)}{P(M)} = \dfrac{0.10}{0.20} = \dfrac{1}{2}$

12.1- Example 2: One card is randomly selected from an ordinary deck of cards.

(a). If a king is drawn, find the probability that the king is a diamond: $P(D|K)$.

(b). If a king is not drawn, find the probability that the card is a diamond: $P(D|K')$.
Solutions:

➤(a).
Step 1: There are 4 kings in an ordinary deck of cards.

Step 2: There is only 1 king that is diamond.

Therefore, $P(D|K) = 1/4$.

➤ (b).
Step 1: There are 48 cards that are not kings in an ordinary deck of cards.

Step 2: An ordinary deck of cards has 13 diamonds of which 12 of them are not kings.

Therefore, $P(D|K') = 12/48 = 1/4$.

12.1 - Example 3: Two cards are drawn at random, without replacement from an ordinary deck of cards.

(a). If the first card drawn is a king, find the probability that the second card drawn is also a king: $P(K_2|K_1)$.

(b). If the first card drawn is a king, find the probability that the second card drawn is not a king: $P(K_2'|K_1)$.

(c). If the first card drawn is not a king, find the probability that the second card drawn is a king: $P(K_2|K_1')$.

(d). If the first card drawn is not a king, find the probability that the second card drawn is also not a king:
$P(K_2'|K_1')$.

(e). If the first card drawn is a king, find the probability that the second card drawn is a queen: $P(Q_2|K_1)$.

Solutions:

There are 52 cards in an ordinary deck of cards of which 4 are kings cards.

➤ **(a).**
Step 1: After the first card is drawn, there are 51 cards remaining in the deck.

Step 2: Since the first card drawn is a king, there are only 3 kings remaining in the deck.

Step 3: Since the remaining deck has only 51 cards and there are only 3 kings remaining in the deck, the probability that the second card drawn is a king is $P(K_2|K_1) = 3/51$.

➤ **(b).**
Step 1: After the first card is drawn, there are 51 cards remaining in the deck.

Step 2: Since the first card drawn is a king, there are 48 non-kings remaining in the deck.

Step 3: Since the remaining deck has only 51 cards and there are 48 non-kings remaining in the deck, the probability that the second card drawn is not a king is $P(K_2'|K_1). = 48/51$.

➤ **(c).**
Step 1: After the first card is drawn, there are 51 cards remaining in the deck.

Step 2: Since the first card drawn is not a king, there are 4 kings remaining in the deck.

Step 3: Since the remaining deck has only 51 cards and there are 4 kings remaining in the deck, the probability that the second card drawn is a king is $P(K_2 |K_1') = 4/51$.

➤ **(d).**
Step 1: After the first card is drawn, there are 51 cards remaining in the deck.

Step 2: Since the first card drawn is not a king, there are 47 non-kings remaining in the deck.

Step 3: Since the remaining deck has only 51 cards and there are 47 non-kings remaining in the deck, the probability that the second card drawn is not a king is $P(K_2' |K_1') = 47/51$.

➤ **(e).**
Step 1: After the first card is drawn, there are 51 cards remaining in the deck.

Step 2: Since the first card drawn is a king, there are 4 queens remaining in the deck.

Step 3: Since the remaining deck has only 51 cards and there are 4 queens remaining in the deck, the probability that the second card drawn is not queen is $(Q_2 |K_1) = 4/51$.

Solved Problems

12.1 - Solved Problem 1: Let **B** be the event that Mr. Jones is stopped for speeding and **E** the event that he is driving under the influence. Assume the chance that he will be stopped for speeding is 0.35 and that the chance he is stopped for speeding and at the same time driving under the influence is 0.25. If he is stopped for speeding, find the probability that he is also driving under the influence.

Solution:

Step 1: $P(B) = 0.35$.

Step 2: $P(E \cap B) = 0.25$

Step 3: $P(E \mid B) = \dfrac{P(E \cap B)}{P(B)} = \dfrac{0.25}{0.35} = \dfrac{5}{7}$

12.1- Solved Problem 2: One card is randomly selected from an ordinary deck of cards.

(a). If a face card is drawn, find the probability that the card is a queen: $P(Q \mid F)$.

(b). If a face card is not drawn, find the probability that the card is a queen: $P(Q \mid F')$.

Solutions:

➤(a).
Step 1: There are 12 face cards in an ordinary deck of cards.

Step 2: There are only 4 queens in the deck.

Therefore, $P(Q \mid F) = 4/12 = 1/3$.

➤(b).
Since a non-face card is selected, and a queen is a face card, $P(Q \mid F') = 0/40 = 0$.

12.1 - Solved Problem 3: Two cards are drawn at random, without replacement from an ordinary deck of cards.

(a). If the first card drawn is a face card, find the probability that the second card drawn is also a face card: $P(F_2 | F_1)$.

(b). If the first card drawn is a face card, find the probability that the second card drawn is not a face card: $P(F_2' | F_1)$.

(c). If the first card drawn is not a face card, find the probability that the second card drawn is a face card:

$P(\mathbf{F}_2 \,|\mathbf{F}_1{}')$.

(d). If the first card drawn is not a face card, find the probability that the second card drawn is also not a face card: $P(\mathbf{F}_2{}' \,|\mathbf{F}_1{}')$.

(e). If the first card drawn is not a face card, find the probability that the second card drawn is a queen :
$P(\mathbf{Q}_2 \,|\mathbf{F}_1{}')$.

Solutions:

There are 52 cards in an ordinary deck of cards of which 12 are face cards.

➤(a).
Step 1: After the first card is drawn, there are 51 cards remaining in the deck.

Step 2: Since the first card drawn is a face card, there are only 11 face cards remaining in the deck.

Step 3: Since the remaining deck has only 51 cards and there are only 11 face cards remaining in the deck, the probability that the second card drawn is a face card is $P(\mathbf{F}_2|\mathbf{F}_1) = 11/51$.

➤(b).
Step 1: After the first card is drawn, there are 51 cards remaining in the deck.

Step 2: Since the first card drawn is a face card, there are 40 (52 - 12) non-face cards remaining in the deck.

Step 3: Since the remaining deck has only 51 cards and there are 40 non-face cards remaining in the deck, the probability that the second card drawn is not a face card is $P(\mathbf{F}_2{}'|\mathbf{F}_1) = 40/51$.

➤(c).
Step 1: After the first card is drawn, there are 51 cards remaining in the deck.

Step 2: Since the first card drawn is not a face card, there are 12 face cards remaining in the deck.

Step 3: Since the remaining deck has only 51 cards and there are 12 face cards remaining in the deck, the probability that the second card drawn is a face card is $P(\mathbf{F}_2 \,|\mathbf{F}_1{}') = 12/51$.

➤(d).
Step 1: After the first card is drawn, there are 51 cards remaining in the deck.

Step 2: Since the first card drawn is not a face card, there are 39 non-face cards remaining in the deck.

Step 3: Since the remaining deck has only 51 cards and there are 39 non-face cards remaining in the deck, the probability that the second card drawn is not a face card is $P(F_2' | F_1') = 39/51$.

➤(e).

Step 1: After the first card is drawn, there are 51 cards remaining in the deck.

Step 2: Since the first card drawn is not a face card, there are 4 queens remaining in the deck.

Step 3: Since the remaining deck has only 51 cards and there are 4 queens remaining in the deck, the probability that the second card drawn is not queen is $(Q_2 | F_1') = 4/51$.

Unsolved Problems with Answers

12.1 - Problem 1: Mr. Jones was recently murdered. Let **A** be the event that Mrs. Jones is arrested for the murder and **G** the event that she is guilty. Assume the chance that she will be arrested is 0.75 and the chance she is arrested and also guilty is 0.20. Find the probability that if she is arrested, then she is guilty.

Answer:

4/15

⇑ *Refer back to* **12.1 - Example 1 & 12.1 - Solved Problem 1**.

12.1- Problem 2: One card is randomly selected from an ordinary deck of cards.

(a). If a diamond is drawn, find the probability that the card is a face card: $P(F|D)$.

(b). If a diamond is not drawn, find the probability that the card is a face card: $P(F|D')$.

Answers:

➤(a). 3/13

➤(b). 3/13

⇑ *Refer back to* **12.2 - Example 2 & 12.1 - Solved Problem 2**.

12.1 - Problem 3: Two cards are drawn at random, without replacement from an ordinary deck of cards.

(a). If the first card drawn is a diamond, find the probability that the second card drawn is also a diamond: $P(D_2|D_1)$.

(b). If the first card drawn is a diamond, find the probability that the second card drawn is not a diamond: $P(D_2'|D_1)$.

(c). If the first card drawn is not a diamond, find the probability that the second card drawn is a diamond card:

$P(D_2 | D_1')$.

(d). If the first card drawn is not a diamond, find the probability that the second card drawn is also not a diamond: $P(D_2' | D_1')$.

(e). If the first card drawn is a diamond, find the probability that the second card drawn is a heart : $P(H_2 | D_1)$.

Answers:

➤(a). 12/51

➤(b). 39/51

➤(c). 13/51

➤(d). 38/51

➤(e). 13/51

⇑ *Refer back to* **12.1 - Example 3 & 12.1 - Solved Problem 3**.

12.2 - A Formula for P(E∩B)

From the formula $P(E|B) = \dfrac{P(E \cap B)}{P(B)}$,   we can derive

$$P(A \cap B) = P(A)P(B|A)$$
or

by multiplying both sides by $P(B)$.

This formula can be generalized as

$$P(E_1 \cap E_2 \cap E_3 \cap ... \cap E_n) =$$

$$P(E_1)P(E_2|E_1)P(E_3|E_1 \cap E_2)P(E_4|E_1 \cap E_2 \cap E_3)...P(E_n|E_1 \cap E_2 \cap E_3.... \cap E_{n-1}).$$

12.2 - Example 1: Assume two cards are drawn from an ordinary deck of cards without replacement. Find the probability that both cards drawn are kings.

Solution:

Step 1: K_1 : the event that a king is drawn on the first drawing.

Step 2: K_2 : the event that a king was also drawn on the second drawing.

Step 3: $K_1 \cap K_2$: the event that both cards drawn are kings.

Step 4: $\#K_1 = 4$

$$P(K_2 \mid K_1) = \frac{3}{51}.$$

Thus, $P(K_1 \cap K_2) = P(K_1)P(K_2 \mid K_1) = (\frac{4}{52})(\frac{3}{51}) = \frac{12}{2652} = \frac{1}{221}.$

12.2 - Example 2: Assume three cards are drawn from an ordinary deck of cards without replacement.

Find the probability that

(a). all three cards drawn are kings.

(b). at least one king is drawn.

Solutions:

➤(a).
K_1: King is drawn on the first drawing.
K_2: King is drawn on the second drawing.
K_3: King is drawn on the third drawing.

$E = K_1 \cap K_2 \cap K_3$: all three cards drawn are kings.

Using the above formula, we have

$$P(E) = P(K_1)P(K_2 \mid K_1)P(K_3 \mid K_1 \cap K_2) = (\frac{4}{52})(\frac{3}{51})(\frac{2}{50}) = \frac{24}{132600}.$$

➤(b).
Let **E** be the event that at least one king is drawn.

Step 1: $E = K_1 \cup K_2 \cup K_3$

Since we don't have a simple formula for the union of more than 3 sets (see Lesson 6, supplementary problem 12),

we first compute

Step 2: $E' = (K_1 \cup K_2 \cup K_3)' = K_1' \cap K_2' \cap K_3'$, no kings are drawn.

Step 3: $P(E') = P(K_1')P(K_2'|K_1')P(K_3'|K_1' \cap K_2') = (\frac{48}{52})(\frac{47}{51})(\frac{46}{50}) = \frac{103776}{132600}$.

Step 4: $P(E) = 1 - P(E') = = 1 - \dfrac{103776}{132600} \approx 0.22$

12.2 - Example 3: An urn contains 10 white, 15 blue and 25 red marbles. Assume three marbles are selected from the urn without replacement. Find the probability that all three marbles selected are red.

Solution:

R_1: The event that the marble drawn is red.

R_2: The event that the second marble drawn is red.

R_3: The event that the third marble drawn is red.

E: The event that all three marbles drawn are red is $E = R_1 \cap R_2 \cap R_3$.

Step 1: The probability that the first marble drawn is red is

$P(R_1) = \dfrac{25}{50}$. .

Step 2: Given that the first marble drawn is red, the probability that the second marble drawn is also red equals is

$P(R_2|R_1) = \dfrac{24}{49}$.

Step 3: Given that the first and second marbles drawn are red, the probability that the third marble drawn is also red is

$P(R_3|R_1 \cap R_2) = \dfrac{23}{48}$.

Step 4: Therefore, the probability that the first, second and third marbles are all red is

$$P(E) = P(R_1 \cap R_2 \cap R_3) = \quad (\frac{25}{50})(\frac{24}{49})(\frac{23}{48}) \; = \; \frac{13800}{117600} \; = \; \frac{23}{196}.$$

12.2 - Example 4: Ms. Jones math class has 10 female and 5 male students. She selects three at random to attend a special math lecture. Find the probability that

(a). all three students selected are females.

(b). two male and one female was selected.

(c). at least one female was selected.

Solutions:

➤ (a).
Step 1: F_1: the event that the first student selected is a female.
F_2: the event that the second student selected is a female.
F_3: the event that the third student selected is a female.

E: the event that all three students selected are females.

Step 2: $E = F_1 \cap F_2 \cap F_3$

$$P(E) = P(F_1 \cap F_2 \cap F_3) = P(F_1)P(F_2|F_1)P(F_3|F_1 \cap F_2) = (\frac{10}{15})(\frac{9}{14})(\frac{8}{13}) \; = \; \frac{24}{91}$$

➤ (b).
Step 1: E: the event that two male and one female was selected.

$$E = (F_1 \cap F_2' \cap F_3') \cup (F_1' \cap F_2 \cap F_3') \cup (F_1' \cap F_2' \cap F_3)$$

Step 2: $P(E) =$

$$P\{(F_1 \cap F_2' \cap F_3') \cup (F_1' \cap F_2 \cap F_3') \cup (F_1' \cap F_2' \cap F_3)\} = P(F_1 \cap F_2' \cap F_3') + P(F_1' \cap F_2 \cap F_3') + P(F_1' \cap F_2' \cap F_3) =$$

$$(\frac{10}{15})(\frac{5}{14})(\frac{4}{13}) \; + \; (\frac{5}{15})(\frac{10}{14})(\frac{4}{13}) \; + \; (\frac{5}{15})(\frac{4}{14})(\frac{10}{13}) \; = \frac{20}{91}$$

➤ (c).
Step 1: E: the event that at least one female was selected.

E': the event that no female was selected.

$$E' = F_1' \cap F_2' \cap F_3'$$

Step 2: $P(E') = P(F_1' \cap F_2' \cap F_3') = (\frac{5}{15})(\frac{4}{14})(\frac{3}{13}) = \frac{2}{91}$

$P(E) = 1 - \frac{2}{91} = \frac{89}{91}$

Solved Problems

12.2 - Solved Problem 1: Assume two cards are drawn from an ordinary deck of cards without replacement. Find the probability that the first card drawn is a king and the second card drawn is an ace.

Solution:

Let K_1 be the event that a king is drawn on the first drawing and A_2 the event that an ace was drawn on the second drawing.

Thus, $P(K_1 \cap A_2) = P(K_1)P(A_2|K_1) = (\frac{4}{52})(\frac{4}{51}) = \frac{16}{2652} = \frac{4}{663}$.

12.2 Solved Problem 2: Assume three cards are drawn from an ordinary deck of cards without replacement. Find the probability that

(a). all three cards are face cards.

(b). at least one face card is drawn.

Solutions:

➤(a).
F_1: Face card is drawn on the first drawing.
F_2: Face card is drawn on the second drawing.
F_3: Face card is drawn on the third drawing.

$E = F_1 \cap F_2 \cap F_3$: all three cards drawn are face cards.

Using the above formula, we have $P(E) = P(F_1)P(F_2|F_1)P(F_3 | F_1 \cap F_2) =$

$(\frac{12}{52})(\frac{11}{51})(\frac{10}{50}) = \frac{1320}{132600}$.

➤(b).
Let E be the event that at least one face card is drawn.

Step 1: $E = F_1 \cup F_2 \cup F_3$

Since we don't have a simple formula for the union of more than 3 sets (see Lesson 6, supplementary problem 12),

we first compute

Step 2: $E' = (F_1 \cup F_2 \cup F_3)' = F_1' \cap F_2' \cap F_3'$, no face cards are drawn.

Step 3: $P(E') = P(F_1')P(F_2'|F_1')P(F_3'|F_1' \cap F_2') = (\frac{40}{52})(\frac{39}{51})(\frac{38}{50}) = \frac{59280}{132600}$

Step 4: $P(E) = 1 - P(E') = = 1 - (\frac{40}{52})(\frac{39}{51})(\frac{38}{50}) = \frac{73320}{132600} \approx 0.55$

12.2 - Solved Problem 3: An urn contains 10 white, 15 blue and 25 red marbles. Assume three marbles are selected from the urn without replacement. Find the probability that the first marble selected is red, the second marble selected is blue and third marble selected is white.

Solution:

R_1: The event that the marble drawn is red.

B_2: The event that the second marble drawn is blue.

W_3: The event that the third marble drawn is white.

E: The event that the first marble selected is red, the second marble selected is blue and third marble selected is white: $E = R_1 \cap B_2 \cap W_3$.

Step 1: The probability that the first marble drawn is red is

$P(E) = \frac{25}{50}$.

Step 2: Given that the first marble drawn is red, the probability that the second marble drawn is blue is

$P(B_2|R_1) = \frac{15}{49}$.

Step 3: Given that the first marble drawn is red, the second marble drawn is blue, the probability that the third marble drawn is white is

$P(W_3|R_1 \cap B_2) = \frac{10}{48}$.

Step 3: Therefore the probability of **E** is

$$P(E) = P(\mathbf{R_1} \cap \mathbf{B_2} \cap \mathbf{W_3}) = (\frac{25}{50})(\frac{15}{49})(\frac{10}{48}) = \frac{75}{2352}.$$

12.2 - Solved Problem 4: In a survey of a famous wine club, 15 members prefer red wine, 10 prefer white wine and 5 prefer neither. Assume four members were randomly selected to taste a new shipment of wines. Find the probability that

(a). all four members selected prefer red wine.

(b). only three members prefer red wine.

(c). at least one member prefers neither red nor white wine.

Solutions:

➤ **(a).**
Step 1: $\mathbf{R_1}$: first member selected prefers red wine.

$\mathbf{R_2}$: second member selected prefers red wine.

$\mathbf{R_3}$: third member selected prefers red wine.

$\mathbf{R_4}$: fourth member selected prefers red wine.

E: the event that all four members selected prefer red wine.

$$\mathbf{E} = \mathbf{R_1} \cap \mathbf{R_2} \cap \mathbf{R_3} \cap \mathbf{R_4}$$

Step 2: $P(E) = P(\mathbf{R_1} \cap \mathbf{R_2} \cap \mathbf{R_3} \cap \mathbf{R_4}) = (\frac{15}{30})(\frac{14}{29})(\frac{13}{28})(\frac{12}{27}) = \dfrac{32760}{657720}$

➤ **(b).**
Step 1: E: the event only three members prefer red wine.

$$\mathbf{E} = (\mathbf{R_1'} \cap \mathbf{R_2} \cap \mathbf{R_3} \cap \mathbf{R_4}) \cup (\mathbf{R_1} \cap \mathbf{R_2'} \cap \mathbf{R_3} \cap \mathbf{R_4}) \cup (\mathbf{R_1} \cap \mathbf{R_2} \cap \mathbf{R_3'} \cap \mathbf{R_4}) \cup (\mathbf{R_1} \cap \mathbf{R_2} \cap \mathbf{R_3} \cap \mathbf{R_4'})$$

Step 2: $P(E) = P\{(\mathbf{R_1'} \cap \mathbf{R_2} \cap \mathbf{R_3} \cap \mathbf{R_4}) \cup (\mathbf{R_1} \cap \mathbf{R_2'} \cap \mathbf{R_3} \cap \mathbf{R_4}) \cup (\mathbf{R_1} \cap \mathbf{R_2} \cap \mathbf{R_3'} \cap \mathbf{R_4}) \cup (\mathbf{R_1} \cap \mathbf{R_2} \cap \mathbf{R_3} \cap \mathbf{R_4'})\} =$

$P(\mathbf{R_1'} \cap \mathbf{R_2} \cap \mathbf{R_3} \cap \mathbf{R_4}) + P(\mathbf{R_1} \cap \mathbf{R_2'} \cap \mathbf{R_3} \cap \mathbf{R_4}) + P(\mathbf{R_1} \cap \mathbf{R_2} \cap \mathbf{R_3'} \cap \mathbf{R_4}) + P(\mathbf{R_1} \cap \mathbf{R_2} \cap \mathbf{R_3} \cap \mathbf{R_4'}) =$

$$(\frac{15}{30})(\frac{15}{29})(\frac{14}{28})(\frac{13}{27}) + (\frac{15}{30})(\frac{15}{29})(\frac{14}{28})(\frac{13}{27}) +$$

$$(\frac{15}{30})(\frac{14}{29})(\frac{15}{28})(\frac{13}{27}) + (\frac{15}{30})(\frac{14}{29})(\frac{13}{28})(\frac{15}{27}) = \frac{16380}{65772}$$

➤ **(c).**

Step 1: N$_1$: first member selected prefers neither red nor white wine.

N$_2$: second member selected prefers neither red nor white wine.

N$_3$: third member selected prefers neither red or nor white wine.

N$_4$: fourth member selected prefers neither red nor white wine.

E: the event that at least one prefer neither red nor white wine.

E': the event that none prefer neither red nor white wine.

$$E' = N_1' \cap N_2' \cap N_3' \cap N_4'$$

Step 2: $P(E') = P(N_1' \cap N_2' \cap N_3' \cap N_4') = (\frac{25}{30})(\frac{24}{29})(\frac{23}{28})(\frac{22}{27}) = \frac{303600}{657720}$

$$P(E) = 1 - \frac{303600}{657720} = \frac{2951}{5481}$$

Unsolved Problems with Answers

12.2 - Problem 1: Assume two cards are drawn from an ordinary deck of cards without replacement. Find the probability that the first card is a king and the second card is an ace or queen.

Answer:

8/663

⇑ *Refer back to* **12.2 - Example 1 & 12.2 - Solved Problem 1**.

12.2 - Problem 2: Assume three cards are drawn from an ordinary deck of cards without replacement. Find the probability that

(a). all three cards are clubs.

(b). at least one club is drawn.

Answer:

➤(a). 1716/132600

➤(b). 77766/132600

⇑ *Refer back to* **12.2 - Example 2 & 12.2 - Solved Problem 2**.

12.2 - Problem 3: An urn contains 10 white, 15 blue and 25 red marbles. Assume three marbles are selected from the urn without replacement. Find the probability that the first is red, the second is blue and third is red.

Answer:

15/196

⇑ *Refer back to* **12.2 - Example 3 & 12.2 - Solved Problem 3.**

12.2 - Problem 4: The members of a political action committee consist of 8 Democrats and 10 Republicans. Three members made a speech. Find the probability that

(a). all three were Democrats.

(b). only two were Democrats.

(c). at least one was a Republican.

Answers:

➤ (a). $\dfrac{7}{102}$

➤ (b). $\dfrac{35}{102}$

➤ (c). $\dfrac{95}{102}$

⇑ *Refer back to* **12.2 - Example 4 & 12.2 - Solved Problem 4.**

12.3 - Writing the Event E in Terms of Other Events.

Under certain conditions, one event can be expressed in terms of other events.

The following formulas are important:

Boolean algebra formulas:

1. $E = (E \cap B) \cup (E \cap B')$

2. If $E = B_1 \cup B_2 \cup ... \cup B_n$ then $E' = B_1' \cap B_2' \cap ... \cap B_n'$

Probability formulas:

1a. $P(A \cap B) = P(B)P(A|B)$

1b. $P(A \cap B) = P(A)P(B|A)$

2a. $P(A \cup B) = P(A) + P(B) - P(A \cap B)$

2b. If $A \cap B = \varphi$ then $P(A \cup B) = P(A) + P(B)$

3a. $P(A_1 \cup A_2) = 1 - P(A_1' \cap A_2')$

3b. $P(A_1 \cup A_2 \cup ... \cup A_n) = 1 - P(A_1' \cap A_2' \cap ... \cap A_n')$

4. If $B_i \cap B_j = \varphi$ $(1 \leq i,j \leq n)$ and $E = B_1 \cup B_2 \cup ... \cup B_n$ then

$P(E) = P(B_1) + P(B_2) + ... + P(B_n)$

12.3 - Example 1: Two urns are sitting on a table. Urn 1 has 3 blue marbles and 7 white marbles. Urn 2 has 5 blue marbles and 10 white marbles. A coin is tossed once. If heads occurs a marble is selected at random from urn 1 and if tails occurs a marble is selected at random from urn 2. Find the probability that the marble selected is white.

Solution:

U_1: the event that urn 1 is selected.

U_2: the event that urn 2 is selected.

W: the event that a white marble is selected.

$W \cap U_1$: the event that a white is selected **AND** urn 1 is selected.

$W \cap U_2$: the event that a white is selected **AND** urn 2 is selected.

$W = (W \cap U_1) \cup (W \cap U_2)$.

To find $P(W)$, we do the following steps:

Step1: $P(W) = P(W \cap U_1) + P(W \cap U_2)$

Step2: $P(W \cap U_1) = P(U_1)P(W|U_1) = (\frac{1}{2})(\frac{7}{10}) = \frac{7}{20}$.

Step3: $P(W \cap U_2) = P(U_2)P(W|U_2) = (\frac{1}{2})(\frac{10}{15}) = \frac{1}{3}$.

Step4: $P(W) = P(W \cap U_1) + P(W \cap U_2) = \frac{7}{20} + \frac{1}{3} = \frac{41}{60}$.

12.3 - Example 2: Two cards are drawn at random without replacement from an ordinary deck of cards.

(a). Find the probability that the second card drawn is a king.

(b). Find the probability that the first card is a king or the second card is a king.

(c). Find the probability that a king and queen are drawn.

(d). Find the probability that at least one king or one queen is drawn.

Solutions:

➤ **(a).**
The chance of drawing a king on the second drawing depends on whether a king had been drawn on the first drawing.

K_1: the event that a king is drawn on the first draw.

K_2: the event that a king is drawn on the second drawing.

$K_1 \cap K_2$: the event that a king is drawn on the first **AND** second drawing.

$K_1' \cap K_2$: the event that a king is **NOT** drawn on the first **AND** a king is drawn on the second drawing.

$K_2 = (K_1 \cap K_2) \cup (K_1' \cap K_2)$: a king is drawn on first and on the second or a king is not drawn on the first but a king is drawn on the second.

$$P(K_2) = P(K_1 \cap K_2) + P(K_1' \cap K_2) = (\frac{4}{52})(\frac{3}{51}) + (\frac{48}{52})(\frac{4}{51}) = \frac{204}{2652} = \frac{4}{52}$$

➤ **(b).**
K_1: the event that a king is drawn on the first draw.

K_2: the event that a king is drawn on the second draw.

E: the event that a king is drawn on the first or second drawing.

$E = K_1 \cup K_2$

$$P(E) = P(K_1 \cup K_2) = P(K_1) + P(K_2) - P(K_1 \cap K_2) = \frac{4}{52} + \frac{4}{52} - P(K_1)P(K_2|K_1) =$$

$$\frac{4}{52} + \frac{4}{52} - \left(\frac{4}{52}\right)\left(\frac{3}{51}\right) = \frac{408}{(52)(51)} - \frac{12}{(52)(51)} = \frac{33}{221}$$

Alternative solution:

$$E' = (K_1 \cup K_2)' = K_1' \cap K_2'$$

$$P(E') = P(K_1' \cap K_2') = P(K_1')P(K_2'|K_1') = (48/52)(47/51)$$

$$P(E) = 1 - P(E') = 1 - (48/52)(47/51) = 1 - 2256/2652 = 396/2652 = 33/221$$

➤ **(c).**
E: the event that a king and queen are drawn.

Since order is not required we must consider all possibilities: $E = (K_1 \cap Q_2) \cup (Q_1 \cap K_2)$

$$P(E) = P(K_1 \cap Q_2) + P(Q_1 \cap K_2) = P(K_1)P(Q_2|K_1) + P(Q_1)P(K_2|Q_1) =$$

$$\left(\frac{4}{52}\right)\left(\frac{4}{51}\right) + \left(\frac{4}{52}\right)\left(\frac{4}{51}\right) + \frac{32}{2652}$$

➤ **(d).**
E: at least 1 king or 1 queen is drawn.

$$E = (K_1 \cup Q_2) \cup (Q_1 \cup K_2)$$

E': no king and no queen are drawn.

$$E' = (K_1' \cap Q_2') \cap (Q_1' \cap K_2') = (K_1' \cap Q_1') \cap (K_2' \cap Q_2')$$

$$P(E') = P[(K_1' \cap Q_1') \cap (K_2' \cap Q_2')] = P(K_1' \cap Q_1')P[(K_2' \cap Q_2')|(K_1' \cap Q_1')] =$$

$$\left(\frac{44}{52}\right)\left(\frac{43}{51}\right) = \frac{1892}{2652} = \frac{473}{663}$$

$$P(E) = 1 - P(E') = 1 - \frac{473}{663} = \frac{190}{663}$$

12.3 - Example 3: Two cards are drawn at random without replacement from an ordinary deck of cards.

(a). Find the probability that the first card is a diamond and the second card is a king.

(b). Find the probability that the first card is a diamond or the second card is a king.

Solutions:

➤ **(a).**

D_1: the event that a diamond is drawn on the first draw.

K_1: the event that a king is drawn on the first draw.

K_2: the event that a king is drawn on the second draw.

$E = D_1 \cap K_2$: the event that a diamond is drawn on the first drawing and a king on the second drawing.

$$D_1 = (D_1 \cap K_1) \cup (D_1 \cap K_1')$$

$$E = D_1 \cap K_2 = [(D_1 \cap K_1) \cup (D_1 \cap K_1')] \cap K_2 = [(D_1 \cap K_1) \cap K_2] \cup [(D_1 \cap K_1') \cap K_2]$$

$$P(E) = P(D_1 \cap K_2) = P([(D_1 \cap K_1) \cap K_2] \cup [(D_1 \cap K_1') \cap K_2]) = P[(D_1 \cap K_1) \cap K_2] + P[(D_1 \cap K_1') \cap K_2] =$$

$$P(D_1 \cap K_1)P[K_2 | (D_1 \cap K_1)] + P(D_1 \cap K_1')P[K_2 | (D_1 \cap K_1')] = (\frac{1}{52})(\frac{3}{51}) + (\frac{12}{52})(\frac{4}{51}) = \frac{1}{52}$$

➤ **(b).**

D_1: the event that a diamond is drawn on the first draw.

K_2: the event that a king is drawn on the second draw.

E: the event that a diamond is drawn on the first or a king is selected on second drawing.

$$E = D_1 \cup K_2$$

$P(K_2) = 4/52$ (See **12.3 - Example 2**).

$$P(E) = P(D_1 \cup K_2) = P(D_1) + P(K_2) - P(D_1 \cap K_2) = \frac{13}{52} + \frac{4}{52} - \frac{1}{52} = \frac{16}{52}$$

12.3 - Example 4: Two urns are sitting on a table. Urn 1 has three blue marbles and 7 white marbles. Urn 2 has 4 blue marbles and 10 white marbles. A marble is selected from urn 1 and placed in urn 2. Next, a marble is selected from urn 2. Find the probability that the marble selected from urn 2 is white.

Solution:

The chance of selecting a white marble from urn 2 depends on whether a white marble was selected from urn 1 and placed in urn 2.

W_1: the event that a white marble is drawn from urn 1 and placed in urn 2.

W_2: the event that a white is drawn from urn 2.

$W_1 \cap W_2$: the event that a white marble is drawn from urn 1 and urn 2.

$W_1' \cap W_2$: the event that a white marble is **NOT** drawn from urn 1 and a white marble is drawn from urn 2.

$W_2 = (W_1' \cap W_2) \cup (W_1 \cap W_2)$, the event that a white is drawn from urn 1 and urn 2 **OR** the event that a white is **NOT** drawn from urn 1 **AND** a white is selected from urn 2

$$P(W_1 \cap W_2) = P(W_1)P(W_2 | W_1) = (\frac{7}{10})(\frac{11}{15})$$

$$P(W_1' \cap W_2) = P(W_1')P(W_2 | W_1') = (\frac{3}{10})(\frac{10}{15})$$

Therefore,

$$P(W_2) = P(W_1 \cap W_2) + P(W_1' \cap W_2) = (\frac{7}{10})(\frac{11}{15}) + (\frac{3}{10})(\frac{10}{15}) = \frac{77}{150} + \frac{30}{150} = \frac{107}{150}.$$

12.3 - Example 5: In a state where cars have to be tested for emissions of pollutants, 25% of all cars emit excessive amounts of pollutants. When tested, 99% of all cars that emit excessive amounts of pollutants will fail, but 17% of the cars that do not emit excessive amounts of pollutants will also fail. A car is selected at random. What is the probability that the car fails the test?

Solution:

Let **E** be the event that a car emits excessive amounts of pollutants.

Let **F** be the event that the car will fail the test.

$P(E)$ = the probability that a car will emit excessive pollutants = 0.25

$P(F|E)$ = the probability that all cars that emit excessive amounts of pollutants will fail = 0.99

$P(F|E')$ = the probability that a car that does not emit excessive pollutants fails the test = 0.17

We now need to solve for $P(F)$.

Step 1: F is the event that a car fails the test **and** it emits excessive pollutants **or** a car fails the test and it does not emit excessive pollutants: $F = (F \cap E) \cup (F \cap E')$.

Step 2: $P(F) = P(F \cap E) + P(F \cap E') = P(E)P(F|E) + P(E')P(F|E') = 0.25(0.99) + 0.75(0.17) = 0.375$.

Solved Problems

12.3 - Solved Problem 1: Two urns are sitting on a table. Urn 1 has five blue marbles, 10 red marbles and 15 white marbles. Urn 2 has 5 blue marbles and 10 white marbles. A coin is tossed twice. If heads occur twice, a marble is selected at random from urn 1; otherwise, a marble is selected at random from urn 2. Find the probability that the marble selected is white.

Solution:

Let $\mathbf{U}_1$ be the event that urn 1 is selected and let $\mathbf{U}_2$ be the event that urn 2 is selected. Let $\mathbf{W}$ be the event that a white marble is selected.

Here, $P(\mathbf{U}_1) = 1/4$.

To find the $P(\mathbf{W})$ we do the following steps:

Step 1: $P(\mathbf{W}) = P(\mathbf{W} \cap \mathbf{U}_1) + P(\mathbf{W} \cap \mathbf{U}_2)$

Step 2: $P(\mathbf{W} \cap \mathbf{U}_1) = P(\mathbf{U}_1)P(\mathbf{W}|\mathbf{U}_1) = (\frac{1}{4})(\frac{15}{30}) = \frac{1}{8}$

Step 3: $P(\mathbf{W} \cap \mathbf{U}_2) = P(\mathbf{U}_2)P(\mathbf{W}|\mathbf{U}_2) = (\frac{3}{4})(\frac{10}{15}) = \frac{1}{2}$

Step 4: $P(\mathbf{W}) = P(\mathbf{W} \cap \mathbf{U}_1) + P(\mathbf{W} \cap \mathbf{U}_2) = \frac{1}{8} + \frac{1}{2} = \frac{5}{8}$

12.3 - Solved Problem 2: Two cards are drawn at random without replacement from an ordinary deck of cards.

(a). Find the probability that the second card drawn is a diamond.

(b). Find the probability that the first card is a diamond or the second card is a diamond.

(c). Find the probability that a diamond and a club are drawn.

(d). Find the probability that at least one diamond or one club is drawn.

Solutions:

➤ **(a).**
The chance of drawing a diamond on the second drawing depends on whether a diamond had been drawn on the first drawing.

$\mathbf{D}_1$: the event that a diamond is drawn on the first draw.

$\mathbf{D}_2$: the event that a diamond is drawn on the second drawing.

$D_1 \cap D_2$: the event that a diamond is drawn on the first **AND** second drawing.

$D_1' \cap K_2$: the event that a diamond is **NOT** drawn on the first **AND** a diamond is drawn on the second drawing.

$D_2 = (D_1 \cap D_2) \cup (D_1' \cap D_2)$: a diamond is drawn on first and on the second or a diamond is not drawn on the first but a diamond is drawn on the second.

$$P(D_2) = P(D_1 \cap D_2) + P(D_1' \cap D_2) = (\frac{13}{52})(\frac{12}{51}) + (\frac{39}{52})(\frac{13}{51}) = \frac{663}{2652} = \frac{13}{52}$$

➤ **(b).**
D_1: the event that a diamond is drawn on the first draw.

D_2: the event that a diamond is drawn on the second draw.

E: the event that a diamond is drawn on the first or second drawing.
$E = D_1 \cup D_2$

$$P(E) = P(D_1 \cup D_2) = P(D_1) + P(D_2) - P(D_1 \cap D_2) = \frac{13}{52} + \frac{13}{52} - (\frac{13}{52})(\frac{12}{51}) = \frac{1170}{2652}$$

Alternative solution:

$$E' = (D_1 \cup D_2)' = D_1' \cap D_2'$$

$$P(E') = P(D_1' \cap D_2') = P(D_1')P(D_2'|D_1') = (39/52)(38/51)$$

$$P(E) = 1 - P(E') = 1 - (39/52)(38/51) = 1 - 1482/2652 = 1170/2652$$

➤ **(c).**
E: the event that a diamond and club are drawn.

Since order is not required we must consider all possibilities: $E = (D_1 \cap C_2) \cup (C_1 \cap D_2)$

$$P(E) = P(D_1 \cap C_2) + P(C_1 \cap D_2) = P(D_1)P(C_2|D_1) + P(C_1)P(D_2|C_1) =$$

$$(\frac{13}{52})(\frac{13}{51}) + (\frac{13}{52})(\frac{13}{51}) + \frac{338}{2652}$$

➤ **(d).**
E: at least 1 diamond or club is drawn.

$$E = (D_1 \cup C_2) \cup (C_1 \cup D_2)$$

E′: no diamond and no club are drawn.

$$E' = (D_1' \cap C_2') \cap (C_1' \cap D_2') = (D_1' \cap C_1') \cap (D_2' \cap C_2')$$

$$P(E') = P[(D_1'\cap C_1')\cap(D_2'\cap C_2')] = P(D_1'\cap C_1')P[(D_2'\cap C_2')|(D_1'\cap C_1')] = (\frac{26}{52})(\frac{25}{51}) = \frac{650}{2652}$$

$$P(E) = 1 - P(E') = 1 - \frac{650}{2652} = \frac{2002}{2652}$$

12.3 - Solved Problem 3: Two cards are drawn at random without replacement from an ordinary deck of cards.

(a). Find the probability that the first card is a king and the second card is a diamond.

(b). Find the probability that the first card is a king or the second card is a diamond.

Solutions:

➤ **(a).**
K_1: the event that a king is drawn on the first draw.

D_1: the event that a diamond is drawn on the first draw.

D_2: the event that a king is drawn on the second draw.

$E = K_1\cap D_2$: the event that a king is drawn on the first drawing and a diamond on the second drawing.

$$K_1 = (K_1\cap D_1)\cup(K_1\cap D_1')$$

$$E = K_1\cap D_2 = [(K_1\cap D_1)\cup(K_1\cap D_1')]\cap D_2 = [(K_1\cap D_1)\cap D_2]\cup[(K_1\cap D_1')\cap D_2]$$

$$P(E) = P(K_1\cap D_2) = P([(K_1\cap D_1)\cap D_2]\cup[(K_1\cap D_1')\cap D_2]) = P[(K_1\cap D_1)\cap D_2] + P[(K_1\cap D_1')\cap D_2] =$$

$$P(K_1\cap D_1)P[D_2|(K_1\cap D_1)] + P(K_1\cap D_1')P[D_2|(K_1\cap D_1')] = (\frac{1}{52})(\frac{12}{51}) + (\frac{3}{52})(\frac{13}{51}) = \frac{1}{52}$$

➤ **(b).**
K_1: the event that a king is drawn on the first draw.

D_2: the event that a diamond is drawn on the second draw.

E: the event that a king is drawn on the first drawing or a diamond on the second drawing.

$$E = K_1\cup D_2$$

$P(D_2) = 13/52$ (**See 12.3 - Solved Problem 2**).

$$P(E) = P(K_1\cup D_2) = P(K_1) + P(D_2) - P(K_1\cap D_2) = \frac{4}{52} + \frac{13}{52} - \frac{1}{52} = \frac{4}{13}$$

12.3 - Solved Problem 4: Two urns are sitting on a table. Urn 1 has 3 red marbles and 7 white marbles. Urn 2 has 6 red marbles and 3 white marbles. A marble is selected from urn 1 and placed in urn 2. Next a marble is selected from urn 2. Find the probability that the marble selected from urn 2 is red.

Solution:

Let $\mathbf{R}_1$ be the event that a red is drawn from urn 1 and placed in urn 2. Let $\mathbf{R}_2$ be the event that a red is drawn from urn 2. Then,

$$\mathbf{R}_2 = (\mathbf{R}_1 \cap \mathbf{R}_2) \cup (\mathbf{R}_1' \cap \mathbf{R}_2)$$

$$P(\mathbf{R}_1) = 3/10$$

$$P(\mathbf{R}_2 | \mathbf{R}_1) = 7/10$$

$$P(\mathbf{R}_1') = 7/10$$

$$P(\mathbf{R}_2 | \mathbf{R}_1') = 6/10$$

$$P(\mathbf{R}_1 \cap \mathbf{R}_2) = P(\mathbf{R}_1)P(\mathbf{R}_2 | \mathbf{R}_1) = (3/10)(7/10) = 21/100$$

$$P(\mathbf{R}_1' \cap \mathbf{R}_2) = P(\mathbf{R}_1')P(\mathbf{R}_2 | \mathbf{R}_1') = (7/10)(6/10) = 42/100$$

$$P(\mathbf{R}_2) = P(\mathbf{R}_1 \cap \mathbf{R}_2) + P(\mathbf{R}_1' \cap \mathbf{R}_2) = 21/100 + 42/100 = 63/100$$

12.3 - Solved Problem 5: According to the Arizona Chapter of the American Lung Association, 7.0% of the population has lung disease. Of those people that have lung disease, 90% smoke and of those not having lung disease, 25.3% are smokers. Suppose a person is selected at random from the population. Find the probability that the person selected is a smoker.

Solution:

Let **S** be the event that the person selected is a smoker.

Let **D** be the event that the person has a lung disease.

From the problem, P(**D**) is the probability that the person selected has lung disease: P(**D**) = 0.07.

P(**S**|**D**) is given that the person has a lung disease, the probability the person smokes: P(**S**|**D**) = 0.90.

P(**S**|**D'**) is given that the person does not have a lung disease, the probability the person smokes:

$P(S|D') = 0.253$.

Now $S = (S \cap D) \cup (S \cap D')$ and

$P(S) = P(S \cap D) + P(S \cap D') = P(D)P(S|D) + P(D')P(S|D') = (0.070)(0.90) + (0.93)(0.253) \approx 0.30$.

Unsolved Problems with Answers

12.3 - Problem 1: Two urns are sitting on a table. Urn 1 has 5 blue marbles, 10 red marbles and 15 white marbles. Urn 2 has 10 blue marbles, 10 red marbles and 10 white marbles. A card is selected from an ordinary deck of cards. If a king occurs, a marble is selected at random from urn 1; otherwise a marble is selected at random from urn 2. Find the probability that the marble selected is white or red.

Answer:

53/78

⇑ *Refer back to* **12.3 - Example 1 & 12.3 - Solved Problem 1.**

12.3 - Problem 2: Two cards are drawn at random without replacement from an ordinary deck of cards.

(a). Find the probability that the second card drawn is a face card.

(b). Find the probability that the first card is a face card or the second card is a face card.

(c). Find the probability that a face card and ace are drawn.

(d). Find the probability that at least one face card or one ace is drawn.

Answers:

➤ **(a).** 12/52

➤ **(b).** 1092/2652

➤ **(c).** 96/2652

➤ **(d).** 1392/2652

⇑ *Refer back to* **12.3 - Example 2 & 12.3 - Solved Problem 2**.

12.3 - Problem 3: Two cards are drawn at random without rreplacement from an ordinary deck of cards.

(a). Find the probability that the first card is a diamond and the second card is a face card.

(b). Find the probability that the first card is a diamond or the second card is a face card.

.
Answers:

➤ **(a).** 3/52

➤ **(b).** 22/52

⇑ *Refer back to* **12.3 - Example 3 & 12.3 - Solved Problem 3.**

12.3 - Problem 4: Two urns are sitting on a table. Urn 1 has 3 red marbles and 7 white marbles. Urn 2 has 4 blue marbles, and 10 white marbles. A marble is selected from urn 1 and placed in urn 2. Next a marble is selected from urn 2. Find the probability that the marble selected from urn 2 is white.

Answer:

107/150

⇑ *Refer back to* **12.3 - Example 4 & 12.3 - Solved Problem 4.**

12.3 - Problem 5: Two large toy companies wish to sell talking teddy bears. One company, Toys International estimates there is a 75% chance their teddy bear will make a profit for the company provided the competing toy company does not introduce a talking teddy bear on the market and a 35% chance it will be profitable if the competing company introduces such a toy on the market. Further, it estimates there is a 60% chance the competing company will introduce the toy. Find the probability that the teddy bear is profitable.

Answer:

0.51

⇑ *Refer back to* **12.3 - Example 5 & 12.3 - Solved Problem 5.**

12.4 - Mutually Independent Events

Two Independent Events

Two events are said to be independent if the occurrence of one event has no affect on the occurrence of the other event. This can be tested in two ways:

Two events **A** and **B** are independent if

1. $P(A|B) = P(A)$

or

2. $P(A \cap B) = P(A)P(B)$.

Events that are not independent are said to be dependent.

It can be shown that if **A** and **B** are independent events then $P(A'|B) = P(A')$.

Pair-wise Independent Events

A sequence of events $A_1, A_2,..., A_n$ is said to be pair-wise independent if

$P\{A_j \cap A_k\} = P(A_j)P(A_k)$ for all j, k $(1 \le j < k \le n)$.

Mutually Independent Events

A sequence of events $A_1, A_2,..., A_n$ is said to be mutually independent if for all combinations
$(1 \le j < k <...\le n)$ the multiplication rules

$P\{A_i \cap A_k\} = P(A_i)P(A_k)$

$P\{A_i \cap A_j \cap A_k\} = P(A_i)P(A_j)P(A_k)$

::

$P\{A_1 \cap A_2 \cap...\cap A_n\} = P(A_1)P(A_2)...P(A_n)$

Although there are examples of pair wise independent sequence of events that are not mutually independent (see supplementary problems 34 and 35), we shall assume, unless otherwise stated, that all independent events are mutually independent.

12.4 - Example 1: A card is drawn from an ordinary deck of cards. Show that the event a (a). king is drawn and the event a diamond is drawn are independent events.

(b). king is drawn and the event queen is drawn are dependent events.

Solutions:

➤ **(a).**
K: The event a king is drawn.
D: The event a diamond is drawn.
K∩D: A king of diamonds is drawn.

$P(K∩D) = 1/52$

$P(K) = 4/52 = 1/13$

$P(D) = 13/52 = 1/4$

$P(K)P(D) = (1/13)(1/4) = 1/52 = P(K∩D)$, which shows that the two events are independent.

➤ **(b).**
K: The event a king is drawn.
Q: The event a queen is drawn.

$P(Q|K) = 0 ≠ P(Q)$

Therefore, the two events are dependent.

12.4 - Example 2: Two cards are drawn from an ordinary deck of cards without replacement.

(a), Show that the event a face card is drawn on the first drawing and the event a diamond card is drawn on the second drawing are independent.

(b). Show that the event a king is drawn on the first drawing and the event a face card is drawn on the second drawing are dependent.

Solutions:

➤ **(a).**
F_1: the event a face card is drawn on the first drawing.

D_2: the event a diamond card is drawn on the second drawing.

$P(D_1) = 13/52 = 1/4$

$P(F_2) = 12/52$ (**See 12.3 - unsolved problem 2**)

$D_1 = (D_1 \cap F_2) \cup (D_1 \cap F_2')$

$P(D_1 \cap F_2) = P\{[(D_1 \cap F_1) \cup (D_1 \cap F_1')] \cap F_2\} = P\{[(D_1 \cap F_1) \cap F_2] \cup [(D_1 \cap F_1') \cap F_2]\} =$

$P\{(D_1 \cap F_1) \cap F_2\} + P\{(D_1 \cap F_1') \cap F_2\} = P(D_1 \cap F_1)P(F_2 | D_1 \cap F_1) + P(D_1 \cap F_1')P(F_2 | D_1 \cap F_1') =$

$(3/52)(11/51) + (10/52)(12/51) = 153/2652 = 3/52$

$P(D_1)P(F_2) = (13/52)(12/52) = 3/52$

Therefore, $P(D_1 \cap F_2) = P(D_1 P(F_2)$.

which shows independence of the two events.

➤ **(b).**
K_1: the event a king is drawn on the first drawing.
F_2: the event a face is drawn on the second drawing.

$P(F_2 | K_1) = 11/51 \neq P(F_2) = 12/52$.

Therefore, the these two events are dependent.

12.4 - Example 3: A recent study of criminal arrests based on gender showed the following data:

	Males	Females	Total
Guilty	150	225	375
Innocent	50	75	125
Total	200	300	500

From this group, a person is selected at random. Let **M** be the event a male is selected and **G** the event that the person is guilty. Find out if these two events are independent.

Solution:

We will use the test:

If the events **M** and **G** are independent then $P(G | M) = P(G)$: Given that a male is selected the probability that he is guilty is the same as the probability that a person (male or female) selected at random is guilty.

Since #(G∩M) = 150, #M = 200, #G = 375,

Step 1: $P(G|M) = \dfrac{P(G\cap M)}{P(M)} = \dfrac{\frac{150}{500}}{\frac{200}{500}} = \dfrac{150}{200} = \dfrac{3}{4}.$

Step 2: $P(G) = \dfrac{375}{500} = \dfrac{3}{4}.$

Step 3: $P(G|M) = P(G) = \dfrac{3}{4}.$

Therefore, **G** and **M** are independent.

12.4 - Example 4: A survey was made of families with three children. One such family is picked at random. Let **B** be the event that the family has both sexes and **G** the event that at most one girl is in the family. Find out if these two events are independent.

Solution:

Assume **S** is the sample space in order of birth.

S = {(g,g,b), (g,b,g), (b,g,g), (g,g,g), (b,b,g), (b,g,b), (g,b,b), (b,b,b)}

B: the event that the family has at least one girl and one boy = {(b,b,g), (b,g,b), (g,b,b), (g,g,b), (g,b,g), (b,g,g)}.

G: the event that the family has one girl or no girls = {(b,b,b), (b,b,g), (b,g,b), (g,b,b)}.

B∩G: the event that the family has exactly one girl and two boys = {(b,b,g), (b,g,b), (g,b,b)}.

To test if the events **B** and **G** are independent we use $P(B\cap G) = P(B)P(G)$.

Step 1: $P(B\cap G) = \dfrac{\#(B\cap G)}{\#S} = \dfrac{3}{8}.$

Step 2: $P(B) = \dfrac{\#B}{\#S} = \dfrac{6}{8}.$

Step 3: $P(G) = \dfrac{\#G}{\#S} = \dfrac{4}{8}.$

Step 4: $P(B)P(G) = (\dfrac{6}{8})(\dfrac{4}{8}) = \dfrac{3}{8}.$

Therefore, $P(B\cap G) = P(B)P(G)$.

Thus, **B** and **G** are independent.

12.4 - Example 5: Mr. Hope is a famous baseball handicapper. He claims that his chance of predicting a winning game is 0.60. Assume he bets on 3 games. Find the probability that he wins 2 games.

Solution:

Let $\mathbf{W}_1$, $\mathbf{W}_2$, $\mathbf{W}_3$ represent the events that a win occurs on the first, second, and third games respectively. It is reasonable to assume that these events are mutually independent since one selection should not affect another selection.

T: The event that he wins on exactly two games.

The following are the ways he can win exactly two games:

He wins the first and second games but loses the third game: $\mathbf{W}_1 \cap \mathbf{W}_2 \cap \mathbf{W}_3'$.

He wins the first and third games but loses the second game: $\mathbf{W}_1 \cap \mathbf{W}_2' \cap \mathbf{W}_3$.

He loses the first game but wins the second and third game: $\mathbf{W}_1' \cap \mathbf{W}_2 \cap \mathbf{W}_3$.

The event that he wins 2 games is the union of the three above events:

$$\mathbf{T} = (\mathbf{W}_1 \cap \mathbf{W}_2 \cap \mathbf{W}_3') \cup (\mathbf{W}_1 \cap \mathbf{W}_2' \cap \mathbf{W}_3) \cup (\mathbf{W}_1' \cap \mathbf{W}_2 \cap \mathbf{W}_3).$$

Since these three events are disjoint we have

$$P(\mathbf{T}) = P(\mathbf{W}_1 \cap \mathbf{W}_2 \cap \mathbf{W}_3') + P(\mathbf{W}_1 \cap \mathbf{W}_2' \cap \mathbf{W}_3) + P(\mathbf{W}_1' \cap \mathbf{W}_2 \cap \mathbf{W}_3)$$

Since $\mathbf{W}_1$, $\mathbf{W}_2$, $\mathbf{W}_3$ are mutually independent events, we have

$$P(\mathbf{W}_1 \cap \mathbf{W}_2 \cap \mathbf{W}_3') = P(\mathbf{W}_1)P(\mathbf{W}_2)P(\mathbf{W}_3')$$

$$P(\mathbf{W}_1 \cap \mathbf{W}_2' \cap \mathbf{W}_3) = P(\mathbf{W}_1)P(\mathbf{W}_2')P(\mathbf{W}_3)$$

$$P(\mathbf{W}_1' \cap \mathbf{W}_2 \cap \mathbf{W}_3) = P(\mathbf{W}_1')P(\mathbf{W}_2)P(\mathbf{W}_3)$$

We can now write $P(\mathbf{T}) = P(\mathbf{W}_1 \cap \mathbf{W}_2 \cap \mathbf{W}_3') + P(\mathbf{W}_1 \cap \mathbf{W}_2' \cap \mathbf{W}_3) + P(\mathbf{W}_1' \cap \mathbf{W}_2 \cap \mathbf{W}_3) =$

$P(\mathbf{W}_1)P(\mathbf{W}_2)P(\mathbf{W}_3') + P(\mathbf{W}_1)P(\mathbf{W}_2')P(\mathbf{W}_3) + P(\mathbf{W}_1')P(\mathbf{W}_2)P(\mathbf{W}_3) =$

$(0.60)(0.60)(0.4) + (0.6)(0.4)(0.6) + (0.4)(0.6)(0.6) = 3(0.6)^2(0.4) = 0.432.$

12.4 - Example 6: At a local college, a survey was taken in a class of 52 foreign language students. The following was the result of this survey:

1 students speaks German(**G**),French(**F**) and Italian (**I**).
2 students speaks French and German.
2 students speaks Italian and German.

13 students speaks Italian and French.

4 students speaks German

26 students speaks French

26 students speaks Italian

A student is randomly selected. Show that the events **G, F, I** are mutually independent.

Solution:

From the Venn diagram constructed from the data above we find:

#**G** = 4

#**F** = 26

#**I** = 26

#**S** = 52, the cardinality of the sample space **S**.

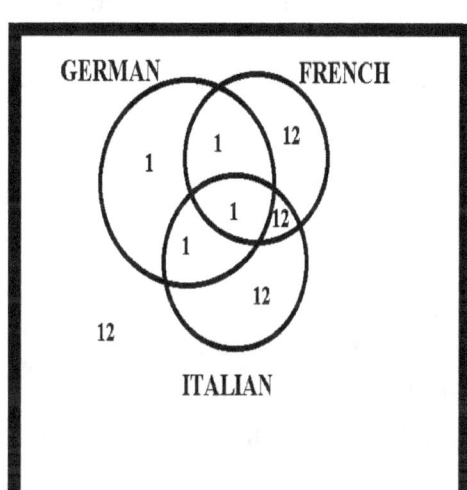

#(**F**∩**G**) = 2

#(**I**∩**G**) = 2

#(**F**∩**I**) = 13

#(**F**∩**G**∩**I**) = 1

$$P(\mathbf{F} \cap \mathbf{G}) = \frac{2}{52} = \frac{1}{26}$$

$$P(\mathbf{F})P(\mathbf{G}) = (\frac{26}{52})(\frac{4}{52}) = \frac{1}{26}$$

Therefore, $P(\mathbf{F} \cap \mathbf{G}) = P(\mathbf{F})P(\mathbf{G})$

$$P(\mathbf{I} \cap \mathbf{G}) = \frac{2}{52} = \frac{1}{26}$$

$$P(\mathbf{I})P(\mathbf{G}) = (\frac{26}{52})(\frac{4}{52}) = \frac{1}{26}$$

Therefore, $P(\mathbf{I} \cap \mathbf{G}) = P(\mathbf{I})P(\mathbf{G})$

$$P(\mathbf{I} \cap \mathbf{F}) = \frac{13}{52} = \frac{1}{4}$$

$$P(\mathbf{I})P(\mathbf{F}) = (\frac{26}{52})(\frac{26}{52}) = \frac{1}{4}$$

Therefore, $P(\mathbf{I} \cap \mathbf{F}) = P(\mathbf{I})P(\mathbf{F})$

$$P(\mathbf{I} \cap \mathbf{F} \cap \mathbf{G}) = \frac{1}{52}$$

$$P(I)P(F)P(G) = (\frac{26}{52})(\frac{26}{52})(\frac{4}{52}) = \frac{1}{52}$$

Therefore, $P(I \cap F \cap G) = P(I)P(F)P(G)$

We conclude that the three events are mutually independent.

Solved Problems

12.4 - Solved Problem 1: A card is drawn from an ordinary deck of cards. Show that the event

(a). face card is drawn and the event a diamond is drawn are independent events.

(b). face card is drawn and the event a king is drawn are dependent events.

Solutions:

➤ **(a).**
F: The event a face card is drawn.
D: The event a diamond is drawn.
F∩D: A face card that is also a diamond is drawn.

$P(F \cap D) = 3/52$

$P(F) = 12/52 = 3/13$

$P(D) = 13/52 = 1/4$

$P(F)P(D) = (3/13)(1/4) = 3/52 = P(F \cap D)$, which shows that the two events are independent.

➤ **(b).**
F: The event a face card is drawn.
K: The event a king is drawn.
F∩K: A face card that is also a king is drawn.

$P(F|K) = 1$

$P(F) = 12/52 = 3/13 \neq 1$

which shows that the two events are dependent.

12.4 - Solved Problem 2: Two cards are drawn from an ordinary deck of cards without replacement.

(a). Show that the event a diamond is drawn on the first drawing and the event a king is drawn on the second drawing are independent.

(b). Show that the event a king is drawn on the first drawing and the event a king is drawn on the second drawing are dependent.

Solutions:

➤ (a).
D_1: the event a diamond is drawn on the first drawing.
K_2: the event a king is drawn on the second drawing.

$P(D_1) = 13/52 = 1/4$
$P(K_2) = 4/52 = 1/13$ (See **3.3 Example 3.**)

$D_1 = (D_1 \cap K_1) \cup (D_1 \cap K_1')$

$P(D_1 \cap K_2) = P\{[(D_1 \cap K_1) \cup (D_1 \cap K_1')] \cap K_2\} = P\{[(D_1 \cap K_1) \cap K_2] \cup [(D_1 \cap K_1') \cap K_2]\} =$

$P\{(D_1 \cap K_1) \cap K_2\} + P\{(D_1 \cap K_1') \cap K_2\} = P(D_1 \cap K_1)P(K_2 | D_1 \cap K_1) + P(D_1 \cap K_1')P(K_2 | D_1 \cap K_1') =$

$$(\frac{1}{52})(\frac{3}{51}) + (\frac{12}{52})(\frac{4}{51}) = \frac{1}{52} = (\frac{1}{4})(\frac{1}{13}) = P(D_1)P(K_2),$$

which shows independence of the two events.

➤ (b).
K_1: the event a king is drawn on the first drawing.
K_2: the event a king is drawn on the second drawing.

$P(K_2 | K_1) = 3/51 \neq P(K_2) = 4/52$

12.4 - Solved Problem 3: A recent study of criminal arrests based on gender showed the following data:

	Males	Females	Total
Guilty	150	225	375
Innocent	50	75	125
Total	200	300	500

From this group, a person is selected at random. Let **F** be the event a female is selected and **I** the event that the person is innocent. Find out if these two events are independent.
Solution:

Since #($\mathbf{I} \cap \mathbf{F}$) = 75, #$\mathbf{F}$ = 300, #$\mathbf{I}$ = 125,

Step 1: $P(\mathbf{I}|\mathbf{F}) = \dfrac{P(\mathbf{I} \cap \mathbf{F})}{P(\mathbf{F})} = \dfrac{\frac{75}{500}}{\frac{300}{500}} = \dfrac{75}{300} = \dfrac{1}{4}.$

Step 2: $P(\mathbf{I}) = \dfrac{125}{500}.$

Therefore, **F** and **I** are independent.

12.4 - Solved Problem 4: A survey was made of families with three children. One such family is picked at random. Let **B** be the event that the family has both sexes and **G** the event that at least two girls are in the family. Find out if these two events are independent. **S** is the sample space in order of birth.

Solution:

S = {(g,g,b), (g,b,g), (b,g,g), (g,g,g), (b,b,g), (b,g,b), (g,b,b), (b,b,b)}

B = {(b,b,g), (b,g,b), (g,b,b), (g,g,b), (g,b,g), (b,g,g)}.

G = {(g,g,g), (g,g,b), (g,b,g), (b,g,g)}

B∩**G** = {(g,g,b), (g,b,g), (b,g,g)}.

Step 1: $P(\mathbf{B} \cap \mathbf{G}) = \dfrac{3}{8}.$

Step 2: $P(\mathbf{B}) = \dfrac{\#\mathbf{B}}{\#\mathbf{S}} = \dfrac{6}{8}.$

Step 3: $P(\mathbf{B}) = \dfrac{\#\mathbf{G}}{\#\mathbf{S}} = \dfrac{4}{8}.$

Step 4: $P(\mathbf{B})P(\mathbf{G}) = (\dfrac{6}{8})(\dfrac{4}{8}) = \dfrac{3}{8}.$

Thus **B** and **G** are independent.

12.4 - Solved Problem 5: Mr. Hope is a famous baseball handicapper. He claims that his chance of predicting a winning game is 0.60. Assume he bets on 3 games. Find the probability that he wins at least 2 games.

Solution:

Let W_1, W_2, W_3 represent the events that a win occurs on the first, second, and third games respectively. It is reasonable to assume that these events are mutually independent. Let **T** be the event that he wins at least two games:

$$T = (W_1 \cap W_2 \cap W_3') \cup (W_1 \cap W_2' \cap W_3) \cup (W_1' \cap W_2 \cap W_3) \cup (W_1 \cap W_2 \cap W_3)$$

$$P(T) = P(W_1 \cap W_2 \cap W_3') + P(W_1 \cap W_2' \cap W_3) + P(W_1' \cap W_2 \cap W_3) + P(W_1 \cap W_2 \cap W_3) =$$

$$P(W_1)P(W_2)P(W_3') + P(W_1)P(W_2')P(W_3) + P(W_1')P(W_2)P(W_3) + P(W_1)P(W_2)P(W_3) =$$

$(0.6)(0.6)(0.4) + (0.6)(0.4)(0.6) + (0.4)(0.6)(0.6) + (0.6)(0.6)(0.6) = 0.144 + 0.144 + 0.144 + 0.216 = 0.648.$

12.4 - Solved Problem 6: From 12.4 - Example 5, show that the events **G′**, **F′**, **I′** are mutually independent.

Solution:

From the Venn diagram constructed from the data in Example 4.5, we find:
#**G′** = 48
#**F′** = 26
#**I′** = 26
#**S** = 52, the cardinality of the sample space **S**.

$\#(F' \cap G') = \#(F \cup G)' = 24$

$\#(I' \cap G') = \#(I \cup G)' = 24$

$\#(F' \cap I') = \#(F \cup I)' = 13$

$\#(F' \cap G' \cap I') = \#(F \cup G \cup I)' = 12$

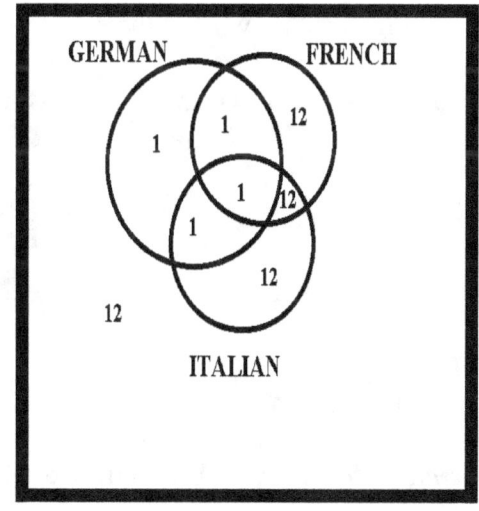

$P(F' \cap G') = \dfrac{24}{52} = \dfrac{6}{13}$

$P(F')P(G') = (\dfrac{26}{52})(\dfrac{48}{52}) = \dfrac{6}{13}$

Therefore, $P(F' \cap G') = P(F')P(G')$.

$P(I' \cap G') = \dfrac{24}{52} = \dfrac{6}{13}$

$P(I')P(G') = (\dfrac{26}{52})(\dfrac{48}{52}) = \dfrac{6}{13}$

Therefore, $P(\mathbf{I}' \cap \mathbf{G}') = P(\mathbf{I}')P(\mathbf{G}')$.

$$P(\mathbf{I}' \cap \mathbf{F}') \;=\; \frac{13}{52} \;=\; \frac{1}{4}$$

$$P(\mathbf{I}')P(\mathbf{F}') = \; \left(\frac{26}{52}\right)\left(\frac{26}{52}\right) \;=\; \frac{1}{4}$$

Therefore, $P(\mathbf{I}' \cap \mathbf{F}') = P(\mathbf{I}')P(\mathbf{F}')$.

$$P(\mathbf{I}' \cap \mathbf{F}' \cap \mathbf{G}') \;=\; \frac{12}{52} \;=\; \frac{3}{13}$$

$$P(\mathbf{I}')P(\mathbf{F}')P(\mathbf{G}') = \; \left(\frac{26}{52}\right)\left(\frac{26}{52}\right)\left(\frac{48}{52}\right) \;=\; \frac{3}{13}$$

Therefore, $P(\mathbf{I}' \cap \mathbf{F}' \cap \mathbf{G}') = P(\mathbf{I}')P(\mathbf{F}')P(\mathbf{G}')$.

We conclude that the three events are mutually independent.

Unsolved Problems with Answers

12.4 - Problem 1: A card is drawn from an ordinary deck of cards. Show that the event

(a). a king is not drawn and the event a diamond is drawn are independent events.

(b). a king is not drawn and a queen is drawn are dependent events.

Answers:

➤ **(a).**
$P(\mathbf{K}')P(\mathbf{D}) = \; 3/13 = P(\mathbf{K}' \cap \mathbf{D})$

➤ **(b).**
$P(\mathbf{K}'|\mathbf{Q}) = 1 \neq P(\mathbf{K}')$

⇑ *Refer back to* **12.4 - Example 1 & 12.4 - Solved Problem 1**.

12.4 - Problem 2: Two cards are drawn from an ordinary deck of cards without replacement. Show that the event a

(a). face card is drawn on the first drawing and the event a diamond is not drawn on the second drawing are independent.

(b). king is drawn on the first drawing and the event a queen is not drawn on the second drawing are dependent .

Answers:

➤ (a).
$P(F_1)P(D_2') = 9/52 = P(F_1 \cap D_2')$

➤ (b).
$P(Q_2'|K_1) = 47/51 \neq P(Q_2') = 48/52$

⇑ *Refer back to* **12.4 - Example 2 & 12.4 - Solved Problem 2**

12.4 - Problem 3: In September, 1988, the House of Representatives voted on an amendment requiring life imprisonment for drug-related murders. Results of the vote were reported as shown below:

	YEA	NAY	DID NOT VOTE	Total
Democrat	153	83	19	255
Republican	169	0	8	177
Total	322	83	27	432

One person from this group is selected at random. Let **D** be the event that he or she is a Democrat and **Y** the event that this person selected voted for the amendment. Find out if these two events are independent.

Answer:

Not independent.

⇑ *Refer back to* **12.4 - Example 2 & 12.4 - Solved Problem 2.**

.
12.4 - Problem 4: A survey was made of families with two children. One such family is picked at random. Let **B** be the event that the family has both sexes and **G** the event that exactly 2 girls are in the family. Find out if these two events are independent.

Answer:

Not independent.

⇑ *Refer back to* **12.4 - Example 3 & 12.4 - Solved Problem 3.**

12.4 - Problem 5: Mr. Hope is a famous baseball handicapper. He claims that his chance of predicting a winning game is 0.60. Assume he bets on 3 games. Find the probability that he wins only two games in a row.

Answer:

0.288

⇑ *Refer back to* **12.4 - Example 4 & 12.4 - Solved Problem 4**.

12.4 - Problem 6: From 12.4 - Example 5, show that the events **G, F′, I′** are mutually independent.

Answers:

$$P(G) = \frac{1}{13} \ , \ \ P(F') = \frac{1}{2} \ , \ \ P(I') = \frac{1}{2} \ , \ \ P(G \cap F') = \frac{1}{26} \ , \ \ P(G \cap I') = \frac{1}{26} \ , \ \ P(I' \cap F') = \frac{1}{4} \ ,$$

$$P(G \cap I') = \frac{1}{26} = P(G)P(I') = (\frac{1}{13})(\frac{1}{2}) \ , \ \ P(G \cap F') = \frac{1}{26} = P(G)P(F') = (\frac{1}{13})(\frac{1}{2}) \ ,$$

$$P(I' \cap F') = \frac{1}{4} = P(I')P(F') = (\frac{1}{2})(\frac{1}{2}) \ , \ \ P(G \cap I' \cap F') = \frac{1}{52} = P(G)P(I')P(F') = (\frac{1}{13})(\frac{1}{2})(\frac{1}{2}).$$

⇑ *Refer back to* **12.4 - Example 5 & 12.4 - Solved Problem 5**.

Supplementary Problems

1. Two cards are drawn without replacement from an ordinary deck of cards. Find the probability a king and queen are drawn.

2. Two cards are drawn without replacement from an ordinary deck of cards. Find the probability that the first card is a king and the second card is a diamond.

3. An urn contains 5 white marbles, 10 blue marbles and 25 red marbles. Assume three marbles are selected from the urn without replacement. Find the probability that all three marbles are the same color.

4. Assume two cards are drawn from an ordinary deck of cards without replacement. Find the probability that the first card is a king or queen and the second card is a king or queen.
(Hint: $(A \cup B) \cap (C \cup D) = (A \cap C) \cup (A \cap D) \cup (B \cap C) \cup (B \cap D)$.)

5. Assume two cards are drawn from an ordinary deck of cards without replacement. Find the probability that a king or queen are drawn .

6. Mr. Jones received a box containing 100 computer chips, of which five, unknown to him are defective. He decides to select three chips and test them. If at least one chip is defective, he will

return the box.

a. Find the probability that he will not return the box.

b. Find the probability that at least two of the three chips selected are defective.

7. Three urns are sitting on a table. Urn 1 has three blue marbles and 7 white marbles, urn 2 has 5 red marbles and 10 white marbles and urn 3 has 10 red marbles and 10 white marbles. A die is tossed once. If a 1 occurs, a marble is selected at random from urn 1; if a 2 or 3 occurs, a marble is selected at random from urn 2 and if a 4, 5, or 6 occurs, a marble is selected at random from urn 3. Find the probability that the marble selected is white.

8. Assume two cards are drawn from an ordinary deck of cards without replacement. Find the probability that the first card is a king or diamond and the second card is a king.

9. Mr. Hope is a famous baseball handicapper. He claims that his chance of predicting a winning game is 0.60. Assume he bets until he wins.

a. Find the probability that he will stop betting by the third game.

b. Find the probability that he will stop betting on the third game.

10. Three urns are sitting on a table. Urn 1 has three blue marbles and 7 white marbles. Urn 2 has 4 blue marbles, and 10 white marbles and urn 3 has 5 white marbles and 5 blue marbles. A marble is selected from urn 1 and placed in urn 2. Next a marble is selected from urn 2 and placed in urn 3. Finally a marble is selected from urn 3. Find the probability that the marble selected from urn 3 is blue.

11. Two decks of ordinary cards are sitting on a table. A card is randomly selected from the first deck and placed in the second deck. Next a card is selected at random from the second deck and placed in the first deck. Finally a card is randomly selected from the first deck. Find the probability that the card finally selected is a diamond.

12. Two urns are sitting on a table. Urn 1 has 5 red and 5 black marbles and Urn 2 has 7 red marbles and 2 black marbles. A die is tossed once. If a 3 appears a marble is drawn at random from urn 1 and placed in urn 2. Next a marble is selected at random from urn 2. Otherwise a marble is drawn at random from urn 2 and placed in urn 1. Next a marble is selected at random from urn 1. Find the probability that the last marble selected is black.

13. A English class has 60 women and 40 men. Students are selected at random one at a time , without replacement, until two males are selected. Find the probability that this process will stop on the fourth selection.

14. Two non-empty events **A** and **B** are said to be mutually exclusive if $A \cap B = \phi$. Are such events always dependent?

15. Three cards are drawn from an ordinary deck of cards without replacement. Let K_1 = the event a king is drawn on the first drawing, K_2 = the event a king is drawn on the second drawing and K_3 = the event a king is drawn on the third drawing.

a. Find $P(K_1 | K_3)$.

b. Does $P(K_1 | K_3) = P(K_3 | K_1)$?

16. Assume that **A** and **B** are independent events. Show that

a. **A** and **B'** are independent.

(First show $P(B' | A) = 1 - P(B | A)$).

b. **A'** and **B'** are independent.

17. Two urns are sitting on a table. Urn A has 6 white marbles and 4 black marbles. Urn B has 3 white marbles and 7 black marbles. Two marbles are to be drawn from one or both urns under the following rule:

A die is tossed. If a 1 or 2 occurs then 2 marbles are selected from urn A, if a 3, 4, or 5 occurs then 2 marbles are selected from urn B, and if a 6 occurs then one marble is selected from each urn.

Find the probability that both marbles selected are white.

18. Show $P(E | A \cup B) = \dfrac{P(E \cap A) + P(E \cap B) - P(E \cap A \cap B)}{P(A) + P(B) - P(A \cap B)}$.

19. Show Rule 2: $P(A \cup B | E) = P(A | E) + P(B | E) - P(A \cap B | E)$.

20. Show Rule 1: $P(E' | B) = 1 - P(E | B)$.

21. Two cards are drawn, without replacement, from an ordinary deck of cards. Let K_1 be the event that a king is drawn on the first card and K_2 the event that a king is drawn on the second card. Show that

$P(K_2 | K_1') \neq 1 - P(K_2 | K_1)$.

22. Two cards are drawn, without replacement, from an ordinary deck of cards. Let K_1 be the event a king is drawn on the first card, Q_1 the event a queen is drawn on the first card, and J_2 the event

that a jack is drawn on the second card. Show that

$$P(J_2 |\ K_1 \cup Q_1) \neq P(J_2 | K_1) + P(J_2 | Q_1).$$

Assume an experiment generates a sample space **S** with non-empty events **A** and **B** and **A**, **B** ≠ **S**. For problems 23 - 29 select the correct answers.

23. **A** and **S** are independent. (a). true (b). false (c). undetermined.

24. **A** and **A'** are independent. (a). true (b). false (c). undetermined.

25. If **A**∩**B** = ϕ then **A** and **B** are independent. (a). true (b). false (c). undetermined.

26. If **A**⊂**B** then **A** and **B** are independent. (a). true (b). false (c). undetermined.

27. **A**∩**B** ≠ ϕ then **A** and **B** are independent. (a). true (b). false (c). undetermined.

28. If $P(A|B) = P(B|A)$ then $P(A) = P(B)$. (a). true (b). false (c). undetermined.

29. If $P(A) = P(B)$ then $P(A|B) = P(B|A)$ (a). true (b). false (c). undetermined.

30. Show that $P(A \cap B \cap C) = P(A)P(B|A)P(C|A \cap B)$ is always true. (Assume $A \cap B \cap C \neq \phi$).

31. Two cards are drawn from an ordinary deck of cards without replacement. Find the probability that a diamond is drawn on the first drawing and a face card drawn on the second drawing.

32. Assume two cards are drawn, without replacement, from an ordinary deck of cards. If the first card drawn is a king, find the probability that the second card is a diamond.

33. Show that

$$P(E_1 \cap E_2 \cap E_3 ... \cap E_n) = P(E_1)P(E_2|E_1)P(E_3|E_1 \cap E_2)P(E_4|E_1 \cap E_2 \cap E_3)...P(E_n|E_1 \cap E_2 \cap E_3 ... \cap E_{n-1})$$

34. Assume a die is tossed twice. Let E_1 be the event that an even number occurs on the first toss, E_2 be the event that an even number occurs on the second toss, and **B** the event that the sum of the two numbers is odd.

Show that these events are pair- wise independent but not mutually independent.

35. Assume a die is tossed three times. Let $E_{1,2}$ be the event that the first and second toss results in the same number, $E_{1,3}$ be the event that the first and third toss results in the same number, and $E_{2,3}$ be the event that the second and third toss results in the same number. Show that these events are pair-wise independent but not mutually independent.

36. Assume the events **A**, **B**, **C**, **D** are mutually independent. Show that the two events

a. **E** = **A**∩**B** and **F** = **C**∩**D** are independent.

b. **E** = **B**∪**C** and **D** are independent.

37. Assume an experiment results in events A_k (k = 1, 2, 3). Using the notation B_k = A_k or A_k', show that the events A_k are mutually independent if $P(B_1 \cap B_2 \cap B_3) = P(B_1)P(B_2)P(B_3)$. Note: This result can be generalized to any number of events.

38. Cards are drawn from an ordinary deck, without replacement, until 2 diamonds are drawn or 5 cards are drawn whichever occurs first. Find the probability that 5 cards were drawn.

39. Assume 2 cards are drawn without replacement from an ordinary deck of cards. If a king is not drawn on the first drawing, find the probability that a face card is drawn on the second drawing.

40. From 12.4 example1 , we know that if 1 card is drawn from an ordinary deck of cards, the events "a king is drawn" and the event "a diamond is drawn" are independent events. For these two events show that for any third non -trivial event the three events would be no be mutually independent.

41. A fair coin is tossed until 2 heads occur or 4 tosses, which ever occurs first.

a. Find the probability of each elementary sample value of the sample space **S**.

b. Show that the events that a head occurs on the first toss and a head occurs on the fourth toss are NOT independent events.

c. Find the probability that 4 tosses occurred.

d. If 4 tosses occur, find the probability that a head occurred on the first toss.

42. Two cards are drawn at random without replacement from an ordinary deck of cards.

a. Find the probability that one card is a king and the other card is any kind of diamond..
b. Find the number of ways that one card is a king and the other card is any kind of diamond..

43. Three cards are selected from an ordinary deck of cards. Show that $P(K_1 \cap K_2) = P(K_2 \cap K_3) = P(K_1 \cap K3)$.

44. In a game of poker, Mr. Jones selects 5 cards from an ordinary deck of cards.

a. In ordinary English, express the event $E = (K_1 \cup K_2 \cup K_3) \cap (K_4' \cap K_5')$.

b. Find P(**E**).

c. Find the number of ways **E** can happen.

45. From an ordinary deck of cards, 2 hands are placed on the table where each hand has 2 cards.

a. Find the probability that both hands have exactly 1 king.

b. Find the probability that at least 1 hand has exactly 1 king.

46. A fair die is tossed twice.

a. If the sum of the two numbers is 6, find the probability that the first toss resulted in a 2.

b. If the sum of the two numbers is a 5 or 6, find the probability that the first toss resulted in a 2.

47. Two urns contain red and white marbles. Urn A has 6 red and 3 white and urn B has 8 red and 11 white.
A card is drawn from an ordinary deck of cards. If the card is a diamond, a marble is drawn from urn A and placed in urn B; then a marble is drawn from urn B. If the card is not a diamond, a marble is drawn from urn B and placed in urn A; then a marble is drawn from urn A. Find the probability that the second marble drawn is red.

Definition of conditional independence of events: Assume we have events **A**, **B**, **C**. We define the events **A** and **B** to be conditionally independent relative to **C** if P(A∩B | C) = P(A|C)P(B |C). (Assuming P(**C**) >0.)

48. Two urns are sitting on a table. Urn A has 5 red marbles and 5 black marbles. Urn B has 3 red marbles and 7 black marbles. A fair coin is tossed once. If heads appears, then a marble is first selected from urn A and a second marble is selected from urn B. If tails appears , then a marble is first selected from urn B and a second marble is selected from urn A. Let R_1 be the event that the first marble selected is red and B_2 the event that the second marble selected is black. Let **H** be the event that heads is tossed.

a. Show that the events R_1 and B_2 are not independent.

b. Show that the events R_1 and B_2 are conditionally independent relative to **H**.

49. Assume a fair die is tossed twice. Let **A** be the event that the number 2 appears on the first toss, **B** the event that the number 4 appears on the second toss and **C** the event that the sum of the two numbers is 6.

a. Show that the events **A** and **B** are independent.

b. Show that the events **A**, **B** , and **C** are not mutually independent.

c. Show that the events **A** and **B** are not conditionally independent relative to **C**.

50. Assume a fair die is tossed twice. Let T_1 be the event that the number 2 appears on the first toss, T_2 the event that the number 2 appears on the second toss, E_1 the event that an even number appears on the first toss, and E_2 the event that an even number appears on the second toss.

a. Show that the events T_1 and T_2 are independent.

b. Show that the events T_1 and T_2 are conditionally independent relative to $E_1 \cap E_2$.

51. Assume the pair of events **A** , **B** are independent and the pair of events **C,D** are also independent.
If $A \subseteq C$ and $B \subseteq D$, show $P(A \cap B \mid C \cap D) = P(A \mid C)P(B \mid D)$.

52. Assume the events **A, B** are conditionally independent relative to **C**

a. Show **A'** and **B** are conditionally independent relative to **C**.

b. Show **A'** and **B'** are conditionally independent relative to **C.**

53. A fair die is rolled 3 times. Let $A_{i,k}$ be the event that the ith and kth rolls produce the same number. Show that the events $A_{i,k}$ ($1 \le i < k \le 3$) are pair wise independent but not mutually independent.

54. Three cards are drawn, without replacement, from an ordinary deck of cards. Find the probability that the first card drawn is a face card, the second is a diamond and the third card drawn is a king.

55. Three urns sit on a table. urn 1 has 3 red marbles and 7 black, urn 2 has 12 red and 8 black and urn 3 has 5 red and 5 black. An urn is selected at random and 2 marbles are drawn without replacement from the selected urn. If the first marble selected is red, find the probability that the second marble selected is black.

56. A room contains 2 tables. On each table sits 2 urns containing red and white marbles. The following table contains the contents of the urns.

TABLE 1		TABLE 2	
URN 1	URN 2	URN 3	URN 4
12 red marbles 8 white marbles	15 red marbles 5 white marbles	10 red marbles 10 white marbles	5 red marbles 15 red marbles

A table and an urn sitting on the table are selected at random. From the urn, 1 marble is selected. Find the probability the marble selected is red.

57. Assume 3 urns sit on a table. Urn A has 3 red and 7 white marbles; urn B has 7 red and 3 white marbles; and urn C has 5 red and 5 black marbles. Two urns are selected at random and from each urn 1 marble is drawn. Assume the first marble drawn is selected according to the order of the alphabet of the urn's label. (For example, if urns B and C were selected then the first marble would be drawn from urn B and the second from urn C). Show that the event that both marbles selected are red is not independent.

58. Two urns each contain marbles. Urn A contains 3 red marbles and 5 white marbles and urn B contains 7 red marbles and 3 white marbles. A fair die is tossed once. If the number 3 appears, a marble is selected from Urn A otherwise a marble is selected from Urn B. If a red marble is selected, what is the probability that Urn A was selected?

Bayesian analysis gives us a deeper understanding of the relationships between different events. It addresses the following fundamental question: given that certain events occur, what is the probability that other events will also occur.

13.1 - What is Bayes's Theorem?

Bayes's theorem is a formula that allows us to write $P(A|B)$ in terms of $P(B|A)$:

$$P(A|B) = \frac{P(B|A)P(A)}{P(B)}.$$

Important formulas to use with Bayes's formula:

$$B = (A \cap B) \cup (A' \cap B)$$

$$P(B) = P(A \cap B) + P(A' \cap B)$$

$$P(B) = P(A)P(B|A) + P(A')P(B|A')$$

$$P(A|B) = \frac{P(B|A)P(A)}{P(A)P(B|A) + (P(A')P(B|A')}.$$

$$P(A|B') = \frac{[1 - P(B|A)]P(A)}{1 - P(B)}$$

Frequently this is necessary since there are probability problems where we need to compute $P(B|A)$ but can only compute $P(A|B)$. The following example demonstrates this:

13.1 - Example 1: Assume the chance it will rain on Monday is 0.15 and Tuesday is 0.25. If it rains on Monday however, the chance it will rain on Tuesday is 0.40. If it rains on Tuesday, find the probability that it rained the previous Monday.

Solution:

To solve this problem, we need to change the statement: If it rains on Tuesday find the probability that it rained the previous Monday into the form: $P(M|T)$.

Given the formula,

$$P(M|T) = \frac{P(T|M)P(M)}{P(T)}$$

we need to find the following probabilities: $P(T|M)$, $P(M)$, $P(T)$.

$P(T|M) = 0.40$, the probability that given it rains on Monday it will rain on Tuesday.
$P(M) = 0.15$, the probability that it will rain on Monday.
$P(T) = 0.25$, the probability that it will rain on Tuesday.

Bayes's formula allows us to write:

$$P(M|T) = \frac{P(T|M)P(M)}{P(T)} = \frac{(0.40)(0.15)}{0.25} = 0.24 \ .$$

13.1 - Example 2: Two urns sit on a table. Urn 1 has 7 red marbles and 3 black marbles and urn 2 has 4 red marbles and 5 black marbles. A marble is randomly selected from urn 1 and placed in urn 2. A marble is next selected from urn 2. If the marble selected from urn 2 is red, find the probability that the marble selected from urn 1 was also red.

Solution:

This is a typical Bayesian problem. Let R_1 be the event that a red marble is selected from urn 1. Let R_2 be the event that a red marble is selected from urn 2. We need to find $P(R_1|R_2)$.

Using the above formula we get

$$P(R_1|R_2) = \frac{P(R_2|R_1)P(R_1)}{P(R_2)} .$$

We need to find the following probabilities: $P(R_2|R_1)$, $P(R_1)$, $P(R_2)$.

$P(R_2|R_1)$ = the probability that if a red marble is selected from urn 1 and placed in urn 2 then a red marble will be selected from urn 2.

Urn 2 has originally 4 red and 5 black marbles. If a red marble is selected from urn 1 and placed in urn 2, then urn 2 will have 5 red and 5 black marbles. So selecting a marble from urn 2, the probability that this marble is red will be $P(\mathbf{R}_2|\mathbf{R}_1) = 5/10$.

$\mathbf{P(R_1)} = 7/10$, the probability that a red marble is selected from urn 1.

$\mathbf{P(R_2)}$: the probability that a red marble is selected from urn 2.

The event of drawing a red from urn 2 will be influenced by the first drawing from urn 1. Therefore we write:

$\mathbf{R}_2$ equals the event that a red marble is drawn on the first drawing and the second drawing or that a black was drawn from the first urn and a red drawn from the second urn.

Therefore, $\mathbf{R}_2 = (\mathbf{R}_1 \cap \mathbf{R}_2) \cup (\mathbf{R}_1{}' \cap \mathbf{R}_2)$.

$$P(\mathbf{R}_2) = P(\mathbf{R}_1 \cap \mathbf{R}_2) + P(\mathbf{R}_1{}' \cap \mathbf{R}_2) = P(\mathbf{R}_1)P(\mathbf{R}_2|\mathbf{R}_1) + P(\mathbf{R}_1{}')P(\mathbf{R}_2|\mathbf{R}_1{}') =$$

$$\left(\frac{7}{10}\right)\left(\frac{5}{10}\right) + \left(\frac{3}{10}\right)\left(\frac{4}{10}\right) = \frac{47}{100} .$$

Therefore $P(\mathbf{R}_1|\mathbf{R}_2) = \dfrac{P(\mathbf{R}_2|\mathbf{R}_1)P(\mathbf{R}_1)}{P(\mathbf{R}_2)} = \dfrac{\left(\frac{5}{10}\right)\left(\frac{7}{10}\right)}{\frac{47}{100}} = \dfrac{35}{47}.$

13.1 - Example 3: Three urns are sitting on a table. Urn 1 has 5 red marbles and 5 blue marbles; urn 2 has 6 red marbles and 4 blue marbles and urn 3 has 4 red and 11 blue marbles. A coin is tossed twice. If two heads appear, a marble is selected from urn 1; if two tails appear a marble is selected from urn 2; otherwise, a marble is selected from urn 3. Assuming a red marble is selected, find the probability that urn 1 was selected.

Solution:

Let $\mathbf{R}$ equal the event that a red marble is selected and $\mathbf{U}_i$ the event that urn i (i = 1, 2, 3) is selected. We need to find $P(\mathbf{U}_1|\mathbf{R})$ using the Bayes formula

$$P(\mathbf{U}_1|\mathbf{R}) = \frac{P(\mathbf{R}|\mathbf{U}_1)P(\mathbf{U}_1)}{P(\mathbf{R})} .$$

We need to find the following: $P(\mathbf{R}|\mathbf{U}_1)$, $P(\mathbf{U}_1)$, $P(\mathbf{R})$.

$P(\mathbf{R}|\ \mathbf{U}_1)$: the probability that if urn 1 is selected, then a red marble is selected.

Since urn 1 has 5 red and 5 blue then $P(\mathbf{R}|\mathbf{U}_1) = 5/10$.
$P(\mathbf{R}|\mathbf{U}_1) = 1/2$

$P(\mathbf{U_1})$ = the probability that urn 1 is selected.

Since the selecting of urn 1 results from two heads appearing out of two tosses of a coin , then

$P(\mathbf{U_1}) = (1/2)(1/2) = 1/4.$

$P(\mathbf{R})$: the probability that a red marble is selected.

The event of drawing a red marble depends on the urn selected. Therefore, the event of drawing a red is a red is drawn and urn 1 is selected or a red is drawn and urn 2 is selected or a red is drawn and urn 3 is selected. We write this as

$\mathbf{R} = (\mathbf{U_1}\cap\mathbf{R})\cup(\mathbf{U_2}\cap\mathbf{R})\cup(\mathbf{U_3}\cap\mathbf{R}).$

$P(\mathbf{R}) = P(\mathbf{U_1}\cap\mathbf{R}) + P(\mathbf{U_2}\cap\mathbf{R}) + P(\mathbf{U_3}\cap\mathbf{R}) = P(\mathbf{U_1})P(\mathbf{R}|\mathbf{U_1}) + P(\mathbf{U_2})P(\mathbf{R}|\mathbf{U_2}) + P(\mathbf{U_3})P(\mathbf{R}|\mathbf{U_3}) =$

$$\left(\frac{1}{4}\right)\left(\frac{5}{10}\right) + \left(\frac{1}{4}\right)\left(\frac{6}{10}\right) + \left(\frac{1}{2}\right)\left(\frac{4}{15}\right) = \frac{49}{120}$$

$$P(\mathbf{U_1}|\mathbf{R}) = \frac{P(\mathbf{R}|\mathbf{U_1})P(\mathbf{U_1})}{P(\mathbf{R})} = \frac{\left(\frac{1}{4}\right)\left(\frac{5}{10}\right)}{\frac{49}{120}} = \frac{15}{49}$$

Solved Problems

13.1 - Solved Problem 1: It is estimated that at the beginning of a major league baseball game the chance that a player will strike out the first time up at bat is 0.45 and 0.30 the second time up to bat. If he strikes out the first time, then the chance he will strike out the second time is 0.60. Assuming he does not strike out the second time, find the probability that he struck out the first time.

Solution:

Let **A** be the event that he strikes out the first time at bat and **B** the event that he strikes out the second time at bat.

$P(\mathbf{A}) = 0.45$, $P(\mathbf{B}) = 0.30$ and $P(\mathbf{B}|\mathbf{A}) = 0.60$. We need to find

$$P(\mathbf{A}|\mathbf{B'}) = \frac{P(\mathbf{B'}|\mathbf{A})P(\mathbf{A})}{P(\mathbf{B'})} = \frac{(0.40)(0.45)}{0.70} = \frac{18}{70}.$$

13.1 - Solved Problem 2: Two urns sit on a table. Urn 1 has 7 red marbles, 5 white marbles and 3 black marbles and urn 2 has 4 red marbles and 5 black marbles. A marble is randomly selected from urn 1 and placed in urn 2. A marble is next selected from urn 2. If the marble selected from urn 2 is red, find the probability that the marble selected from urn 1 was not red.

Solution:

Let $\mathbf{R}_1$ be the event that a red marble is selected from urn 1 $\mathbf{B}_1$ a blue selected and $\mathbf{W}_1$ a white marble is selected. Let $\mathbf{R}_2$ be the event that a red marble is selected from urn 2. We need to find $P(\mathbf{R}_1' | \mathbf{R}_2)$. Using the above formula we get

$$P(\mathbf{R}_1' | \mathbf{R}_2) = \frac{P(\mathbf{R}_2 | \mathbf{R}_1')P(\mathbf{R}_1')}{P(\mathbf{R}_2)} \; .$$

$P(\mathbf{R}_2 | \mathbf{R}_1') = 4/10$

$P(\mathbf{R}_2 | \mathbf{R}_1) = 5/10$

$P(\mathbf{R}_1') = 8/15$

$P(\mathbf{R}_1) = 7/15$

$$\mathbf{R}_2 = (\mathbf{R}_1 \cap \mathbf{R}_2) \cup (\mathbf{R}_1' \cap \mathbf{R}_2)$$

$$P(\mathbf{R}_2) = P(\mathbf{R}_1 \cap \mathbf{R}_2) + P(\mathbf{R}_1' \cap \mathbf{R}_2) = P(\mathbf{R}_1)P(\mathbf{R}_2 | \mathbf{R}_1) + P(\mathbf{R}_1')P(\mathbf{R}_2 | \mathbf{R}_1')$$

$$= (\frac{7}{15})(\frac{5}{10}) + (\frac{8}{15})(\frac{4}{10}) = \frac{67}{150}$$

Therefore,

$$P(\mathbf{R}_1' | \mathbf{R}_2) = \frac{P(\mathbf{R}_2 | \mathbf{R}_1')P(\mathbf{R}_1')}{P(\mathbf{R}_2)} = \frac{(\frac{4}{10})(\frac{8}{15})}{\frac{67}{150}} = \frac{32}{67}.$$

13.1 - Solved Problem 3: Three urns are sitting on a table. Urn 1 has 5 red and 5 blue marbles; urn 2 has 6 red and 4 blue marbles and urn 3 has 2 red and 8 blue marbles. A die is tossed once. If a 2 appears a marble is selected from urn 1; if 1 or 3 appears a marble is selected from urn 2; otherwise a marble is selected from urn 3. Assuming a red marble is selected, find the probability that urn 1 was selected.

Solution:

Let $\mathbf{R}$ = event that a red marble is selected. Let $\mathbf{U}_i$ be the event that urn i is selected. We need to find $P(\mathbf{U}_1 | \mathbf{R})$ using the Bayes formula

$$P(\mathbf{U}_1 | \mathbf{R}) = \frac{P(\mathbf{R} | \mathbf{U}_1)P(\mathbf{U}_1)}{P(\mathbf{R})}$$

$P(\mathbf{R} | \mathbf{U}_1) = 1/2$

$P(\mathbf{U}_1) = 1/6$

$$\mathbf{R} = (\mathbf{U}_1 \cap \mathbf{R}) \cup (\mathbf{U}_2 \cap \mathbf{R}) \cup (\mathbf{U}_3 \cap \mathbf{R})$$

$$P(\mathbf{R}) = P(\mathbf{U}_1 \cap \mathbf{R}) + P(\mathbf{U}_2 \cap \mathbf{R}) + P(\mathbf{U}_3 \cap \mathbf{R}) = P(\mathbf{U}_1)P(\mathbf{R}|\mathbf{U}_1) + P(\mathbf{U}_2)P(\mathbf{R}|\mathbf{U}_2) + P(\mathbf{U}_3)P(\mathbf{R}|\mathbf{U}_3) =$$

$$(\frac{1}{6})(\frac{1}{2}) + (\frac{2}{6})(\frac{6}{10}) + (\frac{3}{6})(\frac{2}{10}) = \frac{23}{60}$$

Therefore,

$$P(\mathbf{U}_1|\mathbf{R}) = \frac{P(\mathbf{R}|\mathbf{U}_1)P(\mathbf{U}_1)}{P(\mathbf{R})} = \frac{(\frac{1}{2})(\frac{1}{6})}{\frac{23}{60}} = \frac{5}{23}.$$

Unsolved Problems with Answers

13.1 - Problem 1: It is estimated that if a student studies for the final, he or she has a 35% chance of earning at least a B in the course. Also assume that there is a 50% chance that the student will study for the final and a 28% chance that the student will earn at least a B in the course. Find the probability that if the student earned at least a B in the course then he or she studied for the final.

Answer:

0.625

⇑ *Refer back to* **13.1 - Example 1 & 13.1 - Solved Problem 1.**

13.1 - Problem 2: Two urns sit on a table. Urn 1 has 7 red marbles, and 3 black marbles. Urn 2 has 5 black marbles. A marble is randomly selected from urn 1 and placed in urn 2. A marble is next selected from urn 2. If the marble selected from urn 2 is red, find the probability that the marble selected from urn 1 was not red.

Answer:

0

⇑ *Refer back to* **13.1 - Example 2 & 13.1 - Solved Problem 2.**

13.1 - Problem 3: Ms. Jones teaches three sections of Spanish. Section 1 has 25 female and 50 male students. Section 2 has 60 female and 40 male students. Section 3 has 20 female and 30 male students. She holds a contest in each of her sections. The chance that the winner is in section 1 is 75/225, from section 2 is 100\225 and section 3 is 50/225 Assuming the winner is a female, find the probability that the winner is from section 1.

Answer:

⇑ *Refer back to* **13.1 - Example 3 & 13.1 - Solved Problem 3**

13.2 - Real Life Applications

13.2 - Example 1: An anti-smoking organization, in an attempt to show that smoking cigarettes causes cancer, submitted the following statistical data to a Senate committee: 25% of the adult population smokes cigarettes on a regularly basis .Their records also show that of those that contracted cancer, 75% were smokers. They estimated that about 5% of the population has cancer. Assuming that a person smokes cigarettes, find the probability that

(a). he/she will contract cancer.

(b). he/she will not contract cancer.

Solutions:

➤ **(a).**
A: the event that a person smokes cigarettes.
C: The event that a person contracts cancer.

$P(\mathbf{A}) = 0.25$
$P(\mathbf{C}) = 0.05$
$P(\mathbf{A}|\mathbf{C}) = 0.75$, the probability that of those adults that contracted cancer, they are also are smokers.

We need to find $P(\mathbf{C}|\mathbf{A})$, the probability that given a person smokes, he/she will contract cancer.

$$P(\mathbf{C}|\mathbf{A}) = \frac{P(\mathbf{A}|\mathbf{C})P(\mathbf{C})}{P(\mathbf{A})} = \frac{(0.75)(0.05)}{0.25} = 0.15$$

➤ **(b).**
We need to find $P(\mathbf{C}'|\mathbf{A})$, the probability that given a person smokes, he/she will not contract cancer.

$P(\mathbf{C}'|\mathbf{A}) = 1 - P(\mathbf{C}|\mathbf{A}) = 1 - 0.15 = 0.85$

13.2 - Example 2: In a state where cars have to be tested for emissions of pollutants, 25% of all cars emit excessive amounts of pollutants. When tested, 99% of all cars that emit excessive amounts of pollutants will fail, but 17% of the cars that do not emit excessive amounts of pollutants will also fail.

(a). What is the probability that a car which fails the test actually emits excessive amounts of pollutants?

(b). What is the probability that a car which fails the test does not emits excessive amounts of pollutants?

Solutions:

➤ **(a).**
To solve this problem, we need to restate the end of the problem: What is the probability that a car which fails the test actually emits excessive amounts of pollutants: $P(E|F)$?
Given the formula,

$$P(E|F) = \frac{P(F|E)P(E)}{P(F)}.$$

We need to find the following: $P(F|E)$, $P(E)$, $P(F)$.

$P(F|E) = 0.99$, the probability that all cars who emits excessive amounts of pollutants will fail.

$P(E) = 0.25$, the probability that a car will emit excessive pollutants.

$P(F|E') = 0.17$, the probability that a car that does not have excessive pollutants fails the test.

We now need to solve for $P(F)$.

Step 1: F is the event that a car fails the test and it emits excessive pollutants or a car fails the test and it does not emit excessive pollutants: $F = (F \cap E) \cup (F \cap E')$.

Step 2: $P(F) = P(F \cap E) + P(F \cap E') = P(E)P(F|E) + P(E')P(F|E') = 0.25(0.99) + 0.75(0.17) = 0.375$

Step 3: We can now determine the probability from the above:

$$P(E|F) = \frac{P(F|E)P(E)}{P(F)} = \frac{(0.99)(0.25)}{0.375} = 0.66 .$$

➤ **(b).**
E' is the event that the car does not emit excessive amounts of pollutants. Therefore, if the car fails the test, the probability that it does not emit excessive amounts of pollutants is $P(E'|F) = 1 - 0.66 = 0.34$.

13.2 - Example 3: A large publishing firm, recently received a request from an author to publish his manuscript on the life cycle of worms. Past experience has shown that there is only a 10% chance that such a book would be commercially successful. However, past records show that 80% of such books that were successful were extensively promoted and 35% of unsuccessful books of this type were also extensively promoted.

(a). Assuming the book is extensively promoted, find the probability that the book will be successful.

(b). Assuming the book is extensively promoted, find the probability that the book will be a failure.

Solutions:

➤ **(a).**
Looking at the end of the problem, we need to solve:

What is the probability of the book being successful given that it had extensive promotion? Given the formula

$$P(S|E) = \frac{P(E|S)P(S)}{P(E)} \; ,$$

we need to find the following $P(E|S)$, $P(E)$, $P(S)$, where S is the event that the book will be successful and E is the event that the book is extensively promoted.

$P(E|S) = 0.80$, the probability that if the book is successful then it was given extensive promotion.

$P(S) = 0.10$, the probability that the book will be successful.

$P(E)$: the probability that the book was given extensive promotion.

Now the event E, the book was given extensive promotion is dependent on the event of the book being successful. Therefore E equals the event that extensive promotion of the book occurred and the book is successful or the event that extensive promotion was given and the book is not successful. This can be written as: $E = (E \cap S) \cup (E \cap S')$.

$P(E) = P(E \cap S) + P(E \cap S') = P(S)P(E|S) + P(S')P(E|S') = (0.10)(0.80) + (0.90)(0.35) = 0.08 + 0.315 = 0.395$

$$P(S|E) = \frac{P(E|S)P(S)}{P(E)} = \frac{(0.80)(0.10)}{0.395} = 0.202$$

➤ **(b).**
S' is the event that the book will not be successful. Therefore, if the book is extensively promoted, the probability that the book will be a failure is $P(S'|E) = 1 - 0.202 = 0.798$.

13.2 - Example 4: A computer manufacturer purchases ROM memory chips from three different companies: 60% from company A, 25% from company B and 15% from company C. Past records show that 10% of the chips from company A have been found defective, 7% from company B have been defective and 6% from company C has been found defective. This data can be summarized in the following table:

Company	Percentage of chips purchased from each company	Probability of a chip being defective
A	60%	0.10
B	25%	0.07
C	15%	0.06

(a). Assume a chip is tested and found defective. Find the probability that it came from company C.

(b). Assume a chip is tested and found defective. Find the probability that it did not come from company C.

Solutions:

➤ **(a).**

Looking at the end of the problem, we need to solve: What is the probability that a chip came from Company C given that it is defective? Given the formula,

$$P(C|D) = \frac{P(D|C)P(C)}{P(D)}.$$

where **C** is the event that the chip came from company C and D the event that the chip is defective.

We need to find $P(D|C)$, $P(C)$, $P(D)$ where

$P(D|C) = 0.06$, the probability that the chip is defective if it came from company C.

$P(C) = 0.15$,the probability that the chip came from company C.

$P(D)$: the probability that a chip is defective.

The event that a chip is defective is effected by which company manufactured it. Therefore, **D** the event that a chip is defective equals the event that it is defective and was manufactured by company A or defective and manufactured by company B or defective and manufactured by company C. This is written

$\mathbf{D = (D \cap A) \cup (D \cap B) \cup (D \cap C)}.$

$P(D) = P(D \cap A) + P(D \cap B) + P(D \cap C) = P(A)P(D|A) + P(B)P(D|B) + P(C)P(D|C) =$

$(0.60)(0.10) + (0.25)(0.07) + (0.15)(0.06) = 0.0865$

$$P(C|D) = \frac{P(D|C)P(C)}{P(D)} = \frac{(0.06)(0.15)}{0.0865} = 0.10$$

➤ **(b).**

C' is the event that the chip did not come from Company C.

Therefore, if chip is tested and found defective, the probability that it came from company C is

$P(C'|D) = 1 - 0.10 = 0.90$.

Solved Problems

13.2 - Solved Problem 1: In an attempt to show there is a causal relation between attending R-rated films and male teenage violence, a citizens group recently acquired the following statistical data: From the population, 85% of all teenagers attended R-rated films and 7% of all male teenagers have committed a violent crime. From interviewing 1,000 male teenagers that have committed violent crimes, they found that 95% had attended R-rated movies. Assuming that a teenager attends R rated films, find the probability

(a). he will commit a violent crime.

(b). he will not commit a violent crime.

Solutions:

➤ **(a).**
A: the event that a male teenager attends R-rated films.
V: The event that a male teenager will commit a violent crime.

$P(A) = 0.85$
$P(V) = 0.07$

$P(A|V) = 0.95$, the probability that of those male teenagers that have committed violent crimes, they also attend R-rated films.

We need to find $P(V|A)$, the probability that given a teenager attends R-rated films, he will commit a violent crime.

$$P(V|A) = \frac{P(A|V)P(V)}{P(A)} = \frac{(0.95)(0.07)}{0.85} \approx 0.08$$

➤ **(b).**
We need to find $P(V'|A)$, the probability that given a male teenager attends R-rated films , he will not commit a violent crime.

$P(\mathbf{V'}|\mathbf{A}) = 1 - P(\mathbf{V}|\mathbf{A}) = 1 - 0.08 = 0.92$

13.2 - Solved Problem 2: According to the Arizona Chapter of the American Lung Association, 7.0% of the population has lung disease. Of those people that have lung disease, 90% smoke and of those not having lung disease, 25.3% are smokers. Suppose a person is selected at random from the population.

(a). If the person selected is a smoker, find the probability that the person has a lung disease.

(b). If the person selected is a smoker, find the probability that the person does not have a lung disease.

Solutions:

➤ **(a).**
Looking at the end of the problem, we need to solve $P(\mathbf{D}|\mathbf{S})$ where $\mathbf{S}$ is the event that the person smokes and $\mathbf{D}$ the person has lung disease. Since

$$P(\mathbf{D}|\mathbf{S}) = \frac{P(\mathbf{S}|\mathbf{D})P(\mathbf{D})}{P(\mathbf{S})} \, ,$$

we need to find $P(\mathbf{S}|\mathbf{D})$, $P(\mathbf{S})$ and $P(\mathbf{D})$. From the problem, $P(\mathbf{S}|\mathbf{D}) = 0.90$, and $P(\mathbf{D})= 0.070$.

Now $\mathbf{S} = (\mathbf{S} \cap \mathbf{D}) \cup (\mathbf{S} \cap \mathbf{D'})$ and $P(\mathbf{S}) = P(\mathbf{S} \cap \mathbf{D}) + P(\mathbf{S} \cap \mathbf{D'}) = P(\mathbf{D})P(\mathbf{S}|\mathbf{D}) + P(\mathbf{D'})P(\mathbf{S}|\mathbf{D'}) =$

$(0.070)(0.90) + (0.93)(0.253) = 0.29829$.

$$P(\mathbf{D}|\mathbf{S}) = \frac{(0.90)(0.07)}{0.29829} = 0.21$$

➤ **(b).**
$\mathbf{D'}$ is the event that the person does not have a lung disease. Therefore, If the person selected is a smoker, the probability that the person does not have a lung disease is $P(\mathbf{D'}|\mathbf{S}) = 1 - 0.21 = 0.79$

13.2 - Solved Problem 3: Two large toy companies wish to sell talking teddy bears. One company, Toys International estimates there is a 75% chance their teddy bear will make a profit for the company provided the competing toy company does not introduce a talking teddy bear on the market and a 35% chance it will be profitable if the competing company introduces such a toy on the market. Further, it estimates there is a 60% chance the competing company will introduce the toy.

(a). Given that the teddy bear is profitable, find the probability that the competing company also introduced a teddy bear.

(b). Given that the teddy bear is profitable, find the probability that the competing company did not introduce a teddy bear.

Solutions:

➤ **(a).**
Looking at the end of the problem, we need to solve $P(I|S)$ where S is the event that the toy will be profitable and I is the event that the competing firm also introduced the toy.

$$P(I|S) = \frac{P(S|I)P(I)}{P(S)}$$

$P(I) = 0.60,$

$P(S|I) = 0.35$

$S = (S \cap I) \cup (S \cap I')$

$P(S) = P(S \cap I) + P(S \cap I') = P(I)P(S|I) + P(I')P(S|I') = (0.60 \times 0.35) + (0.40 \times 0.75) = 0.21 + 0.30 = 0.51$

$$P(I|S) = \frac{P(S|I)P(I)}{P(S)} = \frac{(0.35)(0.60)}{0.51} = 0.412$$

➤ **(b).**
I' is the event that the competing firm did not introduce. Therefore, if the teddy bear is profitable, the probability that the competing company did not introduce a teddy bear is $P(I'|S) = 1 - 0.412 = 0.588$.

13.2 - Solved Problem 4: From past records, an English teacher estimates that the percentage of her students that will receive an A as a final grade is the following:

Event	Percentage of students that Receive an A	Probability of this Event Occurring
E	10%	0.7
F	15%	0.2
G	20%	0.1

A student's final grade is randomly selected and found to be an A.

(a). Find the probability that 15% of the students will receive an A.

(b). Find the probability that 15% of the students will did not receive an A.

Solutions:

➤ **(a).**

Looking at the end of the problem, we need to solve $P(F|A)$ where F is the event 15% of her students received an A and A the event that the student received a final grade of A. **E** is the event that 10% will receive an A, **F** is the event that 15% will receive an A and **G** is the event that 20% will receive an A.

Bayes's formula gives $P(F|A) = \dfrac{P(A|F)P(F)}{P(A)}$.

From the problem,

$P(A|F) = 0.15$, $P(F) = 0.20$, $A = (A \cap E) \cup (A \cap F) \cup (A \cap G)$.

$P(A) = P(A \cap E) + P(A \cap F) + P(A \cap G) = P(E)P(A|E) + P(F)P(A|F) + P(G)P(A|G) =$

$(0.70)(0.10) + (0.20)(0.15) + (0.10)(0.20) = 0.12$

$P(F|A) = \dfrac{P(A|F)P(F)}{P(A)} = \dfrac{(0.15)(0.20)}{0.12} = 0.25$

➤ **(b).**
F′ is the event that 15% will not receive an A. Therefore, the probability that 15% of the students will not receive an A is $P(F|A) = 1 - 0.25 = 0.75$.

Unsolved problems with Answers

13.2 - Problem 1: To show there is a strong causal relationship between auto fatalities and alcoholic abuse, an anti-alcoholic group submitted the following statistical data to their state legislators: 9% of all auto accidents result in fatalities. Among 10,000 auto fatalities, 35% of the drivers were intoxicated. Their data also showed that during any particular time 5% of all drivers are intoxicated. Assuming that a driver is intoxicated, find the probability that

(a). he/she will get involved in a fatal auto accident.

(b). he/she will not get involved in a fatal auto accident.

Answers:

➤ **(a).** 0.63

➤ **(b).** 0.37

⇑ *Refer back to* **13.2 - Example 1 & 13.2 - Solved Problem 1.**

13.2 - Problem 2: A doctor estimates that about 10% of his patients are diabetics. He recently started to use a new blood test to determine if his patients have diabetes. The manufacturer of this new test claims that it will test 90% positive for those patients that have diabetes and will test incorrectly 15% of the time for those patients that do no have diabetes.

(a). Assume a patient tests positive. Find the probability that he or she has diabetes.

(b). Assume a patient tests positive. Find the probability that he or she does not have diabetes.

Answers:

➤ **(a).** 0.40

➤ **(b).** 0.60

⇑ *Refer back to* **13.2 - Example 2 & 13.2 - Solved Problem 2.**

13.2 - Problem 3: The ABC Shoe Corporation is interested in establishing a new shoe store in a local community based on a survey that shows that there is a 80% chance that sales for new shoes in the area exceeds $500,000. It believes that if sales in the area actually exceeds $500,000 then there is a 70% chance that the new store will be successful. However, if sales in the area are less then $500,000 then there is only a 25% chance their new store will be successful. Assume they decide to open a new store.

(a). If this store is not successful, find the probability that the sales in the area exceed $500,000.

(b). If this store is not successful, find the probability that the sales in the area will not exceed $500,000.

Answers:

➤ **(a).** 0.62

➤ **(b).** 0.38

⇑ *Refer back to* **13.2 - Example 3 & 13.2 - Solved Problem 3.**

13.2 - Problem 4: The XYZ toy company has just distributed to its retail stores a new toy train. The following table is an estimate of the percentage of stores that will return the unsold trains:

Event	Percentage of Stores Returning Unsold trains	Probability of this Event Occurring
A	50%	0.6
B	70%	0.15
C	80%	0.12
D	90%	0.13

A store is selected at random and found to have returned some of these trains.

(a). Find the probability that 70% of the stores will return the unsold trains.

(b). Find the probability that 70% of the stores will not return the unsold trains.

Answers:

➤ (a). 0.17

➤ (b). 0.83

⇑ *Refer back to* **Example 2.4 & solved Problem 2.4.**

Supplementary Problems

1. A recent survey in a music class showed the following preference for classical and jazz.

	Male (M)	Female (F)	Total
Classical (C)	35	40	75
Jazz (J)	50	60	110
Total	85	100	185

A student is selected at random. Verify $P(M|J)$ using Bayes's formula.

2. Under our system of justice, a defendant found not guilty by a jury does not necessary mean that the jury believes the defendant is really innocent. A not guilty decision could result from insufficient evidence. Assume that if a person is truly innocent then there is a 90% chance that the jury will find the defendant not guilty and if guilty of the crime then there is a 70% chance the defendant will be found guilty. Further assume that there is only a 25% chance that the defendant is innocent. If a defendant is found not guilty, find the probability that he or she is really innocent.

3. Sam usually dates two women: Sally and Jane. He usually asks Sally out 60% of the time and Jane out 40% of the time. The chance that Sally will accept is 0.30 and Jane is 0.75. Assuming that Sam had a date last night, find the probability that he was with Sally.

4. A recent report on smoking and lung cancer from a national health agency gives the following information: Each year, approximately 400,000 Americans die from lung cancer. Of this group, approximately 320,000 are smokers. Also, there are approximately 50 million smokers. Assuming the population of the U.S. is 250 million, find the probability that a smoker will die of lung cancer.

5. Under what conditions is $P(A|B) = P(B|A)$?

6. In a recent report issued by a major medical research institute on new-born children acquiring a certain disease, the following data was provided: 5% of all adult males and 7% of all females,

respectively, have this disease. If only the father has the disease, there is a 35% chance the offspring will also have the disease. If only the mother has the disease, there is a 55% chance the offspring will also have the disease. If both parents have the disease, there is a 80% chance the offspring will also have the disease. If neither parents have the disease, there is only a 2% chance the offspring will have the disease. Assume an offspring is found to be born with the disease. Find

a. the probability that the only the father has the disease.

b. the probability that the only the mother has the disease.

c. the probability that both parents have the disease.

d. the probability that neither parents have the disease.

e. the probability that the father has the disease.

7. Three urns sit on a table. Urn A has 3 red marbles and 7 black, urn B has 12 red and 8 black and urn C has 5 red and 5 black. Two marbles are selected randomly from different urns. If the selected marbles are both red, find the probability that urns A and B were selected

8. In an attempt to show there is a causal relation between attending R-rated films and male teenage violence, a citizen group recently acquired the following statistical data: From the population, 7% of all male teenagers have committed a violent crime. From interviewing 1,000 male teenagers , of those that have committed violent crimes, they found that 95% had attended R-rated movies and of those that have not committed violent crimes, 15% have attended R-rated movies. Assuming that a teenager attends R rated films, find the probability he has committed a violet crime.

9. Show $P[C \cap B|A] = P(B|A)P(C|A \cap B)$.

10. Show $P(C|A \cap B) = \dfrac{P(B|C)P(A|B \cap C)P(C)}{P(A \cap B)}$.

11.A room contains 2 tables. On each table sits 2 urns containing red and white marbles. The following table contains the contents of the urns:

TABLE 1		TABLE 2	
URN A	URN B	URN C	URN D
12 red marbles 8 white marbles	15 red marbles 5 white marbles	10 red marbles 10 white marbles	5 red marbles 15 red marbles

A table and an urn sitting on the table is selected at random. From the urn, 1 marble is selected. If the marble selected is red, find the probability that

a. table 2 was selected.

b. urn C was selected.

12. To show a connection between lung disease and smoking, the following data was presented to a governmental agency: of those in the study that had lung disease, 68% were smokers; of those that were smokers, 15% had lung disease. The report also stated that 7% of the population had some form of lung disease. From this data, estimate the percentage of the study that smoke.

13. Two ordinary decks of cards are sitting on a table. Two cards are randomly selected from deck A and randomly placed in deck B. A single card is then drawn at random from deck B. If the card drawn from deck B is a diamond

a. find the probability that at least 1 diamond was selected from deck A.

b. find the probability that a diamond and a club was drawn from deck A.

14. Assume $S = A_1 \cup A_2 \cup ... \cup A_n$ and $A_j \cap A_k = \varphi$ for all $j \neq k$.

Show

$$P(A|B) = \frac{P(B|A)P(A)}{P(A_1)P(B|A_1) + (P(A_1')P(B|A_1') + ... + P(A_n)P(B|A_n) + (P(A_n')P(B|A_n')}.$$

Assume two urns contain the following balls: urn A has 10 red and 5 black balls; urn B has 5 red and 3 black. Two balls are drawn, one by one, from urn A and placed in urn B. Then, two balls are drawn, one by one from urn B.

15. If two black balls were drawn from urn B, find the probability that only one red ball was drawn from urn A.

16. If at least one black ball was drawn from urn B, find the probability that two red balls were drawn from urn A.

14.1 - What is a Random Variable?

A random variable is a rule that assigns to each element of a sample space a numeric value. The symbols for random variables are X, Y, Z, etc.

14.1 - Example 1: A coin is tossed twice. Let the random variable X assign to each element the number of heads. Write out the random variable.

Solution:

The sample space for this experiment is $\mathbf{S}$ = {(h,h), (h,t), (t,h), (t,t)}. X is the random variable representing the number of heads for each possible element of the sample space. Therefore,

X(h,h) = 2, X(h,t) = 1,
X(t,h) = 1, X(t,t) = 0.

The relationship between the assigned values of X and the elements of the sample space can be seen from the following table:

Sample Space	Random Variable X
(h,h)	2
(h,t)	1
(t,h)	1
(t,t)	0

14.1 - Example 2: An urn contains three marbles, each marked with the number 1, 2, 3 respectively. Two marbles are selected at random without replacement. Let the random variable X assign to each element of the sample space the smallest of the two numbers. Write out the random variable.

Solution:

The relationship between the assigned values of X and the list of elements of the sample space can be seen from the following table:

Sample Space	Random Variable X
(1,2)	1
(2,1)	1
(1,3)	1
(3,1)	1
(2,3)	2
(3,2)	2

14.1 - Example 3: An urn contains 4 balls where ball 1 is marked 1, ball 2 is marked 2, ball 3 is marked 3 and ball 4 is marked 4. Three balls are drawn without replacement. We are interested only in the numbers picked. Therefore we assume the order of selection is not important. Let the random variable assigned to the drawing be the average of the three numbers marked on the balls. Write out the random variable.

Solution:

Since the order is not important, the selection, (1,2,3) is the same as (2,1,3), (2,3,1), etc. Therefore, the sample space generated by this experiment $\mathbf{S} = \{(1,2,3), (1,2,4), (1,3,4), (2,3,4)\}$. Let X represent the random variable for the average:

$$X(1,2,3) = \frac{1+2+3}{3} = 2 \;, \qquad X(1,2,4) = \frac{1+2+4}{3} = \frac{7}{3}$$

$$X(1,3,4) = \frac{1+3+4}{3} = \frac{8}{3} \;, \qquad X(2,3,4) = \frac{2+3+4}{3} = 3$$

The relationship between the assigned values of X and the elements of the sample space can be seen from the following table:

Sample Space	Random Variable X
(1,2,3)	2
(1,2,4)	7/3
(1,3,4)	8/3
(2,3,4)	3

Solved Problems

14.1 - Solved Problem 1: A coin is tossed three times. Let the random variable X assign to each element the number of tails. Write out the random variable.

Solution:

S = {(h,h,h), (h,h,t), (h,t,h), (t,h,h), (t,t,h), (h,t,t), (t,h,t), (t,t,t)}

X(h,h,h) = 0, X(h,h,t) = 1, X(h,t,h) = 1, X(t,h,h) = 1,

X(t,t,h) = 2, X(h,t,t) = 2, X(t,h,t) = 2, X(t,t,t) = 3

14.1 - Solved Problem 2: A die and a coin are tossed once. Let the random variable assign to each pair the value of the die. Write out the random variable.

Solution:

The sample space is S = {(h,1), (h,2), (h,3), (h,4), (h,5), (h,6), (t,1,(t,2), (t,3), (t,4), (t,5), (t,6)}.

Therefore,

X(h,1) = 1, X(h,2) = 2, X(h,3) = 3,
X(h,4) = 4, X(h,5) = 5, X(h,6) = 6,
X(t,1) = 1, X(t,2) = 2, X(t,3) = 3,
X(t,4) = 4, X(t,5) = 5, X(t,6) = 6.

14.1 - Solved Problem 3: Four students won a local spelling competition: John, age 16, Mary, age 17, Bill, age 18 and Frankie, age 16. Three of these students were randomly selected. Let the random variable assigned to the drawing be the average age of the three students selected. Write out the random variable.

Solution:

Using the first letter of the students' names, S = {(J,M,B), (J,M,F), (J,F,B), (B,F,M)}. Since we are only interested in the numbers picked, order is not important for the sample space. Let X be the random variable for the average age for each sample:

X(J,M,B) = 17 ,X(J,M,F) = 16.33 ,X(J,F,B) = 16.66, X(B,F,M) = 17 .

Unsolved Problems with Answers

14.1 - Problem 1: A urn contains one red and one white marble. A marble is drawn twice with replacement. Let the random variable X assign to each element the number of red marbles selected. Write out the random variable.

Answer:

X(w,w) = 0 ,X(r,w) = 1, X(w,r) = 1, X(r,r) = 2

⇑ *Refer back to* **14.1 - Example 1 & 14.1 - Solved Problem 1.**

14.1 - Problem 2: Ms. Smith has three children: Billy, age two, Jane, age ten, and Frank, age fifteen. Two of the children are randomly selected. Let the random variable assigned to each pair the oldest age. Write out the random variable. (Assume the order of selection is not important).

Answer:

X(Billy,Jane) = 10, X(Billy,Frank) = 15, X(Jane,Frank) = 15

⇑ *Refer back to* **14.1 - Example 2 & 14.1 - Solved Problem 2.**

14.1 - Problem 3: Mary works for a temporary employment service. Over a five-week period, she worked 10,15,10,20, and 30 hours. A random sample of four weeks is taken. Let the random variable be the average number of hours worked per week over the four-week period. Write out the random variable.

Answer:

X(10,15,10,20) = 13.75, X(10,15,10,30) = 16.25,

X(10,15,20,30) = 18.75, X(10,10,20,30) = 17.50, X(15,10,20,30) = 18.75

⇑ *Refer back to* **14.1 - Example 3 & 14.1 - Solved Problem 3.**

14.2-What is a Probability Distribution of a Random Variable?

There are several types of probability distributions of random variables. In this section, we define discrete distributions; in section 3, cumulative distributions and in lesson 16, a continuous distribution of a random variable.

> A discrete probability distribution is a listing of all possible values of a random variable along with the corresponding probabilities. The notation for the distribution of the random variable X is P{X = x} where x represents all possible numeric values that X equals.

Frequently it is convenient to use a table to represent the distribution.

14.2 - Example 1: A coin is tossed twice. Let the random variable X assign to each element the number of heads. Write out the table that represents the distribution.

Solution: The sample space is **S** = {(h,h), (h,t), (t,h), (t,t)}.Let X be the random variable that assigns to each element of the sample space the number of heads. We now list the events in the sample space that result in the random variable equaling the possible values:

{X = 2}= {(h,h)}, {X = 1} = {(h,t),(t,h)}, {X = 0} = {(t,t)}.

For example, {X = 1} is the event that exactly one head appears. We now compute the probability for each of these events:

P{X = 2}= P{(h,h)} = 1/4, P{X = 1} = P{(h,t),(t,h)}= 2/4, P{X = 0} = P{(t,t)} = 1/4.

We can now represent this probability distribution in the following table:

x	P{X = x}
0	1/4
1	1/2
2	1/4
	Total 1

14.2 - Example 2: A die is tossed twice. Let the random variable assigned to the pair of numbers the sum of these numbers. Write the probability distribution.

Solution:

Let X represent the random variable that assigns to each pair of numbers the sum. For example, {X = 5} = {(1,4), (2,3), (3,2), (4,1)} is the event that the sum of the two number is 5. The random variable X takes on values 2, 3, 4, 5, 6, 7, 8, 9, 10, 11, 12.

Therefore, the probability distribution for X is

{X = 2} = {(1,1)}
{X = 3} = {(1,2), (2,1)}
{X = 4} = {(1,3), (2,2), (3,1)}
{X = 5} = {(1,4), (2,3), (3,2), (4,1)}
{X = 6} = {(1,5), (2,4), (3,3), (4,2), (5,1)}
{X = 7} = {(1,6), (2,5), (3,4), (4,3), (5,2), (6,1)}
{X = 8} = {(2,6), (3,5), (4,4), (5,3), (6,2)}
{X = 9} = {(3,6), (4,5), (5,4), (6,3)}
{X =10} = {(4,6), (5,5), (6,4)}

$\{X =11\} = \{(5,6), (6,5)\}$
$\{X =12\} = \{(6,6)\}$

Since #**S** = 36,

$P\{X = 2\} = 1/36$, $P\{X = 3\} = 2/36$, $P\{X = 4\} = 3/36$, $P\{X = 5\} = 4/36$
$P\{X = 6\} = 5/36$, $P\{X = 7\} = 6/36$, $P\{X = 8\} = 5/36$, $P\{X = 9\} = 4/36$
$P\{X = 10\} = 3/36$, $P\{X = 11\} = 2/36$, $P\{X = 12\} = 1/36$

In table form:

x	2	3	4	5	6	7	8	9	10	11	12	Total
P{X = x}	1/3	2/3	3/3	4/3	5/3	6/3	5/3	4/3	3/3	2/3	1/3	1

14.2 - Example 3: John Wish is a famous gambler. He claims that he has a 60% chance of winning on any college football game. On each game he bets $110. If his team wins, he makes $100. If his team loses, he loses $110. On a particular week, he bets on two separate football games. Write out the probability distribution for the random variable which is the total amount he won over these two football games.

Solution:

Since he will play two games, let W_1 be the event that he wins on the first game and W_2 the event that he wins on the second game. All these events are independent. Let X represent the total amount he wins or loses. There are three possible values that X will equal:

Case 1: he wins both games: $W_1 \cap W_2 = \{X = \$200\}$.

Case 2: he wins one game and loses one game:

$(W_1' \cap W_2) \cup (W_1 \cap W_2') = \{X = -\$10\}$.

Case 3: he losses both games:

$W_1' \cap W_2' = \{X = -\$220\}$.

From these cases we can compute:

$P\{X = \$200\} = P(W_1 \cap W_2) = P(W_1)P(W_2) = (0.6)(0.6) = 0.36$

$P\{X = -\$10\} = P[(W_1' \cap W_2) \cup (W_1 \cap W_2')] = P(W_1' \cap W_2) + P(W_1 \cap W_2') = (0.4)(0.6) + (0.6)(0.4)$
$= 0.48$

$$P\{X = -\$220\} = P[(W_1' \cap W_2') = P[(W_1')P(W_2') = (0.40)(0.40) = 0.16$$

In table form:

x	P{X = x}
-$ 200	0.16
-$10	0.48
$200	0.36
	Total 1

Solved Problems

14.2 - Solved Problem 1: The following game is played: A coin is tossed twice. If the two faces are the same you win $20 otherwise you lose $25. Write out the table that represents the probability distribution.

Solution:

S = {(h,h), (h,t), (t,h), (t,t)}. First we list the events in the sample space that result in the random variable equaling the possible values:

$\{X = \$20\}$ = {(h,h), (t,t)}, $\{X = -\$25\}$ = {(h,t), (t,h)}. Therefore,

$P\{X = \$20\}$ = $P\{(h,h), (t,t)\}$ = 1/2, $P\{X = -\$25\}$ = $P\{(h,t), (t,h)\}$ =1/2

We can now represent this probability distribution in the following table:

x	P{X = x}
-$25	1/2
$20	1/2

14.1 - Solved Problem 2: A die is tossed twice. Assume the following rule for the random variable: If the pair of numbers are equal, assign the single number; otherwise assign the larger of the two numbers. Write the probability distribution.

Solution:

We know #S = 36. The random variable only assigns value 1, 2, 3, 4, 5, 6.

$\{X = 1\}$ = $\{(1,1)\}$

$\{X = 2\} = \{(1,2), (2,1), (2,2)\}$

$\{X = 3\} = \{(1,3), (2,3), (3,3), (3,1), (3,2)\}$

$\{X = 4\} = \{(1,4), (2,4), (3,4), (4,4), (4,1), (4,2), (4,3)\}$

$\{X = 5\} = \{(1,5), (2,5), (3,5), (4,5), (5,5), (5,1), (5,2), (5,3), (5,4)\}$

$\{X = 6\} = \{(1,6), (2,6), (3,6), (4,6), (5,6), (6,6), (6,1), (6,2), (6,3), (6,4), (6,5)\}$

$P\{X = 1\} = 1/36$, $P\{X = 2\} = 3/36$, $P\{X = 3\} = 5/36$, $P\{X = 4\} = 7/36$, $P\{X = 5\} = 9/36$, $P\{X = 6\} = 11/36$

x	1	2	3	4	5	6	TOTAL
$P\{X = x\}$	1/36	3/36	5/36	7/36	9/36	11/36	1

14.2 - Solved Problem 3: John Wish is a famous gambler. He claims that he has a 60% chance of winning on any college football game. On each game he bets $110. If his team wins, he makes $100. If his team loses, he loses $110. On a particular week, he bets on three separate football games. Write out the probability distribution for the random variable which is the total amount he won over these three football games.

Solution:

Since he will play three games, let $\mathbf{W}_1$ be the event that he wins on the first game, $\mathbf{W}_2$ the event that he wins on the second game and $\mathbf{W}_3$ the event that he wins on the third game. All these events are independent.

Let X represent the total amount he wins or loses. There are four possible values that X will equal:

Case 1: he wins all three games: $\mathbf{W}_1 \cap \mathbf{W}_2 \cap \mathbf{W}_3 = \{X = \$300\}$.

Case 2: he wins two games and loses one game:

$(\mathbf{W}_1' \cap \mathbf{W}_2 \cap \mathbf{W}_3) \cup (\mathbf{W}_1 \cap \mathbf{W}_2' \cap \mathbf{W}_3) \cup (\mathbf{W}_1 \cap \mathbf{W}_2 \cap \mathbf{W}_3') = \{X = \$90\}$.

Case 3: he wins one game and loses two:

$(\mathbf{W}_1' \cap \mathbf{W}_2' \cap \mathbf{W}_3) \cup (\mathbf{W}_1 \cap \mathbf{W}_2' \cap \mathbf{W}_3') \cup (\mathbf{W}_1' \cap \mathbf{W}_2 \cap \mathbf{W}_3') = \{X = -\$120\}$.

Case 4: he loses all three games: $(\mathbf{W}_1' \cap \mathbf{W}_2' \cap \mathbf{W}_3') = \{X = -\$330\}$.

From these cases we can compute:

$P\{X = \$300\} = P(\mathbf{W}_1 \cap \mathbf{W}_2 \cap \mathbf{W}_3) = (0.6)(0.6)(0.6) = 0.216$

$P\{X = \$90\} = P[(W_1' \cap W_2 \cap W_3) \cup (W_1 \cap W_2' \cap W_3) \cup (W_1 \cap W_2 \cap W_3')] =$

$P(W_1' \cap W_2 \cap W_3) + P(W_1 \cap W_2' \cap W_3) + P(W_1 \cap W_2 \cap W_3') =$

$(0.4)(0.6)(0.6) + (0.6)(0.4)(0.6) + (0.6)(0.6)(0.4) = 0.432$

$P\{X = -\$120\} = P[(W_1' \cap W_2' \cap W_3) \cup (W_1 \cap W_2' \cap W_3') \cup (W_1' \cap W_2 \cap W_3')] =$

$P[(W_1' \cap W_2' \cap W_3) + P(W_1 \cap W_2' \cap W_3') + P(W_1' \cap W_2 \cap W_3')] =$

$(0.4)(0.4)(0.6) + (0.6)(0.4)(0.4) + (0.4)(0.6)(0.4) = 0.288$

$P\{X = -\$330\} = P(W_1' \cap W_2' \cap W_3') = (0.4)(0.4)(0.4) = 0.064$

In table form:

x	P{X = x}
- $330	0.064
- $120	0.288
$90	0.432
$300	0.216
	Total 1

Unsolved Problems with Answers

14.2 - Problem 1: An urn has 1 green marble, 1 black marble and 1 red marble. Two marbles are randomly selected with replacement. If both marbles are the same color, you win $30; otherwise you lose $20. Write out the table that represents the probability distribution.

Answer:

x	P{X = x}
-$20	6/9
$30	3/9
	Total 1

⇑ *Refer back to* **14.2 - Example 1 & 14.2 - Solved Problem 1.**

14.2 - Problem 2: A die is tossed twice. Assume the following rule for the random variable: If the pair of numbers are equal assign the number zero; otherwise assign the positive number that is their difference. For example, $X(5,5) = 0$, $X(5,2) = 3$, $X(2,5) = 3$. Write the probability distribution.

Answer:

x	P{X = x}
0	6/36
1	10/36
2	8/36
3	6/36
4	4/36
5	2/36
	Total 1

⇑ *Refer back to* **14.2 - Example 2 & 14.2 - Solved Problem 2.**

14.2 - Problem 3: John Wish is a famous gambler. He claims that he has a 60% chance of winning on any college football game. On each game he bets $110. If his team wins, he makes $100. If his team loses, he loses $110. On a particular week, he bets on four separate football games. Write out the probability distribution for the random variable which is the total amount he won over these four football games.

Answer:

x	P{X = x}
-$440	0.0256
-$230	0.1536
-$20	0.3456
$190	0.3456
$400	0.1296
	Total 1

⇑ *Refer back to* **14.2 - Example 3 & 14.2 - Solved Problem 3.**

14.3 - What is A Cumulative Probability Distribution of a Discrete Random Variable?

A cumulative probability distribution of a discrete random variable X is defined as $P\{X \leq x\}$, for all real numbers x:

$$P\{X \leq x\} = P\{X = x_1\} + P\{X = x_2\} + ... + P\{X = x_k\}, \text{ where } x_1 \leq x_2 \leq ... \leq x_k \leq x .$$

14.3 - Example 1: A coin is tossed twice. Let the random variable X assign to each element the number of heads. Write out the probability distribution in the form of $P(\{X \leq x\})$ for x = 0, 1, 2.

Solution:

The following is the sample space generated by the experiment: **S** = {(h,h), (h,t), (t,h), (t,t)}.

Step 1: From section 2, we have the distribution of X:

x	P{X = x}
0	1/4
1	2/4
2	1/4
	Total 1

Step 2:

x	P{X = x}	P{X ≤ x}
0	1/4	1/4
1	2/4	1/4 + 2/4 = 3/4
2	1/4	3/4 + 1/4 = 1
	Total 1	

Which gives the cumulative distribution:

x	P{X ≤ x}
0	1/4
1	3/4
2	1

14.3 - Example 2: A die is tossed twice. Let the random variable assigned to each pair of numbers the sum of the numbers. Write the probability distribution in the form of $P\{X \le x\}$ where x = 2,3,4,5,6,7,8,9,10,11,12.

Solution:

Step 1: From section 2 we have the distribution table for X:

x	2	3	4	5	6	7	8	9	10	11	12	Total
P{X = x}	1/36	2/3	3/3	4/3	5/3	6/3	5/3	4/3	3/3	2/3	1/3	1

Step 2:

x	P{X = x}	P{X ≤ x}
2	1/36	1/36
3	2/36	1/36 + 2/36 = 3/36
4	3/36	3/36 + 3/36 = 6/36
5	4/36	6/36 + 4/36 = 10/36
6	5/36	10/36 + 5/36 = 15/36
7	6/36	15/36 + 6/36 = 21/36
8	5/36	21/36 + 5/36 = 26/36
9	4/36	26/36 + 4/36 = 30/36
10	3/36	30/36 + 3/36 = 33/36
11	2/36	33/36 + 2/36 = 35/36
12	1/36	35/36 + 1/36 = 1
	Total 1	

Which give the cumulative distribution of X:

x	P{X≤ x}
2	1/36
3	3/36
4	6/36
5	10/36
6	15/36
7	21/36
8	26/36

9	30/36
10	33/36
11	35/36
12	3/36

14.3 - Example 3: John Wish is a famous gambler. He claims that he has a 60% chance of winning on any college football game. On each game he bets $110. If his team wins, he makes $100. If his team loses, he loses $110. On a particular week, he bets on two separate football games. Write out the probability distribution for the random variable which is the total amount he won over these two football games using the form $P\{X \le x\}$ where x = -$250, -$150, -$50, $50, $150, $250.

Solution:

Let W_1 = the event that he wins the first game.

Let W_2 = the event that he wins the second game.

Using these two events and their compliments we can write:

Case 1: He wins both games:

$W_1 \cap W_2 = \{X = \$200\}$

$P\{X = \$200\} = P\{W_1 \cap W_2\} = (0.6)(0.6) = 0.36$

Case 2: He wins only one game:

$(W_1 \cap W_2') \cup (W_1' \cap W_2) = \{X = -\$10\}$

$P\{X = -\$10\} = P\{(W_1 \cap W_2') \cup (W_1' \cap W_2)\} = P(W_1 \cap W_2') + P(W_1' \cap W_2)\} = 2(0.6)(0.4) = 0.48$

Case 3: He loses both games:

$(W_1' \cap W_2') = \{X = -\$220\}$

$P\{X = -\$220\} = P(W_1' \cap W_2') = (0.4)(0.4) = 0.16$

Writing the distribution in the form {X = x },

Step 1: P{X = $200} = 0.36

P{X = -$10} = 0.48

P{X = -$220} = 0.16

From these values, we can now write the table:

x	P{X = x}
-$220	0.16
-$10	0.48
$200	0.36
	Total 1

Step 2:

x	P{X ≤ x}	P{X ≤ x}
-$250	P{X ≤ -$250} = 0	0
-$150	P{X ≤ -$150} = P{X = -$220} = 0.16	0.16
-$ 50	P{X ≤ -$50} = P{X = -$220} = 0.16	0.16
$ 50	P{X ≤ $50} = P{X = -$220} + P{X = -$10} = 0.16 + 0.48 = 0.64	0.64
$150	P{X ≤ $150} = P{X = -$220} + P{X = -$10} = 0.16 + 0.48 = 0.64	0.64
$250	P{X ≤ $250} = P{X = -$220} + P{X = -$10} + P{X = 200} = 0.16 + 0.48 + 0.36 = 1	1.00

x	P{X ≤ x}
-$250	0
-$150	0.16
-$50	0.16
$50	0.64
$150	0.64
$250	1.00

14.3 - Example 4: An urn contains red, white and blue marbles. Three marbles are randomly selected. Assume the random variable X is the number of red marbles that are selected with the following probability distribution:

$P\{X \le 0\} = 0.20$

$P\{X \le 1\} = 0.25$

$P\{X \le 2\} = 0.65$

$P\{X \le 3\} = 1.00$

Find $P\{X = 0\}$, $P\{X = 1\}$, $P\{X = 2\}$, $P\{X = 3\}$.

Solution:

Step 1: From the given values, we have the cumulative table for X:

x	$P\{X \le x\}$
0	0.20
1	0.25
2	0.65
3	1

Step 2:

x	$P\{X = x\}$	$P\{X \le x\}$
0	0.20	0.20
1	0.25 - 0.20 = 0.05	0.25
2	0.65 - 0.25 = 0.40	0.65
3	1 - 0.65 = 0.35	1

This gives the distribution of X:

x	$P\{X = x\}$
0	0.20
1	0.05
2	0.40
3	0.35
	Total 1

Solved Problems

14.3 - Solved Problem 1: A statistics class has 4 men and 6 women. Two students are selected at random without replacement. Let the random variable X assign to each element the number of women selected. Write out the probability distribution in the form of $P(\{X \le x\})$ for x = 0, 1, 2.

Solution:

Let F_1 = the event that the first person selected is a woman

Let F_2 = the event that the second person selected is a woman.

No women are selected: $\{X = 0\} = F_1' \cap F_2'$.

One woman is selected: $\{X = 1\} = (F_1 \cap F_2') \cup (F_1' \cap F_2)$.

Two women are selected: $\{X = 2\} = (F_1 \cap F_2)$.

$$P\{X = 0\} = P(F_1' \cap F_2') = P(F_1')P(F_2' \mid F_1') = (\frac{4}{10})(\frac{3}{9}) = \frac{12}{90}$$

$$P\{X = 1\} = P(F_1 \cap F_2') + P(F_1' \cap F_2) = P(F_1)P(F_2' \mid F_1) + P(F_1')P(F_2 \mid F_1') =$$

$$(\frac{6}{10})(\frac{4}{9}) + (\frac{4}{10})(\frac{6}{9}) = \frac{48}{90}$$

$$P\{X = 2\} = P(F_1 \cap F_2) = P(F_1)P(F_2 \mid F_1) = (\frac{6}{10})(\frac{5}{9}) = \frac{30}{90}$$

Step 1:

x	P{X = x}
0	12/90
1	48/90
2	30/90
	Total 1

Step 2:

x	P{X = x}	P{X ≤ x}
0	12/90	12/90
1	48/90	12/90 + 48/90 = 60/90
2	30/90	60/90 + 30/90 = 1
	Total 1	

x	P{X ≤ x}
0	12/90
1	60/90
2	1

14.3 - Solved Problem 2: A die is tossed twice. Let the random variable assign to each pair of numbers the larger of the two numbers or the number if the two numbers are equal . Write the probability distribution in the form of $P\{X \le x\}$ where x = 1, 2, 3, 4, 5, 6.

Solution:

$\{X = 1\} = \{(1,1)\}$

$\{X = 2\} = \{ (1,2), (2,1), (2,2)\}$

$\{X = 3\} = \{ (1,3), (2,3), (3,1), (3,2), (3,3)\}$

$\{X = 4\} = \{ (1,4),(2,4), (3,4),(4,1), (4,2), (4,3), (4,4)\}$

$\{X = 5\} = \{(1,5),(2,5), (3,5), (4,5), (5,1), (5,2),(5,3),(5,4),(5,5)\}$

$\{X = 6\} = \{(1,6), (2,6), (3,6), (4,6), (5,6), (6,1), (6,2), (6,3),(6,4), (6,5), (6,6)\}$

Since #**S** = 36,

$P\{X = 1\} = 1/36$

$P\{X = 2\} = 3/36$

$P\{X = 3\} = 5/36$

$P\{X = 4\} = 7/36$

$P\{X = 5\} = 9/36$

$P\{X = 6\} = 11/36$

Step 1:

x	P{X = x}
1	1/36
2	3/36
3	5/36
4	7/36
5	9/36
6	11/36
	Total 1

Step 2:

x	P{X = x}	P{X ≤ x}
1	1/36	1/36
2	3/36	1/36 + 3/36 = 4/36
3	5/36	4/36 + 5/36 = 9/36
4	7/36	9/36 + 7/36 = 16/36
5	9/36	16/36 + 9/36 = 25/36
6	11/36	25/36 + 11/36 = 1
	Total 1	

x	P{X ≤ x}
1	1/36
2	4/36
3	9/36
4	16/36
5	25/36
6	1

14.3 - Solved Problem 3: John Wish is a famous gambler. He claims that he has a 60% chance of winning on any college football game. On each game he bets $110. If his team wins, he makes $100. If his team loses, he loses $110. On a particular week, he bets on two separate football games. Write out the probability distribution for the random variable which is the total amount he won over these two football games using the form

P{X ≤ x}, where x = -$100, -$50, $0, $50, $100, $200.

Solution:

The winnings for each possible outcome are $200, -$10, -$220. Except for the values of x, this problem is the same as Example 3.3. Therefore, we shall use the following results previously derived:

P{X = $200} = 0.36

P{X = -$10} = 0.48

P{X = -$220} = 0.16

Step 1: We have the table:

x	P{X = x}
-$220	0.16
-$10	0.48
$200	0.36
	Total 1

Step 2:

x	P{X ≤ x}	P{X ≤ x}
-$100	P{X ≤ -$100} = P{X = -$220} = 0.16	0.16
-$50	P{X ≤ -$50} = P{X = -$220} = 0.16	0.16
$ 0	P{X ≤ $0} = P{X = -$220} + P{X = -$10} = 0.16 + 0.48 = 0.64	0.64
$50	P{X ≤ $50} = P{X = -$220} + P{X = -$10} = 0.16 + 0.48 = 0.64	0.64
$100	P{X ≤ $100} = P{X = -$220} + P{X = -$10} = 0.16 + 0.48 = 0.64	0.64
$200	P{X ≤ $200} = P{X = -$220} + P{X = -$10} + P{X = 200} = 0.16 + 0.48 + 0.36= 1	1.00

x	P{X ≤ x}
-$100	0.16
-$50	0.16
$ 0	0.64
$50	0.64
$100	0.64
$200	1.00

14.3 - Solved Problem 4: A random sample of four adults were asked their age. Assume the random variable X is the number of adults in this sample that are over thirty years old with the following probability distribution:

$P\{X \le 0\} = 0.15$

$P\{X \le 1\} = 0.20$

$P\{X \le 2\} = 0.35$

$P\{X \le 3\} = 0.75$

$P\{X \le 4\} = 1.00$

Find P{X = 0}, P{X = 1}, P{X = 2}, P{X = 3}, P{X = 4}.

Solution:

Step 1: From the above, we have the cumulative distribution table of X:

x	P{X ≤ x}
0	0.15
1	0.20
2	0.35
3	0.75
4	1

Step 2:

x	P{X = x}	P{X ≤ x}
0	0.15	0.15
1	0.20 - 0.15 = 0.05	0.20
2	0.35 - 0.20 = 0.15	0.35
3	0.75 - 0.35 = 0.40	0.75
4	1 - 0.75 = 0.25	1

Unsolved problems with Answers

14.3 - Problem 1: Two cards are randomly drawn from an ordinary deck of cards without replacement. Let the random variable X assign to each element the number of clubs selected. Write out the probability distribution in the form of P{X ≤ x } for x = 0, 1, 2.

Answer:

x	P{X ≤ x}
0	1482/2652
1	2496/2652
2	1

⇧ *Refer back to* **14.3 - Example 1 & 14.3 - Solved Problem 1.**

14.3 - Problem 2: One side of a die is painted red. This die is tossed twice. Let the random variable assign to each pair of tosses the number of red sides that are tossed. Write the probability distribution in the form of $P\{X \le x\}$ where x = 0, 1, 2.

Answer:

$P\{X \le 0\} = 25/36$

$P\{X \le 1\} = 35/36$

$P\{X \le 2\} = 1$

⇑ *Refer back to* **14.3 - Example 2 & 14.3 - Solved Problem 2.**

14.3 - Problem 3: John Wish is a famous gambler. He claims that he has a 60% chance of winning on any college football game. On each game he bets $110. If his team wins, he makes $100. If his team loses, he loses $110. On a particular week, he bets on two separate football games. Write out the probability distribution for the random variable which is the total amount he won over these two football games using the form $P\{X \le x\}$, where x = -$300, -$200, -$100, $0, $100, $200, $300.

Answer:

x	$P\{X \le x\}$
-$300	0.00
-$200	0.16
-$100	0.16
$ 0	0.64
$100	0.64
$200	1.00
$300	1.00

⇑ *Refer back to* **14.3 - Example 3 & 14.3 - Solved Problem 3.**

14.3 - Problem 4: A random sample of three adults were asked if they were born in California. Assume the random variable X is the number of adults in this sample that were not born in California.:

$P\{X \le 0\} = 0.15$
$P\{X \le 1\} = 0.15$
$P\{X \le 2\} = 0.75$
$P\{X \le 3\} = 1.0$

Find $P\{X = 0\}$, $P\{X = 1\}$, $P\{X = 2\}$, $P\{X = 3\}$.

Answer:
$P\{X = 0\} = 0.15$, $P\{X = 1\} = 0$, $P\{X = 2\} = 0.60$, $P\{X = 3\} = 0.25$

⇑ *Refer back to* **14.3 - Example 4 & 14.3 - Solved Problem 4.**

Supplementary Problems

1. Assume X is a random variable, that takes on values $x = 0, 1, 2, 3,....$ Show, using sets and set operators that $P\{X = k + 1\} = P\{X \leq k + 1\} - P\{X \leq k\}$.

2. Assume X is a random variable, that takes on values $x = 0, 1, 2, 3, 4, 5$. Also assume that

$P\{X \leq 0\} = 0.2$
$P\{X \leq 1\} = 0.3$
$P\{X \leq 2\} = 0.35$
$P\{X \leq 3\} = 0.50$
$P\{X \leq 4\} = 0.70$
$P\{X \leq 5\} = 1.00$

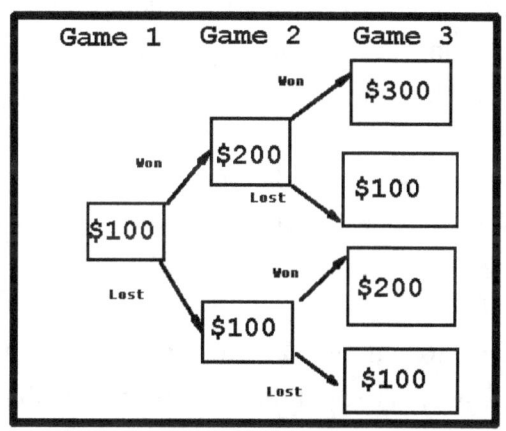

Find:

a. $P\{X \geq 0\}$
b. $P\{X \geq 1\}$
c. $P\{X \geq 2\}$
d. $P\{X = 3\}$
e. $P\{2 \leq X \leq 4\}$
f. $P\{2 < X \leq 4\}$
g $P\{X > 4\}$.

3. John Wish is a famous gambler. He claims that he has a 60% chance of winning on any college football game. $100 bet on a game wins $90 or losses $100. On a particular week, he bets on three separate football games. From the above diagram, he decides to use the following rule for the amount he bets on each of the three games:

For example, assume he losses the first game, but wins the second game. For this case, he bets $100 on the first game, $100 on the second game but $200 on the third game.

Write out the probability distribution for the random variable X which is the total amount he won over these three football games.

4. Mr. Jones, a car salesman, sells at most 5 cars a week. The following distribution table expresses the chance of selling a given number of cars:

x	0	1	2	3	4	5
P{X = x}	0.10	0.25	0.30	0.20	0.10	0.05

Find the following probability that he sells:

a. at least 3 cars.

b. at most 3 cars.

c. three or five cars.

d. between 1 and 4 cars (inclusive).

e. less than 5 cars.

5. A die is tossed twice. Let X_1 be the random variable that assigns the number on the first toss and X_2 be the random variable that assigns the number on the second toss. Define the random variable $X = 2X_1 - X_2$. Write out the probability distribution table for X in the form P(X = k).

6. A die is tossed twice. Let the random variable assigned to each pair of number the sum of the numbers. Write the probability distribution in the form of $P\{X \geq x\}$ where x = 2, 3, 4, 5, 6, 7, 8, 9, 10, 11, 12.

7. A fair die is tossed until 2 heads appear or 5 times, whichever occurs first. Let X equal the number of tosses. Write out the probability distribution $P\{X = x\}$.

8. The conditional probability distribution of two discrete random variables X, Y is defined as

$$P(X = x \mid Y = y) = \frac{P[(X = x) \cap (Y = y)]}{P(Y = y)}, \, P(Y = y) \neq 0.$$

Assume 2 cards are drawn, without replacement, from an ordinary deck of cards. If X is the number of kings and Y the number of queens drawn, find the conditional probability distribution of X and Y.

9. Assume X is a random variable with the following distribution table:

X = x	P{X = x}
-2	1/10
-1	2/10
1	3/10
2	4/10

Find the distribution of the following random variables:

a. X^3

b. X^4

c. $X^4 - X^3$.

The joint distribution of two discrete random variables X,Y is defined as

$p(x_j, y_k) = P\{X = x_j; Y = y_k\} = P[\{X = x_j\} \cap \{Y = y_k\}]$ for all x_j, y_k where $j = 1,...,n$; $k = 1,..., m$.

Using the fact that

$P\{X = x_j\} = p(x_j; y_1) + p(x_j; y_2) + ... + p\{x_j; y_m\}$,

allows user to determine the distribution of X (or Y) from the joint distribution of X,Y.

The following table represents the joint distribution.

X		**Y**			**P{X = x}** (Sum of the rows)
		y_1	y_2 .... y_m		
	x_1	$p(x_1,y_1)$ $p(x_1,y_2)$ ... $p(x_1,y_m)$			$P\{X = x_1\}$
	x_2	$p(x_2,y_1)$ $p(x_2,y_2)$ ... $p(x_2,y_m)$			$P\{X = x_2\}$
	.	.			.
	.	.			.
	.	.			.
	x_n	$p(x_n,y_1)$ $p(x_n,y_2)$ $p(x_n,y_m)$			$P\{X = x_n\}$
P{Y = y} (Sum of the columns)		$P\{Y = y_1\}$ $P\{Y = y_2\}$ ...$P\{Y = y_m\}$			

10. Two cards are drawn from an ordinary deck of cards without replacement. Let the random variable X represent the number of kings and Y the number of queens drawn. For X,Y construct a joint distribution table.

11. Assume the following joint distribution table for 2 random variables:

		Y			X
		0	1	2	3
X	1	2p	0	0	p
	2	6p	6p	6p	0
	3	0	6p	0	0
Y					

Find the distribution of X and Y.

12. A fair die is rolled until the number 5 appears. Let X be the random variable that equals the number of rolls that ends the game.

a. Compute $P\{X = k\}$. (The distribution of X is a special case of the geometric distribution.)

b. Find the smallest number of rolls N for $P\{X \le N\} \ge 0.90$.

c. Show that $P\{X > N\} < 0.10$ can be derived from b.

d. Express in ordinary words the meaning of b.

e. Express in ordinary words the meaning of c.

13. A fair die is rolled until the number 5 appears twice. Let X be the random variable that equals the number of rolls that ends the game.

a. Compute the distribution of X: $P\{X = k\}$ for $k > 1$.

b. Compute $P\{4 < X \le 10\}$.

c. Find the probability that it will take at least 5 rolls to terminate the game.

14. *Chevalier de Mere paradox.* In the 17th century the French gambler, Chevalier de Mere would bet on the following dice games: On rolling a die 4 times he bet that he would get at least 1 ace. (An ace is the "1" of a a six-sided die.) In the other dice game, he would bet that in 24 rolls of a pair of dice, at least 1 double ace would be rolled. (A double ace is pair of dice showing 2 aces.) Chevalier de Mere believed that chance on winning each game was the same.

a. Show Chevalier de Mere was wrong.

b. Chevalier did not understand modern probability theory. How do you think he concluded the chance of winning was the same?

The joint distribution of three discrete random variables X,Y,Z is defined as

$$P\{X = x_j; Y = y_k; Z = z_r\} = P[\{X = x_j\} \cap \{Y = y_k\} \cap (Z = z_r)] \text{ for all } x_j, y_k, z_r.$$

We define pair-wise independents of three discrete random variables X,Y,Z as

$$P\{X = x_j; Y = y_k\} = P\{X = x_j\}P\{Y = y_k\} \text{ for all } x_j, y_k,$$

$$P\{X = x_j; Z = z_k\} = P\{X = x_j\}P\{Z = z_k\} \text{ for all } x_j, z_k,$$

$$P\{Y = y_j; Z = z_k\} = P\{Y = y_j\}P\{Z = z_k\} \text{ for all } y_j, z_k,$$

We define mutually independents of three discrete random variables X,Y,Z as

i. they are pair-wise independent.

ii. $P\{X = x_j; Y = y_k; Z = z_r\} = P\{X = x_j\}P\{Y = y_k\}P\{Z = z_r\}$ for all $x_j, y_k, z_r.$

15. Assume two independent random variables are defined as following:

$$P\{X = 1\} = P\{X = -1\} = 1/2$$

$$P\{Y = 1\} = P\{Y = -1\} = 1/2$$

Define $Z = XY.$

a. Find the distribution of Z.

b. Show that the three random variables are pair-wise independent.

c. Are the three random variables mutually independent?

16. If $m < n$, show

a. $\{X < m\} \subseteq \{X < n\}.$

b. $\{X > n\} \subseteq \{X > m\}$

17. A fair die is tossed until an ace is rolled. If the ace does not occur on the first roll, find the probability that it will take more than three rolls to terminate the game.

18. Assume X_1 and X_2 are independent random variables and both have the same geometric distribution

$$P\{X_j = k\} = q^k p \text{ where } q = 1 - p, k = 0,1,... \text{ and } j = 1,2.$$

a. Define $Z = X_1 + X_2$. Find the distribution of Z.

b. Show that $P\{X_1 = k \mid Z = n\} = \dfrac{1}{n + 1}$ for $0 \le k \le n$, which is a uniform distribution.

19. We have N blank cards each numbered 1 to N respectively. Assume the N cards are randomly shuffled. Define the random variable X_k (k = 1, 2, ... N), where $X_k = 1$ if the card marked k is in the kth position from the top of the deck; If not $X_k = 0$. For example, If card marked 7 is at the 7th position from the top, then $X_7 = 1$.

a. Find the probability distribution of X_k

b. The pair-wise joint distribution of two discrete random variables X_j, X_k is defined as

$p(x_j, x_k) = P\{X_j = x_j; X_k = x_k\} = P[\{X_j = x_j\} \cap \{X_k = x_k\}]$ for all x_j, x_k.

Find the distribution of $p(x_j, x_k)$.

c. Is the distribution of X_k mutually independent?

d. Find $p(1,0,0) = P\{X_j = 1; X_k = 0; X_r = 0\}$.

20. A coin is tossed N times. Assume the probability of heads appearing on each toss is p and tails $1 - p = q$. . For each toss define the sequence of random variables X_k to be the following: If on the kth toss heads occur, $X_k = 1$; otherwise $X_k = 0$. This sequence of random variables is called a Bernoulli trial.

a. Define $S_N = X_1 + X_2 + ... + X_N$. Interpret S_N.

b. Is the random variable sequence S_N mutually independent.

c. Find $P\{S_3 = 2\}$.

d. Does $P\{X_1 = 1 \mid S_2 = 2\} = P\{S_2 = 2 \mid X_1 = 1\}$?

e. Find $P\{S_1 = 1 \mid S_4 = 3\}$.

21. Assume for the two random variables X,Y where $(Y = y_1, y_2, ... y_m)$. Define

$p(x, y) = P[(X = x) \cap (Y = y)]$.

Show $p(x, y_1) + p(x, y_2) + ... + p(x, y_{rm}) = P(X = x)$.

22. A coin is tossed a certain number of times depending on the outcome of a single toss of a fair die. If the rolled die results in the number n ($1 \le n \le 6$), the coin is tossed n times. Let X_j ($1 \le j \le n$) represent the result of tossing the coin on the jth toss where $P(X_j = 1) = p$, $(0 < p < 1)$ and $P(X_j = 0) = 1 - p = q$. Let N represent the result of toss the die once where $P(N = n) = 1/6$; $1 \le n \le 6$.

Define $S = X_1 + X_2 + ... X_n$.

a. What is the meaning of S for this game.

b. Find $P(N = 1|S = 1)$. (Hint: Use Bayes.)

23. Assume we have 2 independent geometric distributions X and Y where $P\{X = k\} = q^{k-1}p = P\{Y = k\}$. Find the distribution of $S = X + Y$.

24. Assume the random variable X has the distribution $P\{X = x_k\}$ where
$P\{X = x_1\} + P\{X = x_2\} + ... + P\{X = x_N\} = 1$.

For $P(A) > 0$, show $P\{X = x_1|A\} + P\{X = x_2|A\} + ... + P\{X = x_N|A\} = 1$

25. A fair coin is tossed until a head occurs or 5 times which ever occurs first. Let X be the random variable that equals the number of tosses to terminate the game. Find the conditional distribution $P(X = k|\mathbf{H})$, where $\mathbf{H}$ is the event that a head is tossed and $(k = 1,2,3,4,5)$.

26. Assume X is a r.v. that has a geometric distribution $P(X = k) = pq^{k-1}$ $(k = 1,2,...)$. Show that
$P(X = n + k| X > n) = P(X = k)$ for $k, n \geq 1$.

27. Assume X is a random variable $P(X = k)$, for $k = 1,2,3....$ If $P(X = n + k| X > n) = P(X = k)$ for $k, n \geq 1$ then X has a geometric distribution $P(X = k) = pq^{k-1}$ $(k = 1,2,...)$.

28. Interpret the meaning of the results of problems 26 and 27.

29. The condition $P(X = n + k| X > n) = P(X = k)$ is called 'the lack of memory' property. Explain in your own words why this is so.

30. John Wish is a famous gambler. He claims that he has a 60% chance of winning on any college football game. On each game he bets $110. If his team wins, he makes $100. If his team loses, he loses $110. On a particular week, he uses the following rule for betting: He bets until he losses a game or he wins three games. Write out the probability distribution for the random variable which is the total amount he won over these three possible football games.

15.1- What is the Expectation of a Random Variable?

The best way to define what the expectation of a Random Variable is by the following example:

Assume Joe's Hamburger Joint sells two types of hamburgers: The Super Deluxe burger and the California Burger. The Super Deluxe sells for $5.65 and the California sells for $4.00. Their records showed that in 1993, they sold 5,850 Super Deluxe burgers and 10,125 California burgers. We ask the question: For a typical customer in 1993, what is the average price of a burger? To find this average, we need the total revenue from all burgers sold. Next, divide the total revenue by the total number of burgers sold:

Total revenue from selling Super Deluxe burgers = ($5.65)(5,850) = $33,052.50 .

Total revenue from selling California burgers = ($4.00)(10,125) = $40,500.00 .

Total revenue from selling burgers = ($5.65)(5,850) + ($4.00)(10,125) = $73,552.50 .

Total number of burgers = 5,850 + 10,125 = 15,975.00 .

The average price per burger is

$$\frac{\text{total revenue}}{\text{total burgers sold}} = \frac{(\$5.65)(5,850) + (\$4.00)(10,125)}{5,850 + 10,125} = \frac{\$73,552.50}{15,975} = \$4.60 \text{ (rounded)}.$$

We can now write,

$$\frac{\text{total revenue}}{\text{total burgers sold}} = \frac{(\$5.65)(5,850) + (\$4.00)(10,125)}{5,850 + 10,125} = (\$5.65)\frac{5850}{15975} + (\$4.00)\frac{10125}{15975}.$$

We are now ready to define a random variable X. Assume the sample space $\mathbf{S}$ = {Super Deluxe burger, California burger}. Let X be the random variable where X(Super Deluxe burger) = $5.65, and X(California burger) = $4.00. Since the total number of burgers consumed in 1993 was

15,975, the   total number of Super Deluxe burgers consumed was 5,850, and the total number of California burgers consumed was 10,125, it follows that

$$P\{X = \$5.65\} = \frac{5850}{15975} \text{ and } P\{X = \$4.00\} = \frac{10125}{15975}.$$

The expectation of the random variable X is defined as

$$E(X) = \$5.65P\{X = \$5.65\} + \$4.00P\{X = \$4.00\} =$$

$$(\$5.65)\frac{5850}{15975} + (\$4.00)\frac{10125}{15975} = \$4.60 \text{ (rounded)}.$$

In table form we can write this process as:

X = x	P{X = x}	xP{X = x}
$5.65	$\dfrac{5850}{15975}$	$(\$5.65)\dfrac{5850}{15975} = \dfrac{\$33,052}{15975}$
$4.00	$\dfrac{10125}{15975}$	$(\$4.00)\dfrac{10125}{15975} = \dfrac{\$40,500}{15975}$
Total 1		$E(X) = \dfrac{\$33,052}{15975} + \dfrac{\$40,500}{15975} = \dfrac{\$73,552.50}{15,975} = \$4.60$

Therefore, the expectation of a random variable is the same as the average of the random variable.

The definition of the expectation of a random variable E(X), where X is equal to a finite number of discrete values $x_1, x_2,..., x_n$ is

$$E(X) = x_1P\{X = x_1\} + x_2P\{X = x_2\} + x_3P\{X = x_3\} + ... + x_nP\{X = x_n\}.$$

15.1 - Example 1: Two cards are randomly selected, without replacement, from an ordinary deck of cards. Find the expected number of diamonds.

Solution:

Step 1: X: The random variable that equals the number of diamonds in a hand containing 2 cards: 0,1,2 .

$(X = 0) = D_1{}'{\cap}D_2{}'$, 0 diamonds.

$(X = 1) = (\mathbf{D_1} \cap \mathbf{D_2}') \cup (\mathbf{D_1}' \cap \mathbf{D_2})$, 1 diamond.

$(X = 2) = \mathbf{D_1} \cap \mathbf{D_2}$, 2 diamonds.

Step 2: $P(X = 0) = P(\mathbf{D_1}' \cap \mathbf{D_2}) = P(\mathbf{D_1}')P(\mathbf{D_2}'|\mathbf{D_1}') = (39/52)(38/51) = (1482)/2652$

$P(X = 1) = P(\mathbf{D_1} \cap \mathbf{D_2}') + P(\mathbf{D_1}' \cap \mathbf{D_2}) = P(\mathbf{D_1})P(\mathbf{D_2}'|\mathbf{D_1}) + P(\mathbf{D_1}')P(\mathbf{D_2}|\mathbf{D_1}') =$
$(13/52)(39/51) + (39/52)(13/51) = (1014)/(2652)$

$P(X = 2) = P(\mathbf{D_1} \cap \mathbf{D_2}) = P(\mathbf{D_1})P(\mathbf{D_2}|\mathbf{D_1})$ $(13/52)(12/51) = (156)/(2652)$

Step : Construct a distribution table:

X = x	P{X = x}	xP{X = x}
0	(1482)/(2652)	0(1482/2652) = 0
1	(1014)/(2652)	1(1014/2652) = (1014)/(2652)
2	(156)/(2652)	2(156/2652) = (312)/(2652)
	Total 1	**E(X)** = 0 + (1014/2652) + (312/2652) = (1326)/(2652) = 0.5

15.1 - Example 2: Mrs. Smith plays the following game: A die is tossed once. If a 1 appears, she wins $10.00, if a 2 or 3 appears, she wins $5.00 and if a 4, 5, or 6 appears, she loses $20.00. Find her expected (average) winning for this game.

Solution:

The random variable X is the amount she wins or loses:

{X = $5.00} = {2, 3}
{X = $10.00} = {1}
{X = -$20.00} = {4, 5, 6}
P{X = $5.00} = P{2, 3} = 2/6
P{X = $10.00} = P{1} = 1/6
P{X = -$20.00} = P{4, 5, 6} = 3/6

X = x	P{X = x}	xP{X = x}
$ 5	2/6	$10/6
$10	1/6	$10/6
-$20	3/6	-$60/6
Total 1		**E(X)** = -$40/6 = -$6.67 loss

15.1 - Example 3: An urn contains 3 red marbles and 1 white marble. Marbles are selected at random without replacement until a white is selected. Find the expected (average) number of marbles selected.

Solution:

Step 1: Let

W_1: the event that a white marble is selected on first drawing.
W_2: the event that a white marble is selected on the second drawing.
W_3: the event that a white marble is selected on the third drawing.
W_4: the event that a white is selected on the fourth drawing.

Step 2: The random variable is the number of selections X:

$\{X = 1\} = W_1$
$\{X = 2\} = W_1' \cap W_2$
$\{X = 3\} = W_1' \cap W_2' \cap W_3$
$\{X = 4\} = W_1' \cap W_2' \cap W_3' \cap W_4$

Step 3: $P\{X = 1\} = P\{W_1\} = \dfrac{1}{4}$

$P\{X = 2\} = P(W_1' \cap W_2) = (\dfrac{3}{4})(\dfrac{1}{3}) = \dfrac{1}{4}$

$P\{X = 3\} = P\{W_1' \cap W_2' \cap W_3\} = (\dfrac{3}{4})(\dfrac{2}{3})(\dfrac{1}{2}) = \dfrac{1}{4}$

$P\{X = 4\} = P\{W_1' \cap W_2' \cap W_3' \cap W_4\} = (\dfrac{3}{4})(\dfrac{2}{3})(\dfrac{1}{2})(\dfrac{1}{1}) = \dfrac{1}{4}$

Step 4:

X = x	P{X = x}	xP{X = x}
1	1/4	1/4
2	1/4	2/4
3	1/4	3/4
4	1/4	4/4
Total 1		E(X) = 10/4 = 2.5 selections

15.1 - Example 4: Mrs. Jones plays a game where depending on three outcomes, she can win $50, $75 or lose $100. Assume she has a 30% chance of winning $50 and a 40% chance of winning $75. With this information complete the following table:

X = x	P{X = x}	xP{X = x}
$50	0.30	$15.00
$75	0.40	$30.00
- $100	p	- $100p
Total 1		E(X) = ?

Solution:

Since the total values in the second column must equal 1, then p = 1 - 0.30 - 0.40 = 0.30. The following is the completed table:

X = x	P{X = x}	xP{X = x}
$50	0.30	$15.00
$75	0.40	$30.00
- $100	0.30	- $100p = -$100(0.3) = - $30.00
Total 1		E(X) = $15.00

Solved Problems

15.1 - Solved Problem 1: An urn has 5 red marbles, 10 blue marbles, and 15 white marbles. Three marbles are selected without replacement. Find the expected (average) number of red marbles selected.

Solution:

The random variable is the number of red marbles selected: $X = 0, 1, 2, 3$.

$\mathbf{R_1}$: the event that red marble occurs on the first drawing.

$\mathbf{R_2}$: the event that a red marble occurs on the second drawing.

$\mathbf{R_3}$: the event that a red marble occurs on the third drawing.

$\{X = 0\} = \mathbf{R_1}'\cap\mathbf{R_2}'\cap\mathbf{R_3}'$

$\{X = 1\} = (\mathbf{R_1}\cap\mathbf{R_2}'\cap\mathbf{R_3}')\cup(\mathbf{R_1}'\cap\mathbf{R_2}\cap\mathbf{R_3}')\cup(\mathbf{R_1}'\cap\mathbf{R_2}'\cap\mathbf{R_3})$

$\{X = 2\} = (\mathbf{R_1}\cap\mathbf{R_2}\cap\mathbf{R_3}')\cup(\mathbf{R_1}\cap\mathbf{R_2}'\cap\mathbf{R_3})\cup\mathbf{R_1}'\cap\mathbf{R_2}\cap\mathbf{R_3})$

$\{X = 3\} = \mathbf{R_1}\cap\mathbf{R_2}\cap\mathbf{R_3}$

$P\{X = 0\} = P(\mathbf{R_1}'\cap\mathbf{R_2}'\cap\mathbf{R_3}') = (\frac{25}{30})(\frac{24}{29})(\frac{23}{28}) = \frac{13800}{24360}$

$P\{X = 1\} = P[\mathbf{R_1}\cap\mathbf{R_2}'\cap\mathbf{R_3}')\cup (\mathbf{R_1}'\cap\mathbf{R_2}\cap\mathbf{R_3}')\cup(\mathbf{R_1}'\cap\mathbf{R_2}'\cap\mathbf{R_3})] =$

$(\frac{5}{30})(\frac{25}{29})(\frac{24}{28}) + (\frac{25}{30})(\frac{5}{29})(\frac{24}{28}) + (\frac{25}{30})(\frac{24}{29})(\frac{5}{28}) = \frac{9000}{24360}$

$P\{X = 2\} = P[(\mathbf{R_1}\cap\mathbf{R_2}\cap\mathbf{R_3}')\cup(\mathbf{R_1}'\cap\mathbf{R_2}\cap\mathbf{R_3})\cup(\mathbf{R_1}\cap\mathbf{R_2}'\cap\mathbf{R_3})] =$

$(\frac{5}{30})(\frac{4}{29})(\frac{25}{28}) + (\frac{25}{30})(\frac{5}{29})(\frac{4}{28}) + (\frac{5}{30})(\frac{25}{29})(\frac{4}{28}) = \frac{1500}{24360}$

$P\{X = 3\} = P[\mathbf{R_1}\cap\mathbf{R_2}\cap\mathbf{R_3}] = (\frac{5}{30})(\frac{4}{29})(\frac{3}{28}) = \frac{60}{24360}$

The following is a table for computing the average number of red marbles:

X = x	P{X = x}	xP{X = x}
0	$\frac{13800}{24360}$	0
1	$\frac{9000}{24360}$	$\frac{9000}{24360}$
2	$\frac{1500}{24360}$	$\frac{3000}{24360}$

X = x	P{X = x}	xP{X = x}
3	$\dfrac{60}{24360}$	$\dfrac{180}{24360}$
Total 1		E(X) = 0.5

15.1 - Solved Problem 2: Roulette is a casino game where the house spins a marble along a wheel which is marked with 38 numbers: 0, 00, 1, 2, 3,..., 36. On each spin of the marble, Mrs. Hope bets $5.00 that the marble will land on an odd number. If she wins, she wins $5.00 otherwise she loses $5.00. Find the expectation of this game for her.

Solution:

The sample space **S** = {0, 00, 1, 2, 3, 4, 5,.., 36}.

The random variable X is the amount that she wins and loses:

{X = $5.00} = {1, 3, 5, 7,..., 35}
{X = -$5.00} = {0, 00, 2, 4, 6, 8,.., 36}

P{X = $5.00} = P{1, 3, 5, 7,..., 35} = 18/38
P{X = -$5.00} = P{{0, 00, 2, 4, 6, 8,..., 36}} = 20/38

X = x	P{X = x}	xP{X = x}
$5	18/38	$90/38
- $5	20/38	- $100/38
Total = 1		E(X) = -$0.26 loss on each game.

15.1 - Solved Problem 3: An urn contains 3 red marbles and 2 white marble. Marbles are selected at random, without replacement until both colors are selected. Find the expected (average) number of marbles selected.

Solution:

Step 1: W$_1$: the event that a white marble is selected on the first drawing.

W$_2$: the event that a white marble is selected on the second drawing.

W$_3$: the event that a white marble is selected on the third drawing.

W_4: the event that a white marble is selected on the fourth drawing.

Step 2: The random variable is the number of selections X:

$$\{X = 2\} = (W_1' \cap W_2) \cup (W_1 \cap W_2')$$

$$\{X = 3\} = (W_1' \cap W_2' \cap W_3) \cup (W_1 \cap W_2 \cap W_3')$$

$$\{X = 4\} = W_1' \cap W_2' \cap W_3' \cap W_4$$

Step 3: $P\{X = 2\} = P\{(W_1' \cap W_2) \cup (W_1 \cap W_2')\} = (\frac{3}{5})(\frac{2}{4}) + (\frac{2}{5})(\frac{3}{4}) = \dfrac{12}{20}$

$P\{X = 3\} = P\{(W_1' \cap W_2' \cap W_3) \cup (W_1 \cap W_2 \cap W_3')\} =$

$(\frac{3}{5})(\frac{2}{4})(\frac{2}{3}) + (\frac{2}{5})(\frac{1}{4})(\frac{3}{3})) = \dfrac{6}{20}$

$P\{X = 4\} = P(W_1' \cap W_2' \cap W_3' \cap W_4) = (\frac{3}{5})(\frac{2}{4})(\frac{1}{3})(\frac{2}{2}) = \dfrac{2}{20}$

Step 4:

X = x	P{X = x}	xP{X = x}
2	12/20	24/20
3	6/20	18/20
4	2/20	8/20
	Total = 1	**E(X)** =50/20 = 2.5 selections

15.1 - Solved Problem 4: In Mrs. Jones third grade class, a survey of students' ages revealed that her students are 8, 9, and 7 years old. 65% of the students are 8 years old, 15% are 9 years old. Find the average age of students in her class.

Solution:

Since the total values in the second column must equal 1, then p = 1 - 0.65 - 0.15 = 0.20 .

The following is the completed table:

X = x	P{X = x}	xP{X = x}
8	0.65	5.20
9	0.15	1.35
7	0.20	1.40

X = x	P{X = x}	xP{X = x}
	Total 1	**E(X) = 7.95** yearly age

Unsolved Problems with Answers

15.1 - Problem 1: An English class has 5 men and 15 women. Three persons are selected without replacement. Find the expected (average) number of women selected.

Answer:

E(X) = 2.25

⇑ *Refer back to* **15.1 - Example 1 & 15.1 - Solved Problem 1.**

15.1 - Problem 2: Roulette is a casino game where the house spins a marble along a wheel which is marked with 38 numbers: 0, 00, 1,2,3,..., 36. On each spin of the marble, Mrs. Hope bets $5.00 on 0 and $10.00 on 00. If the marble lands on 0 she wins $165.00. If the marble lands on 00 she wins $345.00; otherwise, she loses $15.00. Find the expectation of this game for her.
Answer:

E(X) ≈ -$0.79

⇑ *Refer back to* **15.1 - Example 2 & 15.1 - Solved Problem 2.**

15.1 - Problem 3: An urn contains 2 red and 2 white marbles. Marbles are selected 1 at a time, without replacement until 2 whites are selected. Find the expected number of selections.

Answer:

E(X) ≈ 3.33

⇑ *Refer back to* **15.1 - Example 3 & 15.1 - Solved Problem 3.**

15.1 - Problem 4: Ms. White has the following weekly work schedule:

Number hours per day worked	Days
9	Monday
8	Tues & Thursday
7	Wednesday & Friday

Find the average number of hours she works per day.

Answer:

E(X) = 7.8 hours a day.

⇑ *Refer back to* **15.1 - Example 4 & 15.1 - Solved Problem 4.**

15.2 - The expectation of a sum of random variables

Assume we have the sum of sequence of random variables,

$S = X_1 + X_2 + X_3 + ... + X_n.$

It can be shown that the expectation of S equals

$E(S) = E(X_1 + X_2 + X_3 + ... + X_n) = E(X_1) + E(X_2) + E(X_3) + ... + E(X_n).$

15.2 - Example 1: A fair die is tossed 10 times and the total number is recorded. Find the expectation of the sum.

Solution:

Step 1: Let X_k be the resulting outcome on the kth toss of th die (k = 1,2,...,10).

Step 2:

x_k	$P\{X_k = x_k\}$	$x_k P\{X_k = x_k\}$
1	1/6	1/6
2	1/6	2/6
3	1/6	3/6
4	1/6	4/6
5	1/6	5/6
6	1/6	6/6
		E(X_k) = 21/6 = 3.5

$E(S) = E(X_1) + E(X_2) + E(X_3) + ... + E(X_{10}) = 10(3.5) = 35.$

Solved Problems

15.2 - Solved Problem 1: Four cards are drawn, without replacement, from an ordinary deck of cards. Find the expected number of diamonds.

Solution:

Step 1: Let X_k be the number of diamonds drawn on the kth drawing (k = 1,2,3,4).

Step 2:

x_k	$P\{X_k = x_k\}$	$x_k P\{X_k = x_k\}$
0	39/52	0(39/52)
1	13/52	1(13/52)
		$E(X_k)$ = 13/52

$E(S) = E(X_1 + X_2 + X_3 + X_4) = E(X_1) + E(X_2) + E(X_3) + EX_4) = 4(13/52) = 1$ diamond.

Unsolved problems with Answers

15.2 - Problem 1: Mr. Fortune wages on sport events. He believe that there is a 60% chance that he can win a game on football, a 65% chance on baseball and a 70% chance on basketball. For each of these games, he wages $110. If he wins a game, he wins $100; otherwise he loses $110. If he were to wager on each of these games, what would be his average winnings?

Answer:

$79.50

⇑ *Refer back to* **15.2 - Example 1 & 15.2 - Solved Problem 1.**

Supplementary Problems

1. Mr. Jones played the following game: two cards are selected at random without replacement from an ordinary deck of cards. He wins $100 for each king and loses $125 for each queen of hearts or queen of diamonds he draws. Find the expectation (average winnings) of this game.

2. John Wish is a famous gambler. He claims that he has a 60% chance of winning on any college football game. When betting, he has a choice of three strategies:

Strategy 1: He can bet $55.00 on two separate teams. If a team wins, he wins $50; otherwise, he loses $55.

Strategy 2: He can bet $100.00 on a two-team parley. If both teams wins, he wins $260; otherwise he loses $100.00.

Strategy 3: He can bet $110.00 on a single team. If the team wins, he wins $100; otherwise, he loses $110.00.

Find the expected winnings for each strategy.

3. Assume the random variable X has the distribution: $P\{X \le k\} = \dfrac{k^2}{25}$, where

k = 0, 1, 2, 3, 4, 5.

Calculate E(X).

4. The Apple Sports Company owns two retail stores: one in New York City and one in Los Angeles. The following tables show the possible gross revenues from each store:

Total Revenue New York ($1,000) x	P{X = x}	XP{X = x}	Total Revenue Los Angeles ($1,000) y	P{Y = y}	yP{X = y}
$900	0.60	$540	$500	0.70	$350
$1500	0.30	$450	$1100	0.25	$275
$2000	0.10	$200	$1500	0.05	$75
		E(X) = $1190			E(Y) = $700

a. Complete the table for the total revenue for both stores.

b. Show that E(X + Y) = E(X) + E(Y) .

5. Two cards are drawn from an ordinary deck of cards without replacement.

a. Construct the distribution table for the number of diamonds drawn.

b. Find the expected number of diamonds drawn.

6. A game is called a "fair game" if E(X) = 0 where X equals the win/losses resulting from the game.

a. Mr. Smith wages on basketball. Assume he bets $110 on a single game. If he wins, X = $100; otherwise, X = -$110. For this to be a fair game, what must be the probability p that he wins the game.

b. The following table gives the payout when wagering $100 on different parleys[1]:

Number of teams in the parley	Winning
2	$ 260
3	$ 600
4	$1,000
5	$2,000

For each of these parleys, find the value p in the following table that will make the wager a fair game.

Number of teams in the parley	Winning	P
2	$260	
3	$600	
4	$1,000	
5	$2,000	

7. Mr. Fortune wages on sports events. He believes there is a 60% chance that he can win a game on football, a 65% chance on baseball and a 70% chance on basketball. For each of these games, he wages $110. If he wins a game, he wins $100; otherwise he loses $110. During 1996, we wagered on 65 football games, 125 baseball games and 200 basketball games. Find his average winnings.

8. *Define the conditional expectation of X given Y as*

$$E(X = x|Y = y) = x_1 P(X = x_1 |Y = y) + x_2 P(X = x_2 |Y = y) + ... + x_N P(X = x_N |Y = y)$$

Ms. Bullet, a sharpshooter for a local swat team, claims that she can hit a moving target, 1,000 yards away. To test her claim, she shot ten times at such a target and missed eight times. Find the expected number of misses between hits.

9. Show $[E(X)]^2 \le E(X^2)$. (Hint: Expand $E[X - E(X)]^2$)

10. Two discrete random variables are said to be independent, if $P[(X = x_k) \cap (Y = y_r)] =$

---

[1]

    A parley is a preselected number of games where in order to win all the games must win. For example, a person wagering on a three team parley will win only if all three teams win.

$P[(X = x_k)P(Y = y_r)]$ for all k,r. If X and Y are independent, show $E(XY) = E(X)E(Y)$.

11. Ten cards are randomly drawn, without replacement, from an ordinary deck of cards. Find the expected number of diamonds.

12. Assume c is a constant. Show that $E(cX) = cE(X)$.

13. Show that the Schwartz inequality, $[E(XY)]^2 \le E(X^2)E(Y^2)$ is true. (Use the fact that the quadratic polynomial $E[(tX + Y)^2]$ is non-negative.)

14. A die is tossed four times or until two fives appear, whichever occurs first. Find the expected number of tosses.

15. Assume X_k is a finite sequence of random variables where $E(X_k) = \mu$; k = 1,2,...,n.

Define $\overline{X} = \dfrac{X_1 + X_2 + ... + X_n}{n}$.

Show $E(\overline{X}) = \mu$.

16. A fair die is rolled 3 times. Assume X is the random variable that is equal to the largest number rolled.
Find

a. $P\{X \le k\}$; k = 1,2,...,6

b. $P\{X = k\}$; k = 1,2,...,6

c. $E(X)$

17. For any sequence of discrete random variables $X_1, X_2, ..., X_n$, we define the joint distribution of any subset $X_i, X_j,..., X_r$ as

$P\{X_i = x_k, X_j = x_w,..., X_r = x_t\} = P[\{X_i = x_k\} \cap \{X_j = x_w\} \cap ... \cap \{X_r = x_t\}]$.

A sequence of discrete random variables $X_1, X_2, ..., X_n$ is said to be mutually independent if the events

$\{X_1 = x_1\}, \{X_2 = x_2\},..., \{X_n = x_n\}$ are mutually independent (See Lesson 12 for definition).

Three cards are drawn from an ordinary deck of cards. Let X_k (k = 1,2,3) be the total number of diamonds drawn on the kth drawing.

a. Write out the joint distribution of X_1, X_2, X_3.

b. Show this sequence of random variables is not mutually independent.

18. Assume $X = x_k$ ($k = 1,2$) and $Y = y_k$ ($k = 1,2$) are 2 random variables.

a. Show $E(X + Y) = E(X) + E(Y)$.

.

b. Generalize for $X = x_k$ ($k = 1,2,..., n$) and $Y = y_k$ ($k = 1,2,...m$).

c. Show $E(aX + bY) = aE(X) + bE(Y)$, where a and b are constants.

d. Show $E(X_1 + X_2 + ... + X_n) = E(X_1) + E(X_2) + ... + E(X_n)$.

19. Show the following:

 Assume $X = x_k$ ($k = 1,2$) and $Y = y_k$ ($k = 1,2$) are 2 random variables.

a. Show $E(X + Y|Z = z) = E(X|Z = z) + E(Y|Z = z)$.

.

b. Generalize for $X = x_k$ ($k = 1,2,..., n$) and $Y = y_k$ ($k = 1,2,...m$).

c. $E(aX + bY = ax + by \mid Z = z) = aE(X = x \mid Z = z) + bE(Y = y \mid Z = z)$, where a and b are constants.

d. $E(c \mid X = x) = c$, where c is a constant.

e. If X and Y are independent random variables, show $E(X = x \mid Y = y) = E(X)$.

20. Assume X and Y are random variables where ($X = x_k$, $k = 1,2,..,n$) and ($Y = y_k$, $k = 1,2,...,m$).

Show $E[E(X \mid Y)] = E(X)$.

21. A coin is tossed a certain number of tossed depending on the outcome of a single toss of a fair die. If the rolled die results in the number n ($1 \leq n \leq 6$), the coin is tossed n times.
Let X_j ($1 \leq j \leq n$) represent the result of tossing the coin on the jth toss where
$P(X_j = 1) = p$, ($0 < p < 1$) and $P(X_j = 0) = 1 - p = q$. Let N represent the result of toss the die once where $P(N = n) = 1/6$; $1 \leq n \leq 6$. Define $S = X_1 + X_2 + ... X_n$.

a. Find $\mu = E(X_i)$.

b. Show $E(S \mid N) = \mu N$.

c. Show $E(S) = \mu E(N)$.

22. Show the following:

a. If X = c, a constant, Show E(X | Y) = c.

b. If X = $x_1, x_2, \ldots, x_n$ for any random variable X, show E(X |X) = X.

c. If X = $x_1, x_2, \ldots, x_n$ for any random variable X, and Y is a random variable, show E(XY | Y)
=

  YE(X |Y).

We define the random variables X and Y to be almost equal (a.e.) if P(X ≠ Y) = 0.
We write X = Y a.e.

23. If X,Y,Z are random variables, and X = Y a.e. and Y = Z a.e., show X = Z a.e.

24. If the random variables X = $x_1, x_2, \ldots, x_n$ and Y = $y_1, y_2, \ldots, y_m$. If X = Y a.e. and Z is a random variable then E(X|Z) = E(Y| Z).

25. Two random variables are said to be conditionally independent if

P(X = x; Y = y|Z = z) = P(X = x|Z = z)P(Y = y |Z = z) for a values of x, y, z.

If the random variables X = $x_1, x_2, \ldots, x_n$ and Y = $y_1, y_2, \ldots, y_m$ are conditionally independent, show E(XY |Z) = E(X|Z)E(Y|Z).

16.1 - What is the Variance of a Random Variable?

An important application of expectation is a single numeric representation of the set of data that generated it. The closer the data is to the expected value the better the representation. Assume, for example, the following two sets of data:

Example A: 19, 19, 20, 21, 21 has the expectation

$$E(X) = \frac{19 + 19 + 20 + 21 + 21}{5} = \frac{100}{5} = 20.$$

Example B: 10, 15, 20, 25, 30

$$E(X) = \frac{10 + 15 + 20 + 25 + 30}{5} = \frac{100}{5} = 20$$

For both sets of data the mean value is 20; yet the value 20 is a better representation of the data in example A than in example B. The reason is the dispersion of the data away from the mean value 20 is smaller in example A than in example B. We therefore need a way to measure this dispersion.

The traditional way dispersion of data is measured is by the variance of the data or values of the random variable. The symbol is σ^2. The following formula gives us the variance σ^2:

$$\sigma^2 = P\{X = x_1\}[x_1 - E(X)]^2 + P\{X = x_2\}[x_2 - E(X)]^2 + \dots + P\{X = x_n\}(x_n - E(X))^2$$

A shorter version of this formula is

$$\sigma^2 = E(X^2) - [E(X)]^2 \text{ where}$$

$$E(X^2) = x_1^2 \, P\{X^2 = x_1^2\} + x_2^2 \, P\{X^2 = x_2^2\} + \dots + x_n^2 \, P\{X = x_n^2\}$$

16.1 - Example 1: An urn contains 3 red marbles and 1 white marble. Marbles are selected at random without replacement until a white is selected. Find the variance for the number of marbles selected.

Solution:

This example comes from the third example in lesson 15. From this exercise we have the table:

X = x	P{X = x}	xP{X = x}
1	0.25	0.25
2	0.25	0.50
3	0.25	0.75
4	0.25	1.00
Total 1		**E(X) = 2.5 selection**

The following is the table for computing the variance σ^2:

X = x	$X^2 = x^2$	$P\{X^2 = x^2\}$	$x^2P\{X^2 = x^2\}$
1	1	0.25	0.25
2	4	0.25	1.00
3	9	0.25	2.25
4	16	0.25	4.00
			$E(X^2) = 7.5$
		Total 1	$\sigma^2 = E(X^2) - [E(X)]^2 = 7.5 - (2.5)^2 = 1.25$

From the last column we have the variance $\sigma^2 = 1.25$.

16.1 - Example 2: John Wish is a famous gambler. He claims that he has a 60% chance of winning on any college football game. On each game he bets $110. If his team wins, he makes $100. If his team loses, he loses $110. On a particular week, he bets on three separate football games. His expected winnings are computed in the table below. Find the variance of the daily winnings.

X = x	P{X = x}	xP{X = x}
$300	0.216	$ 64.80
90	0.432	38.88
- 120	0.288	- 34.56
- 330	0.064	- 21.12
Total = 1		**E(X) = $48.00 average winnings per day.**

Solution:

The table for computing the variance is

X = x	X² = x²	P{X² = x²}	x²P{X² = x²}
$300	90000	0.216	19440
90	8100	0.432	3499.2
-120	14400	0.288	4147.2
-330	108900	0.064	6969.6
		Total 1	E(X²) = 34056
			σ² = E(X²) - [E(X)]² = $34056 - 48² = $31,752

Solved Problems

16.1 - Solved Problem 1: An urn contains 3 red marbles and 2 white marbles. Marbles are selected at random without replacement until both colors are selected. Find the variance of the number of marbles selected.

Solution:

This problem is from the third problem in lesson 15. From problem 3 we have

X = x	P{X = x}	xP{X = x}
2	36/60	72/60
3	18/60	54/60
4	6/60	24/60
	Total 1	E(X) = 150/60 = 2.5 selections

The table for computing the variance is

X = x	X² = x²	P{X² = x²}	x²P{X² = x²}
2	4	36/60	144/60
3	9	18/60	162/60
4	16	6/60	96/60
		Total 1	E(X²) = 6.7
			σ² = E(X²) - [E(X)]² = 6.7 - 2.5² = 0.45

16.1 - Solved Problem 2: John Wish is a famous gambler. He claims that he has a 60% chance of winning on any college football game. He decides to play two-team parleys: He bets $100 on two teams. If both his teams wins, he wins $215. If at least one of his teams loses , he loses his $100. His expected winnings are computed in the table below. Find the variance.

X = x	P{X = x}	xP{X = x}
$ 215	0.36	$ 77.40
-$100	0.64	- 64.00
	Total 1	**E(X)** = $13.40, average winnings per parley.

Solution:

The table for computing the variance:

X = x	x^2	$P\{X^2 = x^2\}$	$x^2 P\{X^2 = x^2\}$
$215	46225	0.36	16641
-$100	10000	0.64	6400
			$E(X^2)$ = 23041
		Total 1	$\sigma^2 = E(X^2) - [E(X)]^2 = 23041 - 13.40^2 = \$22861.44$

Unsolved Problems with Answers

16.1 - Problem 1: A die is tossed four times or until two fives appears, whichever occurs first. Find the variance on the number of tosses.

Answer:

$\sigma^2 \approx 0.51$

⇑ *Refer back to* **16.1 - Example 1 & 16.1 - Solved Problem 1.**

16.1 - Problem 2: John Wish is a famous gambler. He claims that he has a 60% chance of winning on any college football game. He decides to play three-team parleys: He bets $100 on three teams. If all three teams win, he wins $610. If at least one of his teams loses , he loses his $100. Find his variance on the winnings.

Answer:

$\sigma^2 = \$85,366.31$

⇑ *Refer back to* **16.1 - Example 2 & 16.1 - Solved Problem 2.**

Supplementary Problems

1. Ms. Rich invests in stock options. She is interested in purchasing three 90-day options called option A, B and C. Each option will cost her $500. She feels that there is a 70% chance that option A, a 60% chance that option B and a 55% chance that option C will double in 90 days. The alternative is that the options will expire worthless after 90 days. Find the mean and variance of her return.

2. A pair of dice is tossed once. Find the variance of the sum of the two dice.

3. The Apple Sports Company owns two retail stores: one in New York City and one in Los Angeles. The following tables show the possible gross revenues from each store.

Total Revenue New York ($1,000) x	P{X = x}	XP{X = x}	Total Revenue Los Angeles ($1,000) y	P{Y = y}	yP{X = y}
$900	0.60	$540	$500	0.70	$350
$1500	0.30	$450	$1100	0.25	$275
$2000	0.10	$200	$1500	0.05	$75
		E(X) = $1190			**E(Y) = $700**

a. Find the variance for both stores and the variance of total revenue.

b. Show that $\sigma^2_{X+Y} = \sigma^2_X + \sigma^2_y$.

4. Two cards are drawn from an ordinary deck without replacement.

a. Find the variance on the number of diamonds drawn.

b. Find the variance on the number of clubs drawn.

c. Show that $\sigma^2_{X+Y} \neq \sigma^2_X + \sigma^2_y$.

5. John Wish claims that he has a 60% chance of winning on any college football game. On each game he bets $55. If his team wins, he makes $50. If his team loses, he loses $55. However, if he plays a two team parley and wins, he wins $130 by placing a bet for $50. On a particular day, he has $165 to bet on three separate football games. There are two strategies he is interested in:

Strategy 1: Wager $55 on three separate teams.

Strategy 2: Wager $50 on three separate two-team parleys.

Find the mean and variance for each strategy.

6. Chebyshev's Inequality: Assume that X is a random variable having a mean μ and a variance σ^2. Then for any positive number a, it follows that $P(|X - \mu| \geq a) \leq \dfrac{\sigma^2}{a^2}$.

A machine fills 16.5 ounce bottles with orange juice. If the average fill per bottle is 16.15 ounces and the variance per fill is 0.08 ounces,

a. Find the maximum probability that a bottle will be filled with more than 16.25 ounces or less than 16.05 ounces.

b. Find the maximum probability that a bottle will be filled with more than 16.25 ounces.

7. For any random variable with a finite mean and variance, find the maximum probability that the random variable will be different from its mean by more than

a. two standard deviations (2σ).

b. three standard deviations. (3σ).

σ is called the standard deviation.

8. A game is called a "fair game" if $E(X) = 0$ where X equals the win/losses resulting from the game.

a. Mr. Smith wages on basketball. Assume he bets $110 on a single game. If he wins, X = $100; otherwise
X = -$110. For this to be a fair game, $p \approx 0.523$ the probability that he wins the game. Find the variance.

b. The following table gives the payout when wagering $100 on different parleys[1]:

Number of teams in the parley	Winning
2	$260
3	$600
4	$1,000
5	$2,000

For each of these parleys, p equals the probability of winning each game so that the game is fair

.

---

[1]

A parley is a preselected number of games where in order to win all the games must win. For example, a person wagering on a three team parley will win only if all three teams win.

Number of teams in the parley	Winning	p
2	$ 260	0.527
3	$ 600	0.523
4	$1,000	0.549
5	$2,000	0.544

For each parley, find the variance.

9. Assume X is a random variable and C a constant. Show that $\sigma_{X+C}^2 = \sigma_X^2$.

10. Assume X is a random variable where $\mu = E(X)$ and $\sigma_X^2 > 0$. Define $Z = \dfrac{X-\mu}{\sigma}$. Show that $E(Z) = 0$ and $\sigma_Z^2 = 1$.

11. A game is played with probability p of winning. Each time a player wins, he/she earns 1 point. Find $E(X)$ and σ^2.

For any sequence of discrete random variables $X_1, X_2, ..., X_n$ we define the joint distribution of any subset $X_i, X_j,..., X_r$ as

$P\{X_i = x_k, X_j = x_w,..., X_r = x_t\} = P[\{X_i = x_k\} \cap \{X_j = x_w\} \cap ... \cap \{X_r = x_t\}].$

A sequence of random variables $X_1, X_2, ..., X_n$ is said to be mutually independent if the events $\{X_1 = x_1\}, \{X_2 = x_2\}, ..., \{X_n = x_n\}$ are mutually independent.

12. The definition of variance for a discrete random is

$\sigma^2 = P\{X = x_1\}[x_1 - E(X)]^2 + P\{X = x_2\}[x_2 - E(X)]^2 + ... + P\{X = x_n\}(x_n - E(X))^2.$

a. Show $\sigma^2 = E(X^2) - E(X)^2$.

b. Define $E(X) = \mu$. Derive a formula for $E(X^2)$.

13. Assume the sequence of random variables $X_1, X_2,...,X_n, X_{n+1}$ are mutually independent. random variables.

a. Assume $S_n = X_1 + X_2 + ... + X_n$. Show S_n, X_{n+1} are independent.

b. Assume $\sigma^2(X_k)$ exists and are finite for $k = 1,2,3,..., n+1$. Show

$\sigma^2(S_{n+1}) = \sigma^2(X_1) + ... + \sigma^2(X_{n+1}).$

c. Assume c is a constant. Show $\sigma^2(cX) = c^2\sigma^2(X)$.

d. Assume $\overline{X} = \dfrac{X_1 + X_2 + ... + X_n}{n}$. If $\sigma^2(X_1) = ... = \sigma^2(X_n) = \sigma^2$, show $\sigma^2(\overline{X}) = \dfrac{\sigma^2}{n}$.

e. Define $S^2 = \dfrac{(X_1 - \overline{X})^2 + (X_2 - \overline{X})^2 + ... + (X_n - \overline{X})^2}{n}$.

Assume: $E(X_1) = E(X_2) = ... = E(X_n) = \mu$,

$\sigma^2(X_1) = ... = \sigma^2(X_n) = \sigma^2$.

Show $E(S^2) = \dfrac{n-1}{n}\sigma^2$.

14. A fair die is rolled 3 times. Assume X is the random variable that is equal to the largest number rolled. Find the variance of X. (See supplementary lesson 15, problem 16.)

15. A fair coin is tossed until a head appears or five times, whichever occurs first. Assume X is the random variable that is equal to the number of tosses. Find the mean and variance of X.

16. An urn contains 3 red and 3 back balls. Balls are drawn one at a time, without replacement, until 2 different colors are drawn. Assume X is the random variable that is equal to the number of balls drawn. Find the mean and variance of X.

17. Two urns sit on a table. Urn A has 1 red and 1 white marble; urn B has 1 red and 1 white marble.
A fair coin is tossed. If heads appears a marble is selected from urn A; otherwise from urn B. This process is continued until a red marble is selected. Let X be the random variable equal to the number of tosses of the coin to draw a red marble. Find the mean and the and variance of X.

Covariance of 2 Random Variables. The covariance of X and Y is defined by

$Cov(X,Y) = E[(X - \mu_X)(Y - \mu_Y)]$ where $\mu_X = E(X)$ and $\mu_Y = E(Y)$.

18. Show $Cov(X,Y) = E(XY) - \mu_X\mu_Y$.

19. Assume X, Y are independent. Show $Cov(X,Y) = 0$.

20. A fair coin is tossed twice. Let X_1 be the random variable that equals 1 if a head is tossed the first time or 0 otherwise and X_2 be the random variable that equals 1 if a head is tossed the second time or 0 otherwise. Define $Z = X_1 + X_2$ and $W = X_1 - X_2$.

a. Show $Cov(Z,W) = 0$.

b. Show Z and W are random variables that are dependent.

21. What does the results of problem 19 and 20 show us about the relationship of 2 random variables and their covariance.

22. Two cards are drawn, without replacement, from an ordinary deck of cards. Let X be the random variable equal to the number of diamonds drawn and Y the random variable equal to the number of clubs drawn. Find Cov(X,Y).

23. Show that $\sigma^2_{X+Y+Z} = \sigma_X^2 + \sigma_Y^2 + \sigma_Z^2 + 2[Cov(X,Y) + Cov(X,Z) + Cov(Y, Z)]$

24. Show $Cov(X,Y) = [\sigma^2_{X+Y} - (\sigma_X^2 + \sigma_Y^2)]/2$

25. If X,Y,Z are pair-wise independent, show $\sigma^2_{X+Y+Z} = \sigma_X^2 + \sigma_Y^2 + \sigma_Z^2$.

26. Assume a, b are constants. Show $Cov(X+a, Y+b) = Cov(X,Y)$.

27. Assume X and Y each take on only 2 values each and their distributions are different.

If the Cov(X,Y) = 0, show X and Y are independent.

28. Assume the random variables X_k (k = 1, 2, ...,n) have the same distribution with mean μ and variance

σ^2. Show $\sigma^2_{X1 + X2 + ... + Xn} = n\sigma^2 + n(n-1)Cov(X_1,X_2)$.

29. Assume a box contains 10 balls each respectively numbered 1,2,...,10. Assume r balls are randomly selected, without replacement, and the numbers on the selected balls are added. Find the mean and variance of the total.

30. Assume a fair die is tossed 100 times. Let X_k (k = 1,2,...,100) be the independent random variables that equal the out for each toss. For the following questions, see problem 13 for appropriate definitions.

a. Find $E(\overline{X})$

b. $\sigma^2(\overline{X})$.

c. Find $E(S^2)$.

31. Assume a fair coin is tossed twice. Let X_k (k = 1,2) be the r.v. that equals the number of heads per toss.
Find the probability distribution for the following random variables for the probability sample space **S** = {(h,h),(h,t),(t,h),(t,t)}. See probem 13 for appropriate definitions.

a. X_k

b. $\overline{X}$

c. S^2

d. Confirm $E(S^2) = \dfrac{n-1}{n}\sigma^2$ for $n = 2$.

So far we have used Boolean expressions to solve the probabilities of given events resulting from an experiment. However, it is possible that in some cases these Boolean expressions can be very long and tedious to write out. For example, Assume 7 cards are drawn from an ordinary deck, without replacement. To find the probability that 3 kings, 2 queens are drawn would require a long expression of the union of 210 different possible events! To address this problem, we will develop in this lesson simple counting techniques that will allow us to compute the probabilities of these types of events. We begin with the definition of factorials.

17.1 - What are Factorials?

Factorials are special integer numbers written as n! created by the following system:

$0! = 1$
$1! = 1$
$2! = (2)1 = 2$
$3! = (3)(2)(1) = 6$
$4! = (4)(3)(2)(1) = 24$
$5! = (5)(4)(3)(2)(1) = 120$
$6! = (6)(5)(4)(3)(2)(1) = 720$
$7! = (7)(6)(5)(4)(3)(2)(1) = 5,040$
$8! = (8)(7)(6)(5)(4)(3)(2)(1) = 40,320$
$9! = (9)(8)(7)(6)(5)(4)(3)(2)(1) = 362,880$
$10! = (10)(9)(8)(7)(6)(5)(4)(3)(2)(1) = 3,628,800$

::

$$n! = n(n-1)(n-2)...(2)(1)$$

The above values can also be evaluated by the following system:

$0! = 1$

$1! = 1$

$2! = (2)1 = 2$

$3! = (3)2! = 3(2) = 6$

$4! = (4)3! = 4(6) = 24$

$5! = (5)4! = (5)(24) = 120$

$6! = (6)5! = (6)(120) = 720$

$7! = (7)6! = (7)(720) = 5,040$

$8! = (8)7! = (8)(5,040) = 40,320$

$9! = (9)8! = (9)(40,320) = 362,880$

$10! = (10)9! = (10)(362,880) = 3,628,800.$

17.1 - Example 1: Evaluate $11!$.

Solution:

$11! = (11)10! = 11(3,628,800) = 39,916,800$

17.1 - Example 2: Evaluate $\dfrac{10!}{5!}$.

Solution:

$10! = (10)(9)(8)(7)(6)5!$

Therefore, $\dfrac{10!}{5!} = \dfrac{(10)(9)(8)(7)(6)5!}{5!} = (10)(9)(8)(7)(6) = 30,240.$

17.1 - Example 3: Evaluate $(6!)(4!)$.

Solution:

$6! = (6)[(5)(4)(3)(2)(1)] = (6)5! = (6)(120) = 720$

$4! = (4)3! = (4)[(3)(2)(1)] = (4)6 = 24$

Therefore, $(6!)(4!) = (720)(24) = 17,280$

17.1 - Example 4: Evaluate $\dfrac{10!}{6!4!}$.

Solution:

Step 1: Write $10! = (10)(9)(8)(7)6!$.

Step 2: 6! in the denominator cancels out the 6! in the numerator: $\dfrac{10!}{6!4!} = \dfrac{(10)(9)(8)(7)}{4!}$.

Step 3: Write 4! in the denominator as 4! = (4)(3)(2)(1): $\dfrac{10!}{6!4!} = \dfrac{(10)(9)(8)(7)}{(4)(3)(2)(1)}$.

Step 4: Cancel the denominator where possible: $\dfrac{8}{4} = 2, \dfrac{9}{3} = 3, \dfrac{10}{2} = 5,$

$$\dfrac{10!}{6!4!} = (5)(3)(2)(7) = 210$$

Solved Problems

17.1 - Solved Problem 1: Evaluate 12! .

Solution:

12! = (12)11! = 12(39,916,800) = 479,001,600

17.1 - Solved Problem 2: Evaluate $\dfrac{11!}{7!}$.

Solution:

11! = 11(10)(9)(8)7!

Therefore, $\dfrac{11!}{7!} = \dfrac{(11)(10)(9)(8)7!}{7!} = (11)(10)(9)(8) = 7920$

17.1 - Solved Problem 3: Evaluate (7!)(8!) .

Solution:

7! = (7)(6)[(5)(4)(3)(2)(1)] = (7)(6)5! = (42)(120) = 5040

8! = (8)7! = (8)(5040) = 40,320

Therefore, (7!)(8!) = (5040)(40,320) = 203,212,800

17.1 - Solved Problem 4: Evaluate $\dfrac{15!}{8!7!}$.

Solution:

The 8! in the denominator cancels the 8! in the numerator:

$$\frac{15!}{8!7!} = \frac{(15)(14)(13)(12)(11)(10)(9)8!}{8!7!} = \frac{(15)(14)(13)(12)(11)(10)(9)}{7!}$$

Since $7! = (7)(6)(5)(4)(3)(2)(1)$:

$$\frac{15!}{8!7!} = \frac{(15)(14)(13)(12)(11)(10)(9)}{(7)(6)(5)(4)(3)(2)(1)} = (3)(13)(11)(3)(5) = 6435$$

Unsolved Problems with Answers

17.1 - Problem 1: Evaluate $13!$.

Answer:

6,227,020,800

⇑ *Refer back to* **17.1 - Example 1 & 17.1 - Solved Problem 1**.

17.1 - Problem 2: Evaluate $\dfrac{77!}{74!}$.

Answer:

438,900

⇑ *Refer back to* **17.1 - Example 2 & 17.1 - Solved Problem 2**.

17.1 - Problem 3: Evaluate $(10!)(10!)$.

Answer:

13,168,189,440,000

⇑ *Refer back to* **17.1 - Example 3 & 17.1 - Solved Problem 3**.

17.1 - Problem 4: Evaluate $\dfrac{20!}{10!10!}$.

Answer:

184,756

⇑ *Refer back to* **17.1 - Example 4 & 17.1 - Solved Problem 4**.

17.2 - Counting the Number of Samples

In the past lessons, we have seen many examples of counting the number of elements of sample spaces and samples. There are fundamentally three different ways to sample: sampling with replacement (see Lesson 2), sampling without replacement where **order is important** and sampling without replacement where **order is not important.** In this lesson we study sampling without replacement where order is important and where order is not important.

What is sampling without replacement where order is important?

Sampling without replacement where order is important occurs when each selection is not returned and the number of possible samples is determined by

1. the items selected.

 and

2. given the items selected, each order of selection is different.

This type of sampling is called permutations. The formula for counting the number of sampling is

$$_N P_R = \frac{N!}{(N-R)!} \, ,$$

where N is the number of items to select from and R is the number of items sampled.

17.2 - Example 1: In horse racing, there are three winning positions: first place is called 'win', second place is called 'place', and third place is called 'show'. In a ten horse race, find how many ways three horses can come in win, place or show.

Solution:

Since the prize money is different depending on what order the first three horses come in, this is a sampling where order is important. Since we are selecting from 10 horses, N = 10. And since we are selecting three horses for these three prizes, R = 3. Therefore using the above formula, the number of possible ways that this can happen is

$$_{10}P_3 = \frac{10!}{(10-3)!} = \frac{10!}{7!} = \frac{(10)(9)(8)7!}{7!} = (10)(9)(8) = 720 \text{ different ways three horses can}$$

come in first, second and third positions.

17.2 - Example 2: Ms. Smith's third grade class has 8 girls and 5 boys. On graduation day, she has them line up for group photos. Assume all the girls stand together on the left of the boys. Find the number of ways they can stand together.

Solution:

First, we analyze how many ways the boys can stand together. Here we are selecting five boys out of five to stand together. Since order is important we use the above formula where N = 5 and R = 5. Therefore, the number of ways the five boys can stand together is

$$_5P_5 = \frac{5!}{(5-5)!} = \frac{5!}{0!} = \frac{5!}{1} = (5)(4)(3)(2)(1) = 120 \text{ ways 5 boys can stand together.}$$

Next we analyze how many ways the girls can stand together. Here we are selecting eight girls out of eight to stand together. Since order is important, we use the above formula where N = 8 and R = 8. Therefore, the number of ways the eight girls can stand together is

$$_8P_8 = \frac{8!}{(8-8)!} = \frac{8!}{0!} = \frac{8!}{1} = (8)(7)(6)(5)(4)(3)(2)(1) = 40,320,$$

the number of ways eight girls can stand together.

Now for each arrangement of boys, there are 40,320 number of ways eight girls can stand together. Therefore, the total number of ways the boys and girls can stand together is

$(_5P_5)(_8P_8) =$ (120)(40,320) = 4,838,400 total number of ways the boys and girls can stand together.

What is sampling without replacement where order is NOT important?

Sampling without replacement where order is not important occurs when each selection is not returned and the number of possible samples is determined only by the items selected. This type of sampling is called combinations.

The formula for counting the number of (combinations) samples is

$$\binom{N}{R} = \frac{N!}{R!(N-R)!},$$

where N is the number of items to select from and R is the number of items sampled.

17.2 - Example 3: Five cards are selected without replacement from an ordinary deck of 52 cards. Find the number of 5 card hands.

Solution:

In counting the number of hands, order of the cards is not important. Since there are 52 cards to select from,
N = 52 and since we are selecting 5 cards at random from the deck, R = 5. Using the above formula

$$\binom{52}{5} = \frac{52!}{5!(52-5)!} = \frac{52!}{5!(47)!} = \frac{(52)(51)(50)(49)(48)}{(5)(4)(3)(2)(1)} = 2,598,960.$$

number of five card hands.

17.2 - Example 4: Five cards are selected without replacement from an ordinary deck of 52 cards. Find the number of hands that have exactly 3 kings and 2 queens.

Solution:

Since order is not important, we can think of the five card hand as divided into a left and right part. Assume the left part of the hand contains 3 kings and the right part of the hand containing 2 queens.

Step 1: The number of ways 3 kings can be drawn from 4 kings, where order is not important, is

$$\binom{4}{3} = \frac{4!}{3!(4-3)!} = 4.$$

Step 2: The number of ways 2 queens can be drawn from 4 queens, where order is not important, is

$$\binom{4}{2} = \frac{4!}{2!(4-2)!} = \frac{4!}{2!2!} = 6.$$

Step 3: For each possible hand containing 3 kings there is 6 possible selections of 2 queens. Therefore, the way to compute the total number of hands containing 3 kings and 2 queens is to

multiply $\binom{4}{3}$ by $\binom{4}{2}$.

Performing this multiplication gives

$$\binom{4}{3}\binom{4}{2} = (4)(6) = 24,$$

which is the total number of hands with 3 kings and 2 queens.

Solved Problems

17.2 - Solved Problem 1: Ms. Jones has 25 students in her Spanish class. Each year she holds a spelling contest. The finalists can win 5 different awards. Find the total number of possible winners of these awards.

Solution:

Since the awards are different, this is a sampling where order is important. Since we are selecting from 25 students, N = 25. And since we are selecting five students from these students, R = 5. Therefore, the number of possible ways that this can happen is

$$_{25}P_5 = \frac{25!}{(25-5)!} = \frac{25!}{20!} = (25)(24)(23)(22)(21) = 6,375,600$$

ways of selecting 5 students from 25 students for the 5 different awards.

17.2 - Solved Problem 2: Mrs. Smith's third grade class has 20 girls and 15 boys. She held a spelling contest in which five different prizes were awarded. Three girls and two boys won a prize. Find the number of ways the boys and girls in her class can win.

Solution:

The number of ways 3 girls out of 20 can win three different prizes is

$$_{20}P_3 = \frac{20!}{(20-3)!} = \frac{20!}{17!} = (20)(19)(18) = 6,840 \text{ possible selections.}$$

The number of ways 2 boys out of 15 can win two different prizes is

$$_{15}P_2 = \frac{15!}{(15-2)!} = \frac{15!}{13!} = (15)(14) = 210 \text{ possible selections.}$$

Now for each possible selection of 3 girls, there are 210 possible selections of two boys. Therefore the total possible number of selections is

$$(_{20}P_3)(_{15}P_2) = (6,840)(210) = 1,436,400.$$

17.2 - Solved Problem 3: A coin is tossed 8 times. Find the total number of possible ways 5 heads can occur.

Solution:

The different ways heads can occur, will depend on the location of the five heads. For example, (h,h,t,t,h,h,h,t) is different than (h,t,t,t,h,h,h,h). Therefore we need to find the number of ways of selecting 5 positions out of 8 positions where heads occur and order is not important. This can be computed using $\begin{pmatrix} N \\ R \end{pmatrix}$,

where $N = 8$ and $R = 5$.

Therefore, $\begin{pmatrix} 8 \\ 5 \end{pmatrix} = \dfrac{8!}{5!(8-5)!} = \dfrac{8!}{5!3!} = 56$ total number of possible ways 5 heads can occur.

17.2 - Solved Problem 4: Mr. Fortune likes to wager on football and basketball games. On a given Saturday, 8 games of basketball and 7 games of football are to be played. He decides to wager on 4 basketball games and 5 football games. Find the number of possible selections.

Solution:

Since order is not important, we first compute the total number of possible selections for basketball. Since he is to pick 4 basketball games from 8 games, the number of possible selections is

$$\begin{pmatrix} 8 \\ 4 \end{pmatrix} = \frac{8!}{4!(8-4)!} = \frac{8!}{4!4!} = 70.$$

Next we compute the total number of possible selections for football. Since he is to pick 5 football games out of 7 games, the number of selections is

$$\begin{pmatrix} 7 \\ 5 \end{pmatrix} = \frac{7!}{5!(7-5)!} = \frac{7!}{5!2!} = 21.$$

The total number of selections possible $= (70)(21) = 1{,}470$.

Unsolved Problems with Answers

17.2 - Problem 1: Mrs. Jones has 25 students in her Spanish class. Each year, she holds a spelling contest. The two finalists can win 2 different awards. Find the total number of possible winners of these awards.

Answer:

600

⇑ *Refer back to* **17.2 - Example 1 & 17.2 - Solved Problem 1**.

17.2 - Problem 2: Mrs. Smith has 7 girls and 5 boys in her second grade class. She asks for 3 girls and 3 boys to volunteer to have their pictures taken standing in a row. Find the total number of possible rows of students assuming the boys all stand together and the girls all stand together.

Answer:

25,200

⇑ *Refer back to* **17.2 - Example 2 & 17.2 - Solved Problem 2**.

17.2 - Problem 3: Twenty students are questioned on their approval of the President's foreign policy. Find the total number of possible ways 10 students approve.

Answer:

184,756

⇑ *Refer back to* **17.2 - Example 3 & 17.1 - Solved Problem 3**.

17.2 - Problem 4: Mr. Fortune likes to wager on football, basketball and soccer games. On a given Saturday, 8 games of basketball, 7 games of football and 10 of soccer are to be played. He decides to wager on 4 basketball games, 5 football games and 5 soccer games. Find the number of possible selections.

Answer:

370,440

⇑ *Refer back to* **17.2 - Example 4 & 17.2 - Solved Problem 4**.

17.3 - Probability Applications

Using the above counting techniques allows us to find the probability of events resulting from experiments.

17.3 - Example 1: Ms. Jones has 6 boys and 4 girls in her first grade class. At the end of the year she hires a photographer to take the children's picture. Assuming the children stand together in a straight row for the picture, find the probability that all the boys stand together and all the girls stand together.

Solution:

The sample space is all possible ways the children can line up: **#S** = 10! = 3,628,800.

The event **E** is that all the boys stand together and girls stand together. **#E** is the number of ways this can happen at random.

Step 1: Assume all boys are standing on the left. The number of ways all boys can stand together is 6!. The number of ways all girls can stand together is 4!.

Step 2: From step 1, we multiply (6!)(4!) = 17,280 = # ways all boys are standing together and all girls are standing together where the boys are standing on the left.

Step 3: Assume all boys are standing on the right. The number of ways all boys can stand together is 6!. The number of ways all girls can stand together is 4!

Step 4: From step 1, we multiply (6!)(4!) = 17,280 = # ways all boys are standing together and all girls are standing together where the boys are standing on the right.

Step 5: From step 2 and step 4, **#E** = 17,280 + 17,280 = 34,560.

Step 6: $P(\mathbf{E}) = \dfrac{\#\mathbf{E}}{\#\mathbf{S}} = \dfrac{34,560}{3,628,800}$.

17.3 - Example 2: Five cards are to be selected from an ordinary deck of 52 cards without replacement. Find the probability that the hand contains three kings and 2 queens.

Solution:

The sample space is all possible five-card hands. Since there are 52 cards in the deck and order of selection is not important, we use $\#\mathbf{S} = \dbinom{52}{5} = \dfrac{52!}{5!47!} = 2,598,960.$

The number of ways of getting three kings is the number of ways of drawing three kings from four kings where order is not important. This can be computed using the formula $\dbinom{4}{3} = \dfrac{4!}{3!1!} = 4.$

The number of ways of getting two queens is the number of ways of drawing two queens from 4 queens where order is not important. This can be computed using the formula $\dbinom{4}{2} = \dfrac{4!}{2!2!} = 6.$

For each selection of three kings, there is 6 possible selections of 2 queens. Therefore the total number of ways of selecting a five card-hand with 3 kings and 2 queens is #**E** = (6)(4) = 24.

We thus have $P(\mathbf{E}) = \dfrac{\#\mathbf{E}}{\#\mathbf{S}} = \dfrac{\dbinom{4}{3}\dbinom{4}{2}}{\dbinom{52}{5}} = \dfrac{24}{2,598,960}.$

Solved Problems

17.3 - Solved Problem 1: On a shelf at a local library there are 10 books: 3 are fiction, 5 are non-fiction and 2 are reference books. Assuming the books are filed at random, find the probability that all books of the same category are filed together.

Solution:

The sample space is all possible ways the books can be filed: #**S** = 10! = 3,628,800.

The event **E** is that all books of the same category are filed together #**E** is the number of ways this can happen at random.

Step 1: The number of ways all fiction can be filed together is 3! = 6.

Step 2: The number of ways all non-fiction can be filed together is 5! = 120.

Step 3: The number of ways all reference books can be filed together is 2! = 2.

Step 4: However, each category can be arranged in 3! = 6 different ways.

Step 5: From the above steps, #**E** = (3!)(5!)(2!)(3!)= (6)(120)(2)(6) = 8640.

Step 6: $P(\mathbf{E}) = \dfrac{\#\mathbf{E}}{\#\mathbf{S}} = \dfrac{8640}{3,628,800} \, .$

17.3 - Solved Problem 2: Seven cards are selected from an ordinary deck of 52 cards without replacement. Find the probability that the hand contains four kings, two queens and one ace.

Solution:

The sample space is all possible seven-card hands, #**S**.

Since we are selecting 7 cards, $\#\mathbf{S} = \dbinom{52}{7} = \dfrac{52!}{7!45!} = 133,784,560.$

Step 1: The number of ways of getting 4 kings is $\begin{pmatrix} 4 \\ 4 \end{pmatrix} = 1$.

Step 2: The number of ways of getting 2 queens is $\begin{pmatrix} 4 \\ 2 \end{pmatrix} = 6$.

Step 3: The number of ways of getting 1 ace is $\begin{pmatrix} 4 \\ 1 \end{pmatrix} = 4$.

Step 4: #E $= \begin{pmatrix} 4 \\ 4 \end{pmatrix}\begin{pmatrix} 4 \\ 2 \end{pmatrix}\begin{pmatrix} 4 \\ 1 \end{pmatrix} = (1)(6)(4) = 24$

Step 4: P(E) $= \dfrac{24}{133784560}$

Unsolved Problems with Answers

17.3 - Problem 1: A line of 10 voters is formed at random in front of the voting booth. There are 3 Republicans, 2 are Democrats, 3 are Independent and 2 are Libertarian. Find the probability that voters of the same party are standing together.

Answer:

$$\frac{3,456}{3,628,800}$$

⇑ *Refer back to* **17.3 - Example 1 & 17.3 - Solved Problem 1**.

17.3 - Problem 2: A committee consists of 7 Democrats, 3 Republicans and 5 Independents. Nine members are randomly selected. Find the probability that 3 Democrats, 2 Republicans and 4 are Independent are selected.

Answer:

$$\frac{525}{5005}$$

⇑ *Refer back to* **17.3 - Example 2 & 17.3 - Solved Problem 2**.

Supplementary Problems

1. **The hypergeometric distribution.** Five cards are to selected from an ordinary deck of 52 cards without replacement. Let X be the random variable that equals the number of kings selected. The distribution of X is an example of a hypergeometric distribution. Construct the distribution table for the random variable X.

2. Mrs. Smith has 12 girls and 15 boys in her second grade class. She asks for 5 girls and 5 boys to volunteer to have their pictures taken standing in a row. Find the total number of possible rows of students assuming the boys all stand together.

3. Ms. Jones has 6 boys and 4 girls in her first grade class. At the end of the year she hires a photographer to take the children's picture. Assuming the children are stand together in a straight row for the picture, find the probability that all the boys are standing together.

4. Two decks of cards each have ten cards marked from 1 through 10. Assume one deck is in natural number order and the second deck is randomly shuffled. Matching the cards one at a time, find the probability that

a. a match occurs on the first card.

b. a match occurs on the first and fifth cards.

5. Five married couples purchase seats together at the opera. Assuming seating arrangements are random, find the probability that

a. all the men sit together.

b. each couple sits together.

6. Five people are selected at random. Find the probability that none are born on the same month.

7. Five cards are randomly drawn from an ordinary deck of cards. Find the probability the hand contains 3 aces or 2 kings.

8. Five cards are selected at random without replacement. Find the probability that two cards are kings and remaining three cards are spades.

9. It can be shown that the following formula is true:

$$\binom{N}{0} + \binom{N}{1} + \binom{N}{2} + \ldots + \binom{N}{N} = 2^N .$$

Use this formula to evaluate 2^7.

10. Recently Ms. Wise selected 5 football games to wager on. She decided on the following strategy:

1. Wager on each 5 football games.
2. Wager on every possible 2 team parleys.
3. Wager on every possible 3 team parleys.
4. Wager on every possible 4 team parleys.
5. Wager on every possible 5 team parleys.

Find the number of wages that she will make.

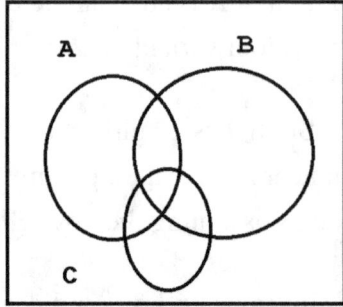

11. Using **A,B,C** , unions, intersections, and compliments for the following Venn diagram, how many non-empty subsets can be generated?

12. Assume 7 cards are drawn at random without replacement from an ordinary deck of cards. We wish to find the probability of the event that 3 kings and 2 queens were selected.

a. Show that to use the technique of Boolean expressions would require 210 different unions of events.

b. Using the number computed in a, find the probability that 3 kings and 2 queens will be selected.

13. In the past, the California Super Lotto™ claimed that the odds to win the maximum prize was 1 chance out of 18,009,460. The game was played under the following simple rule: the player picks 6 different numbers out of 51 numbers. Six different numbers are selected at random from the 51 numbers . The player bets $1.00. If the player picks all 6 numbers, the maximum prize is won.

a. Verify these odds.

b. Assume $1.00 is bet and the maximum prize is 10 million dollars. Assuming no ties, what is the expected gain/loss.?

c. What would the prize have to be to make this a fair game(i.e. E(X) = 0)?

14. Mrs. Jones draws 7 card from an ordinary deck of cards.

a. Find the average number of clubs in her hand.

b. Find the portability that her hand will contain 3 kings and 2 queens or 3 diamonds.

15. A new steak house offers a hamburger sandwich which includes any or all of the following items:
relish, lettuce, tomato, bacon, mushrooms, artichokes, salad dressing, cheese, mayonnaise, pesto sauce.

Compute the number of different sandwich combinations that are possible?

16. A math class has N students (N <365). Assuming 365 days in the year, find

a. the probability that exactly 2 students are born on the same day, and the other N - 2 students are born on different days.

b. the probability that exactly k (k < N) are born on the same day, and the other N - k students are born on different days.

17. The following five card poker hands are the most important for winning the game. Find, for each of these hands the number of ways they can be drawn. (Here, we assume ace can be high and low card.).

a. **Royal Flush:** Ten, Jack, Queen, King, Ace, all of the same suit.

   Example:

b. **Straight Flush:** Five cards all of the same suit that are in sequence.

   Example:

  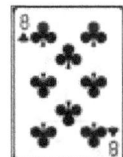 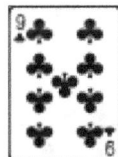 

c. **Four-of-A-Kind**: Four cards of equal rank.

   Example:

    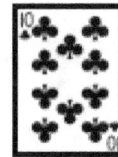

d. **Full House:** Three cards of equal rank, and two different cards of equal rank (Three of a kind and a pair).

> **Example:**

e. **Flush:**

Any five cards of the same suit.

> **Example:**

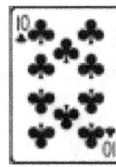    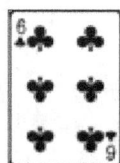

f. **Straight:** Five cards of mixed suits, in sequence .

> **Example:**

  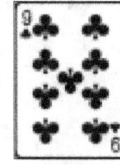  

g. **Two different pairs:** Two different pairs, each of the same kind.

Example:

  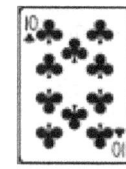  

h. **One pair:** One pair of equal rank. The other three cards are all different.

Example:

18. Two poker players are playing against each other. Assume each player is dealt 5 cards.

a. Find the total number of possible pair of hands.

b. Find the probability that both hands each contain one king.

c. Find the probability that at least 1 hand contains exactly 1 king.

19. Three poker players are playing against each other. Assume each player is dealt 5 cards.

Find the probability that at least 1 hand contains exactly 1 king. (See Lesson 10, Supplementary problem 13 a).

20. Three large urns are sitting on a table. The following game is played. Five balls are tossed at random into the urns. Assume at the end of the game, all five balls are in the urns. Find the probability that no urns are empty.

21. Three large urns are sitting on a table. The following game is played. A player has five balls. A ball is tossed one at a time until all the urns are filled or all the balls are tossed, whichever occurs first. Assuming all the balls fall into an urn, find the expected number of tosses.

22. Assume 4 cards are randomly selected, without replacement, from an ordinary deck of cards. Find the probability that the hand contains 2 kings and 2 diamonds, where a king can be a diamond.

23. *Stirling's Formula.* We can estimate n! for large n, using Stirling's formula:

$n! \sim (2\pi)^{1/2} n^{n+1/2} e^{-n}$ where $\sim$ means that the ratio of the 2 sides approach 1 for large n.

a. Using the error formula, $|n! - (2\pi)^{1/2} n^{n+1/2} e^{-n}|/n!$, find the error for estimating 10! using Stirling's formula.

b. Using Stirling's formula, formulate an estimation for $\dbinom{2n}{n}$.

c. Estimate $\dbinom{20}{10}$.

24. Assume an urn contains n red marbles and N - n black marbles. Assume r marbles are selected at random without replacement. Let k equal the number of red marbles selected ($0 \le k \le r$). The number of ways this can happen is

$$\dbinom{n}{k}\dbinom{N-n}{r-k}.$$

Let X be a discrete random variable that equals the number of red marbles selected. The hypergeometric distribution gives us

$$P(X = k) = \frac{\dbinom{n}{k}\dbinom{N-n}{r-k}}{\dbinom{N}{r}}. \quad k = 0,1,..., r.$$

Show

$$\dbinom{n}{0}\dbinom{N-n}{r-0} + \dbinom{n}{1}\dbinom{N-n}{r-1} + ... + \dbinom{n}{k}\dbinom{N-n}{r-k} + ... + \dbinom{n}{r}\dbinom{N-n}{0} = \dbinom{N}{r}$$

25. Assume 2 urns contain red, white, and blue marbles. Urn A contains 5 red, 5 white and 5 blue marbles. Urn B contains 5 red, 10 white and 5 blue marbles. An urn is selected at random and 5 marbles are selected.

a. Find the probability that 2 red marbles are selected. .

b. If 2 red marbles are selected, find the probability that Urn A was selected.

26. N balls, each numbered 1 to N respectively, sit on a table. Assume r_1 balls are selected without replacement , then r_2 are selected without replacement, etc. This process is continued until all balls are selected where $r_1 + r_2 + ... + r_n = N$ and $r_k > 0$ (k = 0,1,..,n).

a. Show the number of different ways this process can happen for a give selected values of r_k's.

$$\frac{N!}{r_1! r_2! ... r_n!}.$$

b . Assume N = 10, and $r_1 = 4$, $r_2 = 3$, $r_3 = 2$, $r_4 = 1$. Show the total number different selections is 12,600.

27. Assume N urns, each numbered 1 to N respectively, sit on a table. Balls, each numbered 1 to r respectively are tossed into the urns at random.

a. Show that the total number of ways this can happen is N^r.

b. Assume we have 3 urns, and 2 balls numbered 1 and 2. List all the ways the 2 balls can be placed into the 3 urns.

28. The joint hypergeometric distribution for 2 discrete random variables is defined as

$$P(X = x; Y = y) = \frac{\binom{n}{x}\binom{m}{y}\binom{N - n - m}{r - x - y}}{\binom{N}{r}}.$$

Extend this definition to three discrete random variables, X,Y,Z.

29. Assume 2 cards are randomly drawn, without replacement, from an ordinary deck of cards. Let X be the random variable that equals the number of jacks drawn, Y the random variable that equals the number of queens drawn, and Z the random variable that equals the number of kings. drawn. Find the joint distribution of X, Y, Z.

30. Assume we have N urns and N balls each marked 1 to N respectively. Assume we place at random each ball in an such that each urn will only contain 1ball. Show that the total number of possible ways this can happen is N!.

31. Assume we have N urns and r balls each marked 1 to r respectively. Assume we place r_1 marbles in urn 1; r_2 marbles in urn 2, ...; r_N marbles in urn N where $r_1 + r_2 + ... + r_N = r$ and $r_k \geq 0$ for k = 1,2, ..., N.

a. Show the number of ways this can be done is $\dfrac{r!}{r_1! r_2! ... r_N!}$.

b. Assume we have 6 urns and 4 ball each urn marked to 1 to 6 respectively and each ball 1 to 4 respectively. Assume $r_1 = 1$, $r_2 = 2$, $r_3 = 0$, $r_4 = 1$, $r_5 = 0$, $r_6 = 0$. List out the ways these 4 balls can be placed in urns 1,2, 4. Show all the ways the urns can be filled. .

32. *Maxwell-Boltzmann Statistics.* In physics, the Maxwell- Bolzmann statistics deals with the number of ways of placing m distinguishable particles into n distinguishable cells. This is analogous of placing m distinguishable balls into n distinguishable urns. The probability that the n cells contain $r_1,..., r_n$ particles where $r_1 + r_2 + ... + r_n = m$ and $r_k \geq 0$ (k = 1,.. , n) is

$$\frac{m!}{r_1!r_2!...r_n!}n^{-m}.$$

a. Assume n = 2, m = 3. Write out the sample space.

b. Assume m < n. Show that the probability of each cell has at most 1 particle is $_nP_m n^{-m}$

c. If m = 2, n = 3, list the event that each cell has at most 1 particle.

33. *Bose-Einstein Statistics.* In physics, the Bose-Einstein statistics deals with the number of ways of placing m indistinguishable particles into n distinguishable cells. This is analogous of placing m indistinguishable balls into n distinguishable urns. The number of distinguishable arrangement is

$$\binom{n + m - 1}{m}, \quad \text{where each distinguishable arrangement has equal probability.}$$

a. Assume n = 2, m = 3. Write out the sample space.
b. Assume m < n. Show that the probability of each cell has at most 1 particle

is $\dfrac{\binom{n}{m}}{\binom{n + m - 1}{m}}$.

c. If m = 2, n = 4, list the event that each cell has at most 1 particle.

34. *Fermi-Dirac Statistics.* In physics, the Fermi-Dirac statistics deals with the number of ways of placing m indistinguishable particles into n distinguishable cells with the following restrictions:

1. $m \leq n$
2. Each cell can contain at most 1 particle.
3. Each distinguishable arrangement has equal probability.

a. Assume n = 5, m = 3. Write out the sample space.

b. Show that the probability of the event that the cells filled are contiguous

is $(n - m + 1)\left(\dfrac{n}{m}\right)^{-1}$.

35. In a card game 6 players are dealt 4 kings. Show the probability that the kings are dealt evenly (each player is dealt no more than 1 king per player) applying

a. Maxwell- Bolzmann statistics is 5/18.

b Bose-Einstein statistics is 5/42.

36. In a card game 6 players are dealt 4 kings. Let X be the random variable that equals the number of players that were dealt at least 1 king having a Maxwell-Boltzmann statistics distribution. Also let Y be the random variable that equals the number of players that were dealt at least 1 king have the Bose-Einstein statistics.

a. Find the distribution of X.

b. Find the distribution of Y.

37. *The birthday problem.* Assume m people are selected at random. Assuming there are 364 days in the year and m < 364, find the probability

a. that at least 2 people were born on the same day.

b. that at least 3 people were born on the same day.

c. that exactly r people were born on the same day and each of the others were born on different days.

d. Find the number of people that need to randomly selected so that there is at least a 50% chance that at least 2 people were born on the same day.

38. *Another birthday problem.* The ABC Auto Shop specializes in repairing identical models manufactured in Germany. Over a 1 year period, assume m identical repaired automobiles are selected at random. Assuming there are 364 days in the year and m < 364, find the probability that exactly r automobiles were repaired on the same day and each of the others were repaired on different days.

39. An urn contains 6 red balls and 6 white balls. Assume 6 balls are randomly selected without replacement. Find the probability that 3 red marbles were selected.

40. A card game has 6 players. Assume a dealer deals to each player 5 cards from an ordinary deck of cards. If only one of the players was dealt 2 kings, find the probability that all the kings

were dealt from the deck.

41. A fair die is tossed n times. Let X_k (k = 1,2,...,100) be the independent random variables that equal the out for each toss.

Let $\overline{X} = \dfrac{X_1 + X_2 + \ldots + X_n}{n}$.

a. If n = 10, find $P(\overline{X} \le 1.2)$.

b. If n = 100, find $P(\overline{X} \le 1.2)$.

42. The following problem shows the large magnitude of n! for relatively small values of n: A class of 30 mathematics students sit in a class room with 30 chairs. A computer program is to be written that will compute all possible seating arrangements for the 30 students. Assume that the program can compute one million arrangements a second, how long will it take the computer to compute all arrangements?

The binomial sample space is a special sample space generated by a binomial experiment.

18.1 - What is the Binomial Experiment that Generates a Special Sample Space?

The binomial experiment generates a special type of sample space using the following rules:

1. Each element of the sample space is made up from N independent trials.

2. Each trial results in only two possible outcomes which are labeled success or failure.

3. The probability for success on each trial does not change and is denoted as the value p. The probability for failure is $q = 1 - p$.

A sample space generated under these rules is called a binomial sample space.

Also, these types of experiments are called Bernoulli Trials.

18.1 - Example 1: A coin is tossed 3 times. The two outcomes for each toss is heads or tails. Let us define success as head on each toss.

(a). Write out the binomial sample space.

(b). Write out the event that exactly two heads occurs.

(c). Show that this is a binomial sample space.

Solutions:

➤ (a).

Let h be the occurrence of head. Let t be the occurrence of tail.

S = {(h,h,h), (h,h,t), (h,t,h), (t,h,h), (t,t,h), (t,h,t), (h,t,t), (t,t,t)}

➤ **(b).**

E = {(h,h,t), (h,t,h), (t,h,h)}

➤ **(c).**

1. Each toss of the coin is an independent trial. The number of trials is 3.

2. Each trial results in only two possible outcomes: head or tail. We define success as heads on each toss.

3. The probability of success for each tail is 1/2 which is the probability that heads occurs for each toss.

18.1 - Example 2: A die is tossed 4 times. We define success for each toss as to whether the number three occurs.

(a). Write out the binomial sample space.

(b). Write out the event that exactly 1 three occurs.

(c). Show that this is a binomial sample space.

Solutions:

➤ **(a).**
Let 3 indicate the number 3 occurs.

Let f indicate failure: any number from 1 to 6 occurs, except the number 3.

S = {(3,3,3,3)}, (3,3,3,f), (3,3,f,3), (3,f,3,3), (f,3,3,3), (3,3,f,f),(3,f,f,3), (f,f,3,3), (f,3,3,f), (3,f,3,f), (f,3,f,3), (f,f,f,3), (f,f,3,f),(f,3,f,f), (3,f,f,f), (f,f,f,f)}

➤ **(b).**
E = {(3,f,f,f), (f,3,f,f), (f,f,3,f), (f,f,f,3)}

➤ **(c).**
1. Each toss of the die is an independent trial. The number of trials is 4.

2. Each trial results in the success or failure of the occurrence of the number 3.

3. The probability for success, the number 3 occurs, for each trial is 1/6 and the probability for failure f is 5/6 (the number 3 does not occur).

18.1 - Example 3: A statistics class has 3 male students. A survey is made to determine which of these students wears glasses. Assume that the chance a male student is wearing glasses is p.

(a). Write out the binomial sample space.

(b). Write out the event that exactly two male students wear glasses.

(c). Show that this is a binomial sample space.

Solutions:

➤ **(a).**
Let g indicate that a male student wears glasses.
Let f indicate failure: that a male student does not wear glasses.

$S = \{(g,g,g), (g,g,f), (g,f,g), (f,g,g), (f,f,g), (f,g,f), (g,f,f), (f,f,f)\}$

➤ **(b).**
$E = \{(g,g,f), (g,f,g), (f,g,g)\}$

➤ **(c).**
1. Each male student is considered an independent trial. The number of trials is 3.

2. Each trial results in the success or failure of the occurrence of a student wearing glasses.

3. The probability for success is p that a male student is wearing glasses; the probability for failure is
$q = 1 - p$ that the student is not wearing glasses.

18.1 - Example 4: Three cards are drawn without replacement from an ordinary deck of cards. We define success for each drawing as to whether a king is drawn.

(a). Write out the sample space.

(b). Write out the event that no kings occurs.

(c). Show that this is NOT a binomial sample space.

Solutions:

➤ **(a).**
Let k indicate a king is drawn.

Let f indicate failure : that a card other than a king does not occurs.

$S = \{(k,k,k), (k,k,f), (k,f,k), (f,k,k), (k,f,f), (f,k,f), (f,f,k), (f,f,f)\}$

➤ **(b).**
$E = \{(f,f,f)\}$

➤ **(c).**
This sample space is not binomial because the drawings are not independent.

Solved Problems

18.1 - Solved Problem 1: An urn contains 4 red marbles and 6 white marbles. Three marbles are randomly selected with replacement. Let us define success as a white marble is selected.

(a). Write out the binomial sample space.

(b). Write out the event that exactly two white marbles are selected.

(c). Show that this is a binomial sample space.

Solutions:

➤ **(a).**
Let w be the occurrence of a white marble.
Let r be the occurrence of a red marble.

$$\mathbf{S} = \{(w,w,w), (w,w,r), (w,r,w), (r,w,w), (r,r,w), (r,w,r), (w,r,r), (r,r,r)\}$$

➤ **(b).**
$\mathbf{E} = \{(r,w,w), (w,r,w), (w,w,r)\}$

➤ **(c).**
1. Since the sampling is done with replacement, each selection is independent. The number of trials is 3.

2. Each trial results in only two possible outcomes: white or red. We define success as white on each drawing.

3. The probability of success for each trial is 6/10 which is the probability that a white marble is drawn.

18.1 - Solved Problem 2: An urn contains 5 red marbles, 6 white marbles and 4 blue marbles. Four marbles are selected at random with replacement. We define success if a white marble is selected.

(a). Write out the binomial sample space.

(b). Write out the event that exactly 1 white marble occurs.

(c). Show that this is a binomial sample space.

Solutions:

➤ **(a).**

Let w indicate that a white marble is selected.

Let f indicate failure: that a red or blue marble is selected.

S = {(w,w,w,w), (w,w,w,f), (w,w,f,w), (w,f,w,w), (f,w,w,w), (f,f,w,w),

   (f,w,w,f), (f,w,f,w), (w,f,f,w), (w,f,w,f), (w,w,f,f), (f,f,f,w), (f,f,w,f),

   (f,w,f,f), (w,f,f,f), (f,f,f,f)}

➤ **(b).**

E = {(w,f,f,f), (f,w,f,f), (f,f,w,f), (f,f,f,w)}

➤ **(c).**

1. Since selection is made with replacement, the trials are independent. The number of trials is 4.

2. Each trial results in the success or failure on the occurrence of drawing a white marble.

3. The probability for success, that a white marble is drawn for each trial is 6/15 and the probability for failure is 9/15 that a white marble is not selected.

18.1 - Solved Problem 3: A random sample of 3 students was taken from a statistics class to determine which students are majoring in mathematics. We define success for each student that is majoring in mathematics. Assume the chance that a student is majoring in mathematics is p = 0.15.

(a). write out the sample space.

(b). write out the event that all three students are majoring in mathematics.

(c). show that this is a binomial sample space.

Solutions:

➤ **(a).**

Let m indicate that a student is majoring in mathematics.

Let f indicate that a student is not majoring in mathematics.

S = {(m,m,m), (m,m,f), (m,f,m), (f,m,m), (f,f,m), (f,m,f), (m,f,f), (f,f,f)}

➤ **(b).**

E = {(m,m,m)}

➤ **(c).**

1. Each student is considered an independent trial. The number of trials is three.

2. Each trial results in the success if a student is majoring in mathematics.

3. The probability for success is $p = 0.15$ that a student is majoring in mathematics and the probability for failure is $q = 1 - p = 1 - 0.15 = 0.85$ that the student is not majoring in mathematics.

18.1 - Solved Problem 4: An English class has 20 female and 10 male students. Three students are selected at random to read from sections of Hamlet. Assume we define success that a student selected is female.

(a). Write out the sample space.

(b). Write out the event that no females are selected.

(c). Show that this is NOT a binomial sample space.

Solutions:

➤ **(a).**
Let f indicate a female is selected.

$S = \{(f,f,f), (f,f,m), (f,m,f), (m,f,f), (f,m,m), (m,f,m), (m,m,f), (m,m,m)\}$

➤ **(b).**
$E = \{(m,m,m)\}$

➤ **(c).**
This sample space fails to be binomial because each drawing of a student is not an independent trial.

Unsolved problems with Answers

18.1 - Problem 1: QMart Stores has a large number of retail stores. Management estimates that 20% of its retail stores will have a million dollars in sales for the year 1993. Three stores are randomly selected and audited to check if they have a million dollars in sales for 1993.

(a). Write out the binomial sample space.

(b). Write out the event that exactly two stores have a million dollars in sales.

(c). Show that this is a binomial sample space.

Answers:

➤ **(a).**
Let m be the occurrence of a store having a million dollars in sales.

Let f be the occurrence of a store not having a million dollars in sales.

S = {(m,m,m), (m,m,f), (m,f,m), (f,m,m), (m,f,f), (f,m,f), (f,f,m), (f,f,f)}

➤ **(b)**.
E = {(f,m,m), (m,f,m), (m,m,f)}

➤ **(c)**.
1. Since there are a large number of stores to sample from and the sample size is relatively small, it is reasonable to assume the sampling is independent.

2. Each sample results in only two possible outcomes: a store has or does not have a million dollars in sale. We define success as a store that has a million dollars in sales.

3. The probability of success for each sample is $p = 0.20$ which is the proportion of stores that are estimated to have a million dollars in sales. The probability of failure is $1-0.20 = 0.80$.

⇑ *Refer back to* **18.1 - Example 1 & 18.1 - Solved Problem 1**.

18.1 - Problem 2: Ms. Jones teaches 5 sections of German. She has a total of 300 students. The distribution of grades in Ms. Jones's German class is as follows: 20% earned a final grade of A; 25% earned a B; 40% earned a C; 10% earned a D and 5% earned a F. Three of her students are selected at random with replacement. We are interested in those students that earned a grade of B or better is selected.

(a). Write out the binomial sample space.

(b). Write out the event that exactly two students earned a grade of B or better.

(c). Show that this is a binomial sample space.

Answers:

➤ **(a)**.
Let s indicate that a student selected has a grade of B or better.

Let f indicate failure: that a student selected has a grade of C or less.

S = {(s,s,s), (s,s,f), (s,f,s), (f,s,s), (f,f,s), (f,s,f), (s,f,f), (f,f,f)}

➤ **(b)**.
E = {(s,s,f), (s,f,s), (f,s,s)}

➤ **(c)**.
1. Since she has a large number of students and the sample is relatively small, it is reasonable to

assume that the sampling results in independent selections.

2. We define success as a student selected that has a grade of B or better.

3. The probability for success, p = 0.45 and the probability for failure is 0.55 .

⇑ *Refer back to* **18.1 - Example 2 & 18.1 - Solved Problem 2**.

18.1 - Problem 1.3: A large corporation recently hired 3 people in its management training program. A review of their backgrounds was taken to determine if any of the new trainees have studied international law.

(a). Write out the binomial sample space.

(b). Write out the event that none of the trainees have studied international law.

(c). Show that this is a binomial sample space.

Answers:

➤ **(a).**
Let l indicate that a trainee has studied international law.

Let f indicate that a trainee has not studied international law.

S = {(l,l,l), (l,l,f), (l,f,l), (f,l,l), (f,f,l), (f,l,f), (l,f,f), (f,f,f)}

➤ **(b).**
E = {(f,f,f)}

➤ **(c).**
1. Each trainee is considered an independent trial. The number of trials is three.

2. Each trial results in success that a trainee has studied international law or failure that a trainee has not studied international law.

3. The probability for success is p, that a trainee has studied international law, and the probability is
 q = 1 - p, that the trainee has not studied international law.

⇑ *Refer back to* **18.1 - Example 3 & 18.1 - Solved Problem 3**.

18.1 - Problem 4: Mr. Smith has thirty books in his private library. He selects 2 books to read this summer. He is particularly interested in murder mysteries.

(a). Write out the sample space.

(b). Write out the event that both books he selected are murder mysteries.

(c). Show that this is NOT a binomial sample space.

Answers:

➤ **(a).**
Let m indicate a book is a murder mystery and f indicate a book is not a murder mystery.

S = {(m,m), (m,f), (f,m), (f,f)}

➤ **(b).**
E = {(m,m)}

➤ **(c)** .
This sample space fails to be binomial because each drawing of a book is a dependent trial.

⇑ *Refer back to* **18.1 - Example 4 & 18.1 - Solved Problem 4**.

18.2 - Counting the Number of Successes from a Binomial Experiment.

In the next lessons, we study the binomial random variable and the binomial distribution. In order to properly define a binomial distribution, we first need to show how to compute the number of ways a given number of successes can occur from a binomial experiment.

18.2 - Example 1: A coin is tossed 4 times. Find the number of ways exactly two tails can occur.

Solution:

Method 1: Write out the event **E** that exactly 2 tails occurs:

E = {(t,t,h,h), (t,h,h,t), (h,h,t,t), (t,h,t,h), (h,t,t,h), (h,t,h,t)}

Therefore, the number of ways exactly two tails can occur is six since #**E** = 6.

Method 2: Think of each toss of the coin as filling up each of the four slot positions:

_____, _____, _____, _____.

Since we are interested in two tails, two slots are reserved for tails. Therefore, the number of ways of getting 2 tails is equal to the number of ways of selecting 2 slots out of 4 slots where order is not important. Using the counting methods of lesson 8, we have #**E** = $\binom{4}{2}$ = 6.

Throughout this lesson and subsequent lessons on the binomial distribution we will use the following formula:

> Assume a binomial experiment results in n independent trials. The formula for the number of ways that r successes can occur out of the n trials is
>
> $$\binom{n}{r}.$$

18.2 - Example 2: Seven customers are standing at the checkout counter in a local supermarket. Find the number of ways that

(a). three can be females.

(b). six can be males.

(c). none can be females.

Solutions:

For this problem, n = 7.

➤ **(a).**
We define success as female. Therefore r = 3. Using the formula,

$$\binom{n}{r} = \binom{7}{3} = \frac{7!}{3!4!} = 35$$

ways that three customers out of seven are female.

➤ **(b).**
Here we define success as male. Therefore, r = 6. Using the formula

$$\binom{n}{r} = \binom{7}{6} = 7$$

ways that six customers out of seven are male.

➤ **(c).**
We define success as female. Therefore, r = 0 Using the formula

$$\binom{n}{r} = \binom{7}{0} = 1$$

way that no customers out of the seven customers are female.

Solved Problems

18.2 - Solved Problem 1: A die is tossed 5 times. Find the number of ways exactly three 2s' can occur.

Solution:

Method 1: Write out the event **E** that exactly three 2s' occur:

E = {(2,2,2,f,f,), (2,2,f,f,2), (2,f,f,2,2), (f,f,2,2,2),(f,2,2,2,f),

     (2,f,2,f,2), (f,2,f,2,2), (2,2,f,2,f), (f,2,2,f,2), (2,f,2,2,f)}

Therefore, the number of ways exactly three twos's can occur is ten so **#E** = 10.

Method 2: Think of each toss of the die as filling up each of the five slot positions:

_____, _____, _____, _____, _____.

Since we are interested in three 2s', the three slots are reserved for three 2s' Therefore the number of ways of getting three 2s' is equal to the number of ways of selecting 3 slots out of 5 slots where order is not important. Using the counting methods of lesson 8, we have

$$\#E = \binom{5}{3} = 10.$$

18.2 - Solved Problem 2: Ten automobiles are tested for mileage. Find the number of ways that

(a) three cars exceed 30 miles a gallon.

(b). five cars get less than 30 miles a gallon.

(c). two get between 20 and 28 miles a gallon.

Solutions:

For this problem, N = 10.

➤ **(a).**
We define success as a car exceeds 30 miles a gallon. Therefore r = 3. Using the formula

$$\binom{n}{r} = \binom{10}{3} = 120.$$

ways that three cars can exceed 30 miles a gallon.

➤ **(b).**

Here we define success as a car gets less than 30 miles a gallon. Therefore, r = 5 Using the formula

$$\binom{n}{r} = \binom{10}{5} = 252 \text{ ways that five cars out of ten get less than 30 miles per gallon.}$$

➤ **(c).**

We define success as a car getting between 20 and 28 miles a gallon. Therefore r = 2. Using the formula

$$\binom{n}{r} = \binom{10}{2} = 45 \text{ ways that two cars out of ten can get between 20 and 28 miles a gallon.}$$

Unsolved Problems with Answers

18.2 - Problem 1: The Adam's family has three children. Find the number of ways that exactly 2 can be girls.

Answer:

3.

⇑ *Refer back to* **18.2 - Example 1 & 18.2 - Solved Problem 1**.

18.2 - Problem 2: Six bullets are fired at a target. Find the number of ways that

(a) three bullets hit the target.

(b) four miss the target.

Answers:

➤ **(a).** 20

➤ **(b).**15

⇑ *Refer back to* **18.2 - Example 2 & 18.2 - Solved Problem 2**.

Supplementary Problems

The binomial experiment can be generalized to N repeated independent trials where each trial can have one of several outcomes.

1. A die is tossed twice. Write out the multinomial sample space where each toss of the die can result in a 1 or 3.

2. A die is tossed N times. What is the cardinality of this multinomial sample space where each toss of the die can result in a 1, 3 or 5?

3. A die is tossed four times. Write out the event **E** that we get two ones' and two threes'.

4. Assume a multinomial experiment of N independent trials with n different outcomes. Find a formula that will compute the cardinality of the event **E** consisting of $r_1, r_2, r_3, r_4, ..., r_n$ outcomes of each type where $r_1 + r_2 + r_3 + r_4 + ... + r_n = N$.

5. A pair of dice is tossed 10 times and the numbers are summed. Find the number of ways that two elevens, three tens, and four sevens appear.

6. On a given Saturday Mr. Jones wages on 10 football games. How many ways can he win five, lose three and tie two games?

7. If 7 cards are selected, at random, from an ordinary deck of cards, find the number of ways a hand can contain 2 aces, 2 kings and 1 jack?

8. Assume we define $\begin{pmatrix} N \\ k \end{pmatrix} = 0$ for k > N. Show for $0 \le k \le n + m$

$$\begin{pmatrix} n+m \\ k \end{pmatrix} = \begin{pmatrix} n \\ 0 \end{pmatrix}\begin{pmatrix} m \\ k \end{pmatrix} + \begin{pmatrix} n \\ 1 \end{pmatrix}\begin{pmatrix} m \\ k-1 \end{pmatrix} + \begin{pmatrix} n \\ 2 \end{pmatrix}\begin{pmatrix} m \\ k-2 \end{pmatrix} + ... + \begin{pmatrix} n \\ k \end{pmatrix}\begin{pmatrix} m \\ 0 \end{pmatrix}$$

9. Assume we have a Bernoulli trial. Let X be a random variable that has the geometric distribution

$P(X = k) = q^{k-1}p$, where p is the probability of success and q = 1 - p. Show $P(X > k) = q^k$.

19.1 - The Definition of the Binomial Random Variable X

> The binomial random variable is a random variable only defined on a binomial sample space. The binomial random variable assigns to each element of a binomial sample space, the number of successes. The symbol for a binomial random variable is X.

19.1 - Example 1: A coin is tossed twice. Let the binomial random variable X assign to each element the number of heads. Write out the random variable.

Solution:

For this binomial sample space we define head on a trial as success. The sample space is
$S = \{(h,h), (h,t), (t,h), (t,t)\}$.

X is the binomial random variable representing the number of heads for each possible element of the sample space. Therefore, $X(h,h) = 2$, $X(h,t) = 1$, $X(t,h) = 1$, $X(t,t) = 0$.

The relationship between the assigned values of X and the elements of the sample space can be seen from the following table:

Sample Space	X = x
(h,h)	2
(h,t)	1
(t,h)	1
(t,t)	0

19.1 - Example 2: An urn contains red, white and blue marbles. Three marbles are selected with replacement. Let the binomial random variable X assign to each element of the sample space, the number of reds drawn. Write out the random variable.

Solution:

For this binomial sample space we define getting a red on a given trial as success. The binomial sample space is $S = \{(r,r,r), (r,r,f), (r,f,r), (f,r,r), (r,f,f), (f,r,f), (f,f,r), (f,f,f)\}$, where r is a red marble is drawn and f is a red marble is not drawn.

X is the binomial random variable representing the number of reds for each possible element of the binomial sample space. Therefore,

$X(r,r,r) = 3$, $X(r,r,f) = 2$, $X(r,f,r) = 2$, $X(f,r,r) = 2$
$X(r,f,f) = 1$, $X(f,r,f) = 1$, $X(f,f,r) = 1$, $X(f,f,f) = 0$

The relationship between the assigned values of X and the elements of the sample space can be seen from the following table:

Sample Space	X = x
(r,r,r)	3
(r,r,f)	2
(r,f,r)	2
(f,r,r)	2
(r,f,f)	1
(f,r,f)	1
(f,f,r)	1
(f,f,f)	0

19.1 - Example 3: Three students are selected to attend the meeting of the National Bird Watchers Association. Let X assign to each element of the binomial sample space, the number of students that are zoology majors. Write out the random variable.

Solution:

For this binomial sample space we define a student being a zoology major as a success.

The sample space is $S = \{(z,z,z), (z,z,f), (z,f,z), (f,z,z), (f,f,z), (f,z,f), (z,f,f), (f,f,f)\}$, where

z is a zoology major and f is not a zoology major.

X is the binomial random variable representing the number of zoology majors for each possible element of the sample space. Therefore,

$X(z,z,z) = 3$, $X(z,z,f) = 2$, $X(z,f,z) = 2$, $X(f,z,z) = 2$
$X(f,f,z) = 1$, $X(f,z,f) = 1$, $X(z,f,f) = 1$, $X(f,f,f) = 0$.

The relationship between the assigned values of X and the elements of the sample space can be seen from the following table:

Sample Space	X = x
(z,z,z)	3
(z,z,f)	2
(z,f,z)	2
(f,z,z)	2
(f,f,z)	1
(f,z,f)	1
(z,f,f)	1
(f,f,f)	0

Solved Problems

19.1 - Solved Problem 1: A pair of dice are tossed twice. The resulting two numbers are added. Let the binomial random variable X assign to each pair of tosses the number of sevens. Write out the random variable.

Solution:

For this binomial sample space we define the sum 7 as success. The sample space is
S = {(7,7), (7,f), (f,7), (f,f)}.

X is the binomial random variable representing the number of sevens for each possible element of the sample space. Therefore, **X**(7,7) = 2, **X**(7,f) = 1, X(f,7) = 1, **X**(f,f) = 0.

The relationship between the assigned values of X and the elements of the sample space can be seen from the following table:

Sample Space	X = x
(7,7)	2
(7,f)	1
(f,7)	1
(f,f)	0

19.1 - Solved Problem 2: Three pedestrians are questioned about their position on U.S. military in involvement in Europe. Four questions were asked:

Question 1: Should the U.S. military remain in Europe?

Question 2: Should the U.S. completely withdraw its military from Europe?

Question 3: Should the U.S. remain a member of NATO?

Question 4: Should the U.S. allow full membership of Russia in NATO?

Let the binomial random variable X assign to each element of the sample space the number of people that answered yes to question 1. Write out the random variable.

Solution:

The binomial sample space is

$S = \{(y,y,y), (y,y,f), (y,f,y), (f,y,y), (y,f,f), (f,y,f), (f,f,y), (f,f,f)\}$, where

y indicates question 1 was answered 'yes',
f indicates question 1 was answered 'no'.

X is the binomial random variable representing the number of yes responses on question 1. Therefore,

$X(y,y,y) = 3, X(y,y,f) = 2, X(y,f,y) = 2, X(f,y,y) = 2$

$X(y,f,f) = 1, X(f,y,f) = 1, X(f,f,y) = 1, X(f,f,f) = 0$

The relationship between the assigned values of X and the elements of the sample space can be seen from the following table:

Sample Space	X = x
(y,y,y)	3
(y,y,f)	2
(y,f,y)	2
(f,y,y)	2
(y,f,f)	1
(f,y,f)	1
(f,f,y)	1
(f,f,f)	0

19.1 - Solved Problem 3: Three Ramblers are tested for mileage. Let X assign to each element of the binomial sample space, the number of Ramblers that exceed 30 miles per gallon. Write out the random variable.

Solution:

For this binomial sample space, we define a Rambler that gets over 30 miles per gallon as a success. The sample space is S = {(g,g,g), (g,g,f), (g,f,g), (f,g,g), (f,f,g), (f,g,f), (g,f,f), (f,f,f)}, where g is a Rambler get over 30 miles per gallon and f is a Rambler does not get over 30 miles per gallon. X is the binomial random variable representing the number of Rambler that get over 30 miles per gallon for each possible element of the sample space. Therefore,

X(g,g,g) = 3, X(g,g,f) = 2, X(g,f,g) = 2, X(f,g,g) = 2

X(f,f,g) = 1, X(f,g,f) = 1, X(g,f,f) = 1, X(f,f,f) = 0

The relationship between the assigned values of X and the elements of the sample space can be seen from the following table:

Sample Space	X = x
(g,g,g)	3
(g,g,f)	2
(g,f,g)	2
(f,g,g)	2
(g,f,f)	1
(f,g,f)	1
(f,f,g)	1
(f,f,f)	0

Unsolved Problems with Answers

19.1 - Problem 1: Two cards, three times, are selected from an ordinary deck of cards with replacement.. Assume the cards 10 through King each count 10 points and ace counts eleven points. All other cards are given their natural value. Let the binomial random variable X assign to the three trials the number of times the sum of 21 occurs. Write out the random variable.

Answer:

X(21,21,21) = 3, X(21,21,f) = 2, X(21,f,21) = 2, X(f,21,21) = 2,

X(f,f,21) = 1, X(f,21,f) = 1, X(21,f,f) = 1, X(f,f,f) = 0

⇑ *Refer back to* **19.1 - Example 1 & 19.1 - Solved Problem 1**.

19.1 - Problem 2: Ms. Smith's library contains mysteries, love stories, and travel books. Three books were randomly selected from Mrs. Smith's library. She is interested in the number that are mysteries. Write out the random variable.

Answer:

$X(m,m,m) = 3$, $X(m,m,f) = 2$, $X(m,f,m) = 2$, $X(f,m,m) = 2$,

$X(f,f,m) = 1$, $X(f,m,f) = 1$, $X(m,f,f) = 1$, $X(f,f,f) = 0$

⇑ *Refer back to* **19.1 - Example 2 & 19.1 - Solved Problem 2**.

19.1 - Problem 3: Four students check into the health clinic with a fever. The doctor checks each student for strep throat. Write out the random variable.

Answer:

$X(s,s,s,f) = 3$, $X(s,s,f,s) = 3$, $X(s,f,s,s) = 3$, $X(f,s,s,s) = 3$

$X(s,f,s,f) = 2$, $X(s,s,f,f) = 2$, $X(f,f,s,s) = 2$, $X(f,s,f,s) = 2$

$X(f,s,s,f) = 2$, $X(s,f,f,s) = 2$, $X(f,f,f,s) = 1$, $X(f,f,s,f) = 1$

$X(f,s,f,f) = 1$, $X(s,f,f,f) = 1$, $X(f,f,f,f) = 0$,$X(s,s,s,s) = 4$

⇑ *Refer back to* **19.1 - Example 3 & 19.1 - Solved Problem 3**.

Supplementary Problems

1. Assume a coin is tossed four times. Let X be the random variable that defines the number of heads. Define the event of the binomial sample space $\mathbf{E} = \{X > 1\}$. Write out the elements of **E**.

2. Assume five cards are drawn with replacement from an ordinary deck of cards Let X be the random variable that defines the number of kings drawn. Define the event of the binomial sample space
$\mathbf{E} = \{2 \le X \le 4\}$. Write out the elements of **E**.

3. A poll on the Senate race was taken from 10 students at a local college. Let X be the random variable that defines the number of students that believe the incumbent Senator will win the November election. Define the event of the binomial sample space $\mathbf{E} = \{9 \le X\}$. Write out the elements of **E**.

4. Assume a coin is tossed three times. Let X be the random variable that defines the number of heads.
Let $\mathbf{E}_k = \{X \le k\}$ and $\mathbf{F}_k = \{X \ge k\}$ for $k = 0, 1,..., 3$. Write out the events $\mathbf{E}_k$ and $\mathbf{F}_k$.

5. Assume a coin is tossed once and a die is tossed once. Let X be the random variable that defines the number of heads and Y the random variable that defines the number of even numbers that appeared. Let

$E_k = \{Z = X + Y = k\}$ and $F_k = \{Z = X + Y \leq k\}$, for $k = 0, 1, 2, 3$. Write out the events E_k and F_k.

6. From the table, write out the events for $0 \leq k \leq 3$:

Sample Space	X
(y,y,y)	3
(y,y,f)	2
(y,f,y)	2
(f,y,y)	2
(y,f,f)	1
(f,y,f)	1
(f,f,y)	1
(f,f,f)	0

a. $E_k = \{X = k\}$

b. $F_k = \{X \leq k\}$

7. Assume Mr. Jones wages on two football games on Friday, one Saturday and one on Sunday. Let X be the random variable that equals the numbers of games won on Friday, Y the number of games won on Saturday and Z the number of games won on Sunday. Find the events $E_k = \{T = X + Y + Z \leq k\}$ for $k = 0, 1, 2, 3, 4$.

8. Assume X is a binomial random variable defined on a binomial sample space where $\{0 \leq X \leq N\}$.

Let $E_k = \{X \leq k\}$ and $F_k = \{X \geq k\}$.

a. For each integer $0 \leq p, q \leq N$, find $E_p \cap F_q$.

b. For each integer $0 \leq k \leq N$, write $\{X = k\}$ as a Boolean expression of the events E_k.

c. For each integer $0 \leq k \leq N$, write $\{X = k\}$ as a Boolean expression of the events F_k.

d. For each integer $0 \leq k \leq N$, write E_k as a Boolean expression of the events F_k.

20.1 - What is a Binomial Distribution?

The binomial distribution is the probability distribution of the binomial random variable:
$P\{X = k\}; k = 0, 1, 2, 3..., n$. (See lesson 19).

Assume the sample space is generated from a binomial experiment of n independent trials. Assume the probability of success for each trial is p and the probability of failure is $q = 1 - p$. Since the trials are independent,

$$P\{X=k\} = \binom{n}{k} p^k q^{n-k} \text{ where}$$

n = the number of independent trials.

k = the number of successes.

p = probability of success for each trial.

$q = 1 - p$ = probability of failure for each trial.

20.1 - Example 1: Assume a die is tossed 5 times. Write out the binomial distribution, where the random variable is the number of fours that appear.

Solution:

Since there are five tosses of the die: n = 5.

Each toss of the die is independent.

Success on each trial is defined as a four appears: p = 1/6.

Failure on each trial is defined as a four does not appear: q = 1- 1/6 = 5/6.

The number of possible fours range from zero to 5 fours: k = 0, 1, 2, 3, 4, 5.

The distribution of this binomial distribution is

$$P\{X = 0\} = \binom{5}{0} \left(\frac{1}{6}\right)^0 \left(\frac{5}{6}\right)^5 = \frac{3125}{7776}$$

$$P\{X = 1\} = \binom{5}{1} \left(\frac{1}{6}\right)^1 \left(\frac{5}{6}\right)^4 = \frac{3125}{7776}$$

$$P\{X = 2\} = \binom{5}{2} \left(\frac{1}{6}\right)^2 \left(\frac{5}{6}\right)^3 = \frac{1250}{7776}$$

$$P\{X = 3\} = \binom{5}{3} \left(\frac{1}{6}\right)^3 \left(\frac{5}{6}\right)^2 = \frac{250}{7776}$$

$$P\{X = 4\} = \binom{5}{4} \left(\frac{1}{6}\right)^4 \left(\frac{5}{6}\right)^1 = \frac{25}{7776}$$

$$P\{X = 5\} = \binom{5}{5} \left(\frac{1}{6}\right)^5 \left(\frac{5}{6}\right)^0 = \frac{1}{7776}$$

20.1 - Example 2: A large technical university's records reveal that 45% of all students major in Engineering. Ten students were interviewed. Find the probability that

(a). exactly 5 students are majoring in Engineering.

(b). 4 students are not majoring in Engineering.

(c). between 4 and 6 students are majoring in Engineering.

(d). at least 1 student is majoring in Engineering.

(e). at most 9 students are majoring in Engineering.

Solutions:

➤ **(a).**
This is a binomial distribution for the following reasons:

1. Each student interviewed is a trial. Since 10 students are interviewed the number of trials is N = 10.

2. It is reasonable, since the students were selected at random from a large student body, that the selections are independent.

3. Success is defined as a student interviewed is a Engineering major.

4. The probability of a student being an Engineering major is p = 0.45. The probability of a student not being a engineering major is q = 1 - p = 0.55.

5. We wish to compute the probability that 5 of the 10 students interviewed are Engineering majors:

$$P\{X = 5\} = \binom{10}{5} (0.45)^5 (0.55)^5 \approx 0.2340$$

where N = 10, x = 5, p = 0.45, q = 0.55 .

➤ **(b).**
The event "4 students are not majoring in Engineering" is equal to the event that 6 students are majoring in Engineering. Therefore,

$$P\{X=6\} = \binom{10}{6} (0.45)^6 (0.55)^4 \approx 0.1596$$

➤ **(c).**
The event "between 4 and 6 students are majoring in Engineering" can be written as $\{4 \le X \le 6\} = \{X = 4\} \cup \{X = 5\} \cup \{X = 6\}$.

Therefore, $P\{4 \le X \le 6\} = P\{X = 4\} + P\{X = 5\} + P\{X = 6\} =$

$$\binom{10}{4} (0.45)^4 (0.55)^6 + \binom{10}{5} (0.45)^5 (0.55)^5 + \binom{10}{6} (0.45)^6 (0.55)^4 \approx$$

0.2384 + 0.2340 + 0.1596 = 0.6320

➤ **(d).**
The event "at least one student is majoring in Engineering" can be written as $\{1 \le X\} = \{X = 0\}'$. Therefore,

$$P\{1 \le X\} = P\{X = 0\}' = 1 - P\{X = 0\} = \binom{10}{0}(0.45)^0(0.55)^{10} \approx 1 - 0.0025 = 0.9975$$

The event "at most nine students are majoring in Engineering" can be written as $\{X \le 9\} = \{X = 10\}'$. Therefore,

$$P\{X \le 9\} = P\{X = 10\}' = 1 - P\{X = 10\} = 1 - \binom{10}{10}(0.45)^{10}(0.55)^0 \approx 1 - 0.0003 = 0.9997$$

20.1 - Example 3: Each day a large manufacturer of computer disks receives shipments of several boxes of computer chips. Before installing these chips into their computers, a random sample of 10 chips from each box is tested for defects. If among these ten chips two or more chips are found defective, the entire box is returned to the manufacturer. On a certain day, five boxes were received of which 5% of the chips were defective. Find the probability that none of the boxes were returned.

Solution:

Step 1: First we must find the probability that a box of chips will be returned. This is a binomial distribution where X is the number of chips found defective, N = 10, and p = 0.05.

Step 2: E: the event that at least two of the chips tested are found defective.

E′: the event that at most one chip tested is found defective.

$$P(E') = P\{X=0\} + P\{X=1\} = \binom{10}{0}(0.05)^0(0.95)^{10} + \binom{10}{1}(0.05)^1(0.95)^9 \approx 0.60 + 0.32 = 0.92$$

P(E) = 1 - P(E') ≈ 1 - 0.92 = 0.08, the probability that a box will be returned.

Step 3: To find the probability that none of the boxes were returned, let X be the number of boxes returned. Since it is reasonable to assume the return of the boxes are independent of each other, we assume a binomial distribution where N = 5, p = 0.08 and k = 0.

$$P\{X = 0\} = \binom{5}{0}(0.08)^0(0.92)^5 \approx 0.66$$

Solved Problems

20.1 - Solved Problem 1: Assume a die is tossed 8 times. Write out the binomial distribution where the random variable is the number of threes or twos that appear.

Solution:

Since there are eight tosses of the die: n = 8.

Each toss of the die is independent.

Success on each trial is defined as a three or two appears: p = 2/6 = 1/3.

Failure on each trial is defined as neither a two nor a three appears: q = 1 - 1/3 = 2/3.

The number of possible successes range from zero to 8: k = 0, 1, 2, 3, 4, 5, 6, 7, 8.

The distribution of this binomial distribution is

$$P\{X = 0\} = \binom{8}{0}\left(\frac{1}{3}\right)^0\left(\frac{2}{3}\right)^8 = \frac{256}{6561}$$

$$P\{X = 1\} = \binom{8}{1} \left(\frac{1}{3}\right)^1 \left(\frac{2}{3}\right)^7 = \frac{1024}{6561}$$

$$P\{X = 2\} = \binom{8}{2} \left(\frac{1}{3}\right)^2 \left(\frac{2}{3}\right)^6 = \frac{1792}{6561}$$

$$P\{X = 3\} = \binom{8}{3} \left(\frac{1}{3}\right)^3 \left(\frac{2}{3}\right)^5 = \frac{1792}{6561}$$

$$P\{X = 4\} = \binom{8}{4} \left(\frac{1}{3}\right)^4 \left(\frac{2}{3}\right)^4 = \frac{1120}{6561}$$

$$P\{X = 5\} = \binom{8}{5} \left(\frac{1}{3}\right)^5 \left(\frac{2}{3}\right)^3 = \frac{448}{6561}$$

$$P\{X = 6\} = \binom{8}{6} \left(\frac{1}{3}\right)^6 \left(\frac{2}{3}\right)^2 = \frac{112}{6561}$$

$$P\{X = 7\} = \binom{8}{7} \left(\frac{1}{3}\right)^7 \left(\frac{2}{3}\right)^1 = \frac{16}{6561}$$

$$P\{X = 8\} = \binom{8}{8} \left(\frac{1}{3}\right)^8 \left(\frac{2}{3}\right)^0 = \frac{1}{6561}.$$

20.1 - Solved Problem 2: Records kept by a local high school showed that 20% of the graduating students intend to go on to a four-year college, 50% to a two-year college and the remaining students intend not to go to an institution of higher learning. A survey of 10 students are taken. Find the probability that

(a). exactly 4 students intend to go to a four-year college.

(b). 7 students do not intend to go to a two-year college.

(c). between 4 and 7 students will go to a four-year or two-year college.

(d). at least 2 students will go to a two-year college.

(e). at most 7 students will go not go to a two-year or four-year college.

Solutions:

This problem contains three binomial distributions for the following reasons:

1. Each student interviewed is a trial. Since 10 students are interviewed, the number of trials is

N = 10.

2. It is reasonable, since the students were selected at random from a large student body, that the selections are independent.

3. Success is defined depending on the stated problem. However, in each case there are only two choices: success or failure.

4. The probability of a student intending to go to a four-year college is p = 0.20.

The probability of a student intending to go to a two year college is p = 0.50. And the probability of a student not intending to go to an institution of higher learning is p = 1 - 0.20 - 0.50 = 0.30 .

➤ **(a).**
The random variable X is the number of students intending to go to a four year-college. Therefore,

$$P\{X = 4\} = \binom{10}{4} (0.20)^4 (0.80)^6 = 0.088$$

where n = 10, x = 4, p = 0.20 and q = 0.80

➤ **(b).**
X is the random variable representing the number of students intending to go to a two-year college. The event "7 students do not intend to go to a two- year college" can be written as the event 3 intend to go to a two-year college: $\{X = 3\}$. Therefore

$$P\{X = 3\} = \binom{10}{3} (0.50)^3 (0.50)^7 \approx 0.1172.$$

➤ **(c).**
The random variable X is the number of students that intend to go to a four-year or two-year college. The probability of success p = 0.70. The event "between 4 and 7 students intending to go to a four or two-year college" can be written as

$$\{4 \le X \le 7\} = \{X = 4\}\cup\{X = 5\}\cup\{X = 6\}\cup\{X = 7\}.$$

Therefore, $P\{4 \le X \le 7\} = P\{X = 4\} + P\{X = 5\} + P\{X = 6\} + P\{X = 7\} =$

$$\binom{10}{4} (0.70)^4(0.30)^6 + \binom{10}{5} (0.70)^5(0.30)^5 + \binom{10}{6} (0.70)^6(0.30)^4 + \binom{10}{7}(0.70)^7(0.3)^3 =$$

0.0368 + 0.1029 + 0.2001 + 0.2668 = 0.6066.

➤ **(d).**
The random variable is the number of students going to a two year college. Here p = 0.50. The event "at least two students intends to go to two year college" can be written as

$\{2 \le X\} = \{X \le 1\}'$ and $\{X \le 1\} = \{X = 0\} \cup \{X = 1\}$. Therefore,

$P\{2 \le X\} = P\{X \le 1\}' = 1 - P\{X \le 1\} = 1 - P\{\{X = 0\} \cup \{X = 1\}\} = 1 - P\{X = 0\} - P\{X = 1\} =$

$1 - \binom{10}{0}(0.5)^0(0.5)^{10} - \binom{10}{1}(0.5)^1(0.5)^9 \approx 1 - 0.0010 - 0.0098 = 0.98920.$

➤ **(e).**

Here the random variable X is the number of students not intending to go to a two or four-year college. For X we have $p = 0.30$. The event "at most 7 students will go not go to a two-year or four-year college" can be written as $\{X \le 7\} = \{8 \le X\}'$. We can write the event: $\{8 \le X\} = \{X = 8\} \cup \{X = 9\} \cup \{X = 10\}$.

Therefore,

$P(\{X \le 7\}) = P(\{8 \le X\}') = 1 - P\{8 \le X\} = 1 - P(\{X = 8\} \cup \{X = 9\} \cup \{X = 10\}) =$

$1 - P(\{X = 8\}) - P(\{X = 9\}) - P(\{X = 10\}) =$

$1 - \binom{10}{8}(0.3)^8(0.7)^2 - \binom{10}{9}(0.3)^9(0.7)^1 - \binom{10}{10}(.3)^{10}(0.7)^0 \approx 1 - 0.0014 - 0.0001 - 0.00$

$= 0.9985$

20.1 - Solved Problem 3: To decide if new legislation is needed to reduce air pollution in 5 California cities, twenty automobiles from each city was randomly selected and tested for auto emissions. If five or more of these automobiles emit excessive pollutants, then the city is designated for stricter regulations. Assume that 10% of all automobiles in California emit excessive pollutants, find the probability that at least two cities will be required to have stricter regulations.

Solution:

Step 1: First we must find the probability that in a given city five or more of the randomly selected cars will emit excessive pollutants. This is a binomial distribution where X is the number of automobiles that emit excessive pollutants, $N = 20$, and $p = 0.10$.

Step 2: E: the event that five or more of these automobiles emit excessive pollutants.

E′: the event that at most four of these automobiles emit excessive pollutants.

$P(E') = P\{X = 0\} + P\{X = 1\} + P\{X = 2\}P\{X = 3\} + P\{X = 4\} =$

$$\binom{20}{0}(0.10)^0(0.90)^{20} + \binom{20}{1}(0.10)^1(0.90)^{19} + \binom{20}{2}(0.10)^2(0.90)^{18} + \binom{20}{3}(0.10)^3(0.90)^{17}$$

$$+ \binom{20}{4}(0.10)^4(0.90)^{16} \approx 0.12 + 0.27 + 0.29 + 0.19 + 0.09 = 0.96$$

Therefore,

$P(E) = 1 - P(E') \approx 1 - 0.96 = 0.04$, the probability that a city will be designated for stricter regulations.

Step 3: To find the probability that at least two cities will be required to have stricter regulation let X be the number of cities designated for stricter pollution control. Since it is reasonable to assume the cities are independent of each other, we assume a binomial distribution where $N = 5$, $p = 0.04$ and $P\{X \geq 2\}$.

Step 4: $P\{X \geq 2\} = 1 - P\{X \leq 1\} = 1 - P\{X = 0\} - P\{X = 1\} =$

$$1 - \binom{5}{0}(0.04)^0(0.96)^5 - \binom{5}{1}(0.04)^1(0.96)^4 \approx 1 - 0.82 - 0.17 = 0.01$$

Unsolved Problems with Answers

20.1 - Problem 1: A machine fills 16 liquid ounces of cola in 18 ounce bottles. A study shows that the probability of it filling a bottle with more than 16 ounces is $p = 0.10$. A sample of five bottles are checked. For the random variable X, write out the binomial distribution.

Answer:

$P\{X = 0\} \approx 0.5905$
$P\{X = 1\} \approx 0.3281$
$P\{X = 2\} \approx 0.0729$
$P\{X = 3\} \approx 0.0081$
$P\{X = 4\} \approx 0.0005$
$P\{X = 5\} \approx 0.00001$

⇑ *Refer back to* **20.1 - Example 1 & 20.1 - Solved Problem 1**.

20.1 - Problem 2: A new medical test on men has a 80% chance of detecting positive coronary blockage. Fifteen men are tested. Find the probability that

(a). exactly 12 men are tested positive.

(b). 10 men are tested negative.

(c). between 7 and 10 tested negative.

(d). at least 1 man tested positive.

(e). at most 13 men tested negative.

Answers:

➤ **(a)**. 0.2501

➤ **(b)**. 0.0001

➤ **(c)**. 0.0181

➤ **(d)**.1

➤ **(e)**. 1

⇑ *Refer back to* **20.1 - Example 2 & 20.1 - Solved Problem 2**.

20.1 - Problem 3: Mr. Slager teaches a statistics class at a local college. During the semester he gives 8 multiple choice exams. Each exam consists of 10 questions where each question has 4 possible choices. To pass a test, a student must answer at least 70% of the questions correctly. A student decides to randomly guess all the questions. Find the probability that he will fail all of Mr. Slager's exams.

Answer:

0.97

⇑ *Refer back to* **20.1 - Example 3 & 20.1 - Solved Problem 3**.

20.2 - The Mean and Variance of the Binomial Distribution

The Binomial Distribution has simple formulas for the mean (μ) and the variance (σ^2)

The mean of the Binomial Distribution:

$\mu = E(X) = Np$,

where N is the number of trials and p is the probability of success on each trial.

The variance of the Binomial Distribution:

$\sigma^2 = npq$,

where N is the number of trials, p is the probability of success on each trial and q = 1 - p is the probability of failure on each trial.

20.2 - Example 1: A pair of dice are tossed 120 times. Find the expected number of sevens and the variance.

Solution:

Each toss of the pair of dice is a trial. Therefore N = 120. Success is defined for each trial as getting a sum of seven. The chance of getting a seven on a single toss of a pair of dice is p = 6/36 = 1/6. Therefore, the average number of sevens expected on 120 tosses of a pair of dice is $\mu = Np$ = (120)(1/6) = 20 sevens.

The variance formula is $\sigma^2 = npq$ = (120)(1/6)(5/6) = 50/3.

Solved Problem

20.2 - Solved Problem 1: In a large western state, voter registration records show that 45% of voters are registered Republicans. Ten voters are interviewed. Find the average number of Republicans and the variance.

Solution:

The number of trials represents the number of interviews: N = 10.

Success is defined as a voter interviewed is a Republican: p = 0.45 .

Failure is defined as a voter interviewed is not a Republican: q = 1-0.45 = 0.55 .

The average number of Republicans interviewed: $\mu = Np$ = (10)(0.45) = 4.5 .

Variance of the number of Republicans interviewed: $\sigma^2 = npq$ = (10)(0.45)(0.55) = 2.475 .

Unsolved Problems with Answers

20.2 - Problem 1: An editor of a major publishing company estimates that due to typing errors, 5% of words on a manuscript are misspelled. Assume a manuscript consisting of 120,000 words is submitted for publication, find the average number of misspelled words and the variance.

Answer:

$\mu = 6{,}000$, the average misspelled words

$\sigma^2 = 5{,}700$

⇑ *Refer back to* **20.2 - Example 1 & 20.2 - Solved Problem 1**.

Supplementary Problems

1. Mr. Jones loves to play five-card poker. Assume he plays 6 hands. Find the probability that he drew three hands each consisting of exactly two spades.

2. From problem 1, find the minimum number of hands he would have to play to assure the chance is greater than 0.80 that he gets at least two hands each containing exactly two spades.

3. From problem 1, find the number of hands he would have to play to assure the maximum chance of getting four hands each containing two spades.

4. A national auto manufacturer has contracts with two companies that manufacturer transmissions. One company is on the East coast and provides 70% of the transmissions. The other company is on the West coast. Studies show that the percent of defective transmissions from the East coast is 5% and those from the West coast is 7%. Five transmissions are selected at random. Find the probability that at least one of the five selected is defective.

5. Assume a multinomial sample space is generated with more than three binomial random variables with probabilities $p_1, p_2, p_3, ..., p_r$, where

$p_1 + p_2 + p_3 + ... + p_r = 1$. The formula for the multinomial distribution of these binomial variables is

$$P\{X_1 = k_1; X_2 = k_2; ...; X_r = k_r\} = \binom{N}{k_1}\binom{N_1}{k_2}...\binom{N_{r-1}}{k_r} p_1^{k_1} p_2^{k_2} p_3^{k_3}...p_r^{k_r}$$

where $N_1 = N - k_1$, $N_2 = N - k_1 - k_2$, ... $N_{r-1} = N - k_1 - k_2 - ... - k_{r-1}$ and $k_1 + k_2 + ... + k_r = N$.

Assume a die is tossed twelve times. Find the probability that all six numbers appear twice.

6. The voter registration records of a large western state shows the following break-down of voters' political affiliation:

Party	Percent of Registered Voters
Democrat	35%
Republican	40%
Independent	15%
Libertarian	7%
Green	3%

Ten registered voters are interviewed. Find the probability that:

a. Three are Democrats, four are Republicans, two are Libertarians and 1 is Green.

b. Six are Republicans.

c. Three are Democrats.

d. Six are Republicans or three are Democrats.

7. A red and a blue die are both tosses ten times. Find the probability that the red die results in 5 even numbers or the blue die results in 3 odd numbers.

8. A fair die is tossed until the number 3 occurs. Let X equal the number of tosses to terminate this game.

a. Is the distribution of X binomial? Explain.

b. Find the probability that it takes more than 4 tosses to terminate this game.

9. Using the formula

$$\binom{10}{0} a^0 b^{10} + \binom{10}{2} a^2 b^8 + \binom{10}{4} a^4 b^6 + \dots + \binom{10}{10} a^{10} b^0 = \frac{1}{2}\{(a + b)^{10} + (a - b)^{10}\},$$

we can solve the following problems: A fair die is tossed 20 times.

a. Find the probability that an even number of 3s' occur.

b. Find the probability that an odd number of 3s' occur.

10. Each day a large manufacturer of computer disks receives shipments of several boxes of computer chips. Before installing these chips into their computers, a random sample of 10 chips

from each box is tested for defects. If among these ten chips two or more chips are found defective, the entire box is returned to the manufacturer. If a given box has p% chips defective, find p so that the probability the box is returned is 0.01.

11. Each day a large manufacturer of computer disks receives shipments of several boxes of computer chips. Before installing these chips into their computers, a random sample of n chips from each box is tested for defects. If among these ten chips two or more chips are found defective, the entire box is returned to the manufacturer. If a given box has 5% chips defective, find n so that the probability the box is returned is 0.01.

12. A computer randomly sorts numbers 1 to N and prints them in a single row. If any given number is printed in its natural position, then we will say the number is a match. For example if N = 10 and the random sequence printed is 4, 8, $\underline{3}$, 10, 6, 9, $\underline{7}$, 1, 2, 5 then we have two matches as underlined.

Let (X_k = 1) be the event if number k is a match and {X_k = 0} be the event if number k is not a match (k = 1,..., N).

a. Find $P(X_k = 1)$.

b. Find $E(X_k)$.

c. Find $\sigma^2_{X_k}$.

d. Does this sequence of random variables have a binomial distribution? Explain.

e. Let $S = X_1 + ... + X_N$. Interpret S.

f. E(S)

g. Find σ^2_S.

h. Let $\overline{P} = \dfrac{S}{N}$. Find $E(\overline{P})$ and $\sigma^2_{\overline{P}}$.

13. Assume a binomial experiment with N independent trials where p is the probability of success on each trial.

a. Show $\mu = Np$

b. Show $\sigma = \sqrt{Np(1 - p)}$.

14. Assume X and Y are mutually independent binomial distributions where

$$P(X = k) = \binom{n}{k} p^k q^{n-k} \text{ and } P(Y = k) = \binom{m}{k} p^k q^{n-k}.$$

Assume we define $\binom{N}{k} = 0$, for $k > N$.

Show $Z = X + Y$ has a binomial distribution, where $P(Z = k) = \binom{n+m}{k} p^k q^{n+m-k}$, for $0 \le k \le n + m$. (Hint: See Lesson 18, problem 8)

Hypergeometric distribution approximation to the binomial distribution. Assume N balls are in an urn where each ball is marked 1 to N respectively. Assume n balls are red and N - n are black. We randomly selected r balls and define the random variable X to be the number of red marbles selected. Since there are 2 ways to randomly selected the balls, we have the following:

i. Randomly select r balls <u>without</u> replacement. This type of sampling results in the hypergeometric distribution

of X: $P(X = k) = \dfrac{\binom{n}{k}\binom{N-n}{r-k}}{\binom{N}{r}}$, $0 \le k \le r$.

ii. Randomly select r balls <u>with</u> replacement. This type of sampling results in the binomial distribution of X:

$$P(X = k) = \binom{r}{k} p^k q^{r-k}, \text{ where } p = n/N \text{ and } q = 1 - p.$$

For large N, it can be shown that $\dfrac{\binom{n}{k}\binom{N-n}{r-k}}{\binom{N}{r}} \approx \binom{r}{k} p^k q^{r-k}.$

15.
a. Assume 7 cards are randomly selected without replacement from an ordinary deck of cards. Find the probability that the hand will contain at least 1 diamond.

b. Assume 7 cards are randomly selected with replacement from an ordinary deck of cards. Find the probability that the hand will contain at least 1 diamond.

c. Find the difference between the above probabilities.

Assume 7 cards are randomly selected without replacement from multiple deck made up of 5 ordinary deck of cards.

d. Assume 7 cards are randomly selected without replacement from this multiple deck . Find the probability that the hand will contain at least 1 diamond.

e. Assume 7 cards are randomly selected with replacement from this multiple deck . Find the probability that the hand will contain at least 1 diamond.

f. Find the difference between these probabilities.

g. What important conclusion can one come to from the result of question f ?

16. From 2 ordinary deck of cards, 5 cards are dealt from each deck without replacement. Let X be the random variable that equals the number of kings dealt from one of the deck of cards and Y be the random variable that equals the number of kings dealt from the other deck of cards. Write out the distribution of $P(Z = X + Y = k)$ for $0 \le k \le 8$.

17. From problem 16, if the total number of kings drawn was 4, find the probability that 2 cards were drawn from each deck.

18. Assume X and Y are mutually independent binomial distributions. If

$$P(X = k) = \binom{N}{k} p_1^{k} q_1^{r-k}$$

and $P(Y = k) = \binom{M}{k} p_2^{k} q_2^{r-k}$, find the distribution for $Z = X + Y$.

19. *Quality Control Sampling.* The ABC Transmission Corp. has 3 machines that each produce an important part to an automobile transmission system. A quality control study shows that machine A produces 3% defective parts; machine B produces 5% defective parts and machine C produces 1% defective parts. Each hour machine A produces 60 parts; machine B produces 40 parts and machine C produces 50 parts. Assuming that the output of each machine are mutually independent and quality produced by each machine are also independent, find the probability for each hour there is a total of at least 2 defective parts.

20. From problem 18, assume the total output produced at least 2 defective parts. Find the probability that all the defective parts came from machine A .

21. Assume X and Y are independent binomial distributions where
$$P(X = k) = P(Y = k) = \binom{r}{k} p^{k} q^{r-k}.$$

Show that $P(X = k | X + Y = N)$ has a hypergeometric distribution. (Hint: See Problem 14.)

22. Assume r cards are drawn without replacement from an ordinary deck of cards. Find the minimum number r so that there is a 90% chance that at least 2 diamonds are drawn. (Hint: see Problem 15).

23. A journalist at a large college wishes to interview 10 math majors. Assume 5% of the student body are majoring in math, how many students must he sampled from the student body to have a 90% chance to find at least 10 math majors among his/her sample? (Hint: see Problem 22).

Pascal-Negative Binomial Distribution. Assume we have an experiment of independent trials with following rules:

i. The probability of success equal to p for each trial and $q = 1 - p$ for each failure.

ii. The trials terminate once $r > 0$ successes occur.

Let $(X = n)$ be the random variable that equals the numbers of trials n, necessary to terminate the experiment.

The Pascal-negative binomial distribution is $P(X = n) = \begin{pmatrix} n - 1 \\ r - 1 \end{pmatrix} p^r q^{n - r}$ where $n \geq r > 0$.

For the remaining questions, use the Pascal-negative binomial distribution.

24. A fair coin is tossed until 5 heads occur. Find the probability that the coin is tossed 10 times.

25. On an automobile assembly line a robot checks each car's breaking system for defects. It is estimated that there is a 5% chance the robot will detect a defect. If the robot finds 2 defective break systems, the assembly line will be stopped and appropriate adjustments will be made to the manufacturing of the break system.
Find the probability that the assembly line will be shut down on a run of

a. 10 cars.

b. 20 cars .

26. A fair coin is tossed until five heads occur or ten tosses of the coin whichever occurs first.

a. Find the expected number of tosses.

b. Find the probability that five heads will not occur.

27. Explain the difference, if any, between the sample space generated by the Pascal distribution

$$P(X = n) = \binom{n-1}{r-1} p^r q^{n-r} \quad n \geq r > 0,$$ and the sample space generated by the binomial distribution

$$P(Y = k) = \binom{n}{k} p^k q^{n-k}.$$

Since the formula for the binomial distribution is frequently difficult to compute, tables have been developed containing the probability values of the binomial distribution for selected values of p and N for P{X ≥ x}. In this lesson, we present the binomial table (Table A, back of book) for N = 20 and p = 0.05, 0.10, 0.15, ..., 0.75, 0.80, 0.85, 0.90, 0.95 .

21.1 - Using the Binomial Distribution Table.

21.1 - Example 1: Assume p = 0.35. Find P{X ≥ 5}.

Solution:

Step 1: Since p = 0.35, go to the column in bold:

Step 2: Since x = 5, we go along the left column to the value in bold. From this value, we connect to p = 0.35:

Table A: Cumulative binomial distribution: P{X ≥ x} (partial table)

p =	0 .05	0.10	0.15	0.20	0.25	0.30	0.35	0.40	0.45	0.50
x										
0	1.0	1.0	1.0	1.0	1.0	1.0	1.0	1.0	1.0	1.0
1	0.6415	0.8784	0.9612	0.9885	0.9968	0.9992	0.9998	1.0	1.0	1.0
2	0.2642	0.6083	0.8244	0.9308	0.9757	0.9924	0.9979	0.9995	0.9999	1.0
3	0.0755	0.3231	0.5951	0.7939	0.9087	0.9645	0.9879	0.9964	0.9991	0.9998
4	0.0159	0.1331	0.3523	0.5886	0.7748	0.8929	0.9556	0.9840	0.9951	0.9987
5	0.0026	0.0432	0.1702	0.3704	0.5852	0.7625	0.8818	.9490	0.9811	0.9941

Therefore, from the table P{X ≥ 5 } = 0.8818 .

21.1 - Example 2: Assume p = 0.70 . Find P{X ≥ 8}.

Solution:

Step 1: Since p = 0.70, go to the column in bold:

Step 2: Since x = 8, go along the left column to the value in bold. From this value we connect to p = 0.70:

Table A: Cumulative binomial distribution P{X ≥ x} (partial table):

p =	0.55	0.60	0.65	0.70	0.75	0.80	0.85	0.90	0.95
x									
0	1	1	1	1	1	1	1	1	1
1	1	1	1	1	1	1	1	1	1
2	1	1	1	1	1	1	1	1	1
3	1	1	1	1	1	1	1	1	1
4	0.9997	0.9999	1	1	1	1	1	1	1
5	0.9985	0.9997	1	1	1	1	1	1	1
6	0.9936	0.9984	0.9997	1	1	1	1	1	1
7	0.9786	0.9935	0.9985	0.9997	1	1	1	1	1
8	0.9420	0.9790	0.9940	0.9987	0.9998	1	1	1	1

Therefore from the table P{X ≥ 8} = 0.9987 .

21.1 - Example 3: Assume p= 0.35. Find P{5 ≤ X≤ 7}.

Solution:

It is not possible to find P{5 ≤ X ≤ 7} directly from table A. However,

{X ≥ 5} = {5 ≤ X≤ 7}∪{X ≥ 8}.

Therefore, P{X ≥ 5} = P{5 ≤ X ≤ 7} + P{X ≥ 8} and it follows

P{5 ≤ X ≤ 7} = P{X ≥ 5} - P{X ≥ 8}= 0.8818 - 0.3990 = 0.4828 .

21.1 - Example 4: Assume p = 0.35. Find P{X = 5}.

Solution

It is not possible to find P{X = 5} directly from table A. However,

$\{X \geq 5\} = \{X = 5\} \cup \{X \geq 6\}$.

Therefore, $P\{X \geq 5\} = P\{X = 5\} + P\{X \geq 6\}$ and it follows

$P\{X = 5\} = P\{X \geq 5\} - P\{X \geq 6\} = 0.8818 - 0.7546 = 0.1272$.

21.1 - Example 5: Assume $p = 0.35$. Find $P\{X \leq 4\}$.

Solution:

It is not possible to find $P\{X \leq 4\}$ directly from table A.

However, $P\{X \leq 4\} = 1 - P\{X \geq 5\} = 1 - 0.8818 = 0.1182$.

Solved Problems

21.1 - Solved Problem 1: Assume $p = 0.25$. Find $P\{X \geq 7\}$.

Solution:

Step 1: Since $p = 0.25$ go to the column in bold:

Step 2: Since $x = 7$, we go along the left column in bold. From this value we connect to $p = 0.25$

p =	0.05	0.10	0.15	0.20	0.25	0.30	0.35	0.40	0.45	0.50
x										
0	1.0	1.0	1.0	1.0	1.0	1.0	1.0	1.0	1.0	1.0
1	0.6415	0.8784	0.9612	0.9885	0.9968	0.9992	0.9998	1.0	1.0	1.0
2	0.2642	0.6083	0.8244	0.9308	0.9757	0.9924	0.9979	0.9995	0.9999	1.0
3	0.0755	0.3231	0.5951	0.7939	0.9087	0.9645	0.9879	0.9964	0.9991	0.9998
4	0.0159	0.1331	0.3523	0.5886	0.7748	0.8929	0.9556	0.9840	0.9951	0.9987
5	0.0026	0.0432	0.1702	0.3704	0.5852	0.7625	0.8818	0.9490	0.9811	0.9941
6	0.0003	0.0113	0.0673	0.1958	0.3828	0.5836	0.7546	0.8744	0.9447	0.9793
7	0.0000	0.0024	0.0219	0.0867	0.2142	0.3920	0.5834	0.7500	0.8701	0.9423

Therefore, from the table $P\{X \geq 7\} = 0.2142$.

21.1 - Solved Problem 2: Assume $p = 0.25$. Find $P\{X \geq 10\}$.

Solution:

Step 1: Since p = 0.25 go to the column in bold:

Step 2: Since x = 10, we go along the left column in bold. From this value we connect to p = 0.25
.

p=	0 .05	0.10	0.15	0.20	0.25	0.30	0.35	0.40	0.45	0.50
x										
0	1.0	1.0	1.0	1.0	1.0	1.0	1.0	1.0	1.0	1.0
1	0.6415	0.8784	0.9612	0.9885	0.9968	0.9992	0.9998	1.0	1.0	1.0
2	0.2642	0.6083	0.8244	0.9308	0.9757	0.9924	0.9979	0.9995	0.9999	1.0
3	0.0755	0.3231	0.5951	0.7939	0.9087	0.9645	0.9879	0.9964	0.9991	0.9998
4	0.0159	0.1331	0.3523	0.5886	0.7748	0.8929	0.9556	0.9840	0.9951	0.9987
5	0.0026	0.0432	0.1702	0.3704	0.5852	0.7625	0.8818	0.9490	0.9811	0.9941
6	0.0003	0.0113	0.0673	0.1958	0.3828	0.5836	0.7546	0.8744	0.9447	0.9793
7	0.0000	0.0024	0.0219	0.0867	0.2142	0.3920	0.5834	0.7500	0.8701	0.9423
8	0.0000	0.0004	0.0059	0.0321	0.1018	0.2277	0.3990	0.5841	0.7480	0.8684
9	0.0000	0.0001	0.0013	0.0100	0.0409	0.1133	0.2376	0.4044	0.5857	0.7483
10	0.0000	0.0000	0.0002	0.0026	0.0139	0.0480	0.1218	0.2447	0.4086	0.5881

Therefore, from the table, $P\{X \geq 10\} = 0.0139$.

21.1 - Solved Problem 3: Assume p= 0.25 . Find $P\{10 \leq X \leq 12\}$.

Solution:

It is not possible to find $P\{10 \leq X \leq 12\}$ directly from table A. However,

$\{X \geq 10\} = \{10 \leq X \leq 12\} \cup \{X \geq 13\}$.

Therefore, $P\{10 \leq X\} = P\{10 \leq X \leq 12\} + P\{13 \leq X\}$ and it follows

$P\{10 \leq X \leq 12\} = P\{X \geq 10\} - P\{X \geq 13\} = 0.0139 - 0.0002 = 0.0137$.

21.1 - Solved Problem 4: Assume p = 0.25. Find $P\{X = 10\}$.

Solution:

It is not possible to find $P\{X = 10\}$ directly from table A. However,

$\{X \geq 10\} = \{X = 10\} \cup \{X \geq 11\}$

Therefore, $P\{X \geq 10\} = P\{X = 10\} + P\{X \geq 11\}$ and it follows

$P\{X = 10\} = P\{X \geq 10\} - P\{X \geq 11\} = 0.0139 - 0.0039 = 0.01$.

21.1 - Solved Problem 5: Assume $p = 0.25$. Find $P\{X \leq 9\}$.

Solution:

It is not possible to find $P\{X \leq 9\}$ directly from table A.

However, $P\{X \leq 9\} = 1 - P\{X \geq 10\} = 1 - 0.0139 = 0.9861$.

Unsolved Problems with Answers

21.1 - Problem 1: Assume $p = 0.50$. Find $P\{X \geq 15\}$.

Answer:

0.0207

⇑ *Refer back to* **21.1 - Example 1 & 21.1 - Solved Problem 1**.

21.1 - Problem 2: Assume $p = 0.10$. Find $P\{X \geq 2\}$.

Answer:

0.6083

⇑ *Refer back to* **21.1 - Example 2 & 21.1 - Solved Problem 2**.

21.1 - Problem 3: Assume $p = 0.20$. Find $P\{1 \leq X \leq 10\}$.

Answer:

0.9879

⇑ *Refer back to* **21.1 - Example 3 & 21.1 - Solved Problem 3**.

21.1 - Problem 4: Assume $p = 0.40$. Find $P\{X = 15\}$.

Answer:

0.0013

⇑ *Refer back to* **21.1 - Example 4 & 21.1 - Solved Problem 4**.

21.1 - Problem 5: Assume $p = 0.05$. Find $P\{X \leq 10\}$.

Answer:

1.0

⇑ *Refer back to* **21.1 - Example 5 & 21.1 - Solved Problem 5**.

21.2 Real Life Applications

For this section, we use the binomial table, Table A.

21.2 - Example 1: A fair coin is tossed 20 times. Find the probability that

(a). at least 5 heads appear.

(b). exactly 5 heads appears.

(c). less than 5 heads appear.

(d). between 10 and 15 heads appear.

solutions:

➤ **(a).**
Since the coin is fair, we assume that p = 0.50, and since the number of tosses is n = 20, we can use the binomial table. The event, at least 5 heads appear, is $\{X \geq 5\}$ and from the Binomial table where the column is p =0.5 and the row is x = 5, $P\{X \geq 5\} = 0.9941$.

➤ **(b).**
The probability of the event, exactly five head appears, can be written

$P\{X = 5\} = P\{X \geq 5\} - P\{X \geq 6\} = 0.9941 - 0.9793 = 0.0148$.

➤ **(c).**
The event, less than 5 heads appear is $\{X < 5\} = \{X \leq 4\}$. However, to use the table we must write

$P\{X < 5\} = P\{X \leq 4\} = 1 - P\{X \geq 5\} = 1 - 0.9941 = 0.0059$.

➤ **(d).**
The event, between ten and fifteen heads will occur is $\{10 \leq X \leq 15\}$.

$P\{10 \leq X \leq 15\} = P\{X \geq 10 \} - P\{X \geq 16 \} = 0.5881 - 0.0059 = 0.5822$

21.2 - Example 2: According to a national survey, 35% of all registered voters are Republicans. Twenty registered voters are sampled. Find the probability that (a). less than 12 are Republicans.

(b). less than 12 are not Republicans.

(c). more than 11 are not Republicans.

(d). between 12 and 14 are not Republicans.

Solutions:

➤ **(a).**
The event, less than 12 are Republicans is $\{X < 12\} = \{X \leq 11\}$.

To use the Binomial table, we must write $\{X \leq 11\}' = \{X \geq 12\}$.

Using $x = 12$ and $p = 0.35$ in the table,

$P\{X < 12\} = P\{X \leq 11\} = 1 - P\{X \geq 12\} = 1 - 0.0196 = 0.9804$.

➤ **(b).**
The random variable X is defined as the distribution that a voter is not a Republican.

The probability that a person surveyed is not a Republican is $p = 1 - 0.35 = 0.65$.

The event, less than 12 are not Republicans is $\{X < 12\} = \{X \leq 11\}$.

To use the Binomial table, we must write $\{X \leq 11\}' = \{X \geq 12\}$. Using $x = 12$ and $p = 0.65$ in the binomial table,

$P\{X < 12\} = P\{X \leq 11\} = 1 - P\{X \geq 12\} = 1 - 0.7624 = 0.2376$.

➤ **(c).**
The random variable X is defined as the distribution that a voter is not a Republican. The probability that a person surveyed is not a Republican is $p = 1 - 0.35 = 0.65$.

The event, more than 11 are not Republicans is $\{X > 11\} = \{X \geq 12\}$.

From (b). we have $P\{X \leq 11\} = 0.2376$.

$P\{11 < X\} = P\{X \geq 12\} = 1 - P\{X \leq 11\} = 1 - 0.2376 = 0.7624$.

➤ **(d).**
The random variable X is defined as the distribution that a voter is not a Republican.

The probability that a person surveyed is not a Republican is $p = 1 - 0.35 = 0.65$.

The event, that between 12 and 14 are not Republicans is $\{12 \leq X \leq 14\}$.

$$P\{12 \leq X \leq 14\} = P\{X \geq 12\} - P\{X \geq 15\} = 0.7624 - 0.2454 = 0.517 \,.$$

21.2 - Example 3: The MacroStar Computing Company purchases its microchips from a large Northern California computing manufacturer. This company claims that 5% of all the chips it produces are defective. Each day, MacroStar receives boxes each containing 100 microchips. Inspecting each box, MacroStar uses the following decision rule:

A random sample of 20 chips is selected from each box. If two or more chips are defective, the box is rejected and returned to the manufacturing company.

On a given day, 10 boxes are inspected. Find the probability that at most one box is returned.

Solution:

The decision rule states:

Take a sample of 20 chips. If two or more chips are defective then reject the box.

We must first find the probability of getting two or more defective chips given that the chance of any individual chip being defective is 0.05(5%). Therefore we let $N = 20$ and $p = 0.05$. From the binomial tables, we find that the $P\{X \geq 2\} = 0.2642$ which is the probability that a box will be rejected.

Next, we have a new binomial problem stated as the following: If we select $N = 10$ boxes, what's probability that at most one box is returned? Here we let $p = 0.2642$ which is the probability that a box will be rejected. Here X represents the number of boxes rejected.

We need to use the formula $P\{X = x\} = \begin{pmatrix} N \\ x \end{pmatrix} p^x \, q^{N-x} = \begin{pmatrix} 10 \\ x \end{pmatrix} (0.2642)^x (0.7358)^{10-x}$.

To find the probability that at most one box is rejected we need to compute

$$P\{X \leq 1\} = P\{X = 1\} + P\{X = 0\} = \begin{pmatrix} 10 \\ 0 \end{pmatrix} (0.2642)^0 (0.7358)^{10} + \begin{pmatrix} 10 \\ 1 \end{pmatrix} (0.2642)^1 (0.7358)^9 =$$

0.2135, the probability that at most one box will be rejected.

Solved Problems

21.2 - Solved Problem 1: A U.S. Air Force fighter plane fires 20 missiles at different targets. Assuming that the chance that any missile will hit its target is 0.90, find

(a). at least 15 targets are hit.

(b). exactly 15 targets are hit.

(c). less than 15 targets are hit.

(d). between 12 and 17 targets are hit.

Solutions:

➤ **(a).**

Here it is reasonable to assume that the event of any missile hitting its target is independent. We are given that p = 0.90. Since the twenty missiles are fired at targets, we have n = 20 and we can use the binomial table. The probability of the event { X ≥ 15} is the event that at least 15 targets are hit. From the table:

P{ X ≥ 15} = 0.9887 .

➤ **(b).**

The probability of the event exactly fifteen targets are hit can be written

P{X = 15} = P{X ≥ 15} - P{X ≥ 16} = 0.9887 - 0.9568 = 0.0319 .

➤ **(c).**

The event, less than 15 targets are hit is {X < 15} = {X ≤ 14}. However, to use the table we must write

P{X < 15} = P{X ≤ 14} = 1 - P{X ≥ 15} = 1 - 0.9887 = 0.0113 .

➤ **(d).**

The event, between 12 and 17 targets are hit is {12 ≤ X ≤ 17}. Therefore,

P{12 ≤ X ≤ 17} =P{X ≥ 12} - P{X ≥ 18} = 0.9999 - 0.6769 = 0.323 .

21.2 - Solved Problem 2: Mr. Jones believes that he has a 60% chance of picking the winner of a football game. Assume over a month, he selects 20 teams. Find the chance that

(a). he picks the winners of less than half the games.

(b). he loses less than half the games.

(c). he loses more than 15 games.

(d). he loses between 10 and 15 games.

Solutions:

➤ **(a).**

The event, he picks the winner of less than half the games is $\{X < 10\} = \{X \le 9\}$.

To use the Binomial table we must write $\{X \le 9\}' = \{X \ge 10\}$.

Using $x = 10$ and $p = 0.60$ in the table,

$P\{X < 10\} = P\{X \le 9\} = 1 - P\{X \ge 10\} = 1 - 0.8725 = 0.1275$.

➤ **(b).**
The random variable X is defined as the distribution that he loses a game. For this random variable,
$p = 1 - 0.60 = 0.40$.

The event that he loses less than half the games is $\{X < 10\} = \{X \le 9\}$.

Therefore $P\{X < 10\} = P\{X \le 9\} = 1 - P\{X \ge 10\} = 1 - 0.2447 = 0.7553$.

➤ **(c).**
The random variable X is defined as the distribution that he loses a game. For this random variable,
$p = 0.40$.
The event, he loses more than 15 games is $\{X > 15\} = \{X \ge 16\}$.

From the table $P\{X > 15\} = P\{X \ge 16\} = 0.0003$

➤ **(d).**
The event he loses between 10 and 15 games is $\{10 \le X \le 15\}$. For this random variable,
$p = 1 - 0.60 = 0.40$.

$P\{10 \le X \le 15\} = P\{X \ge 10\} - P\{X \ge 16\} = 0.2447 - 0.0003 = 0.2444$

21.2 - Solved Problem 3: The ClearWater Bottling Company each day fills 10,000 bottles with 16 ounces of spring water. Due to the imperfections of the filling machinery, 10% of all bottles are filled with less than 16 ounces. Government regulations require that no more than 3% of all bottles contain less than 16 ounces. To enforce these rules, government inspectors will use the following decision rule:
Select at random 20 bottles from the production line. If one or more of these bottles contains less than 16 ounces, then the production line is shut down.

For five different inspections, find the probability that it will be shut down exactly three times.

Solution:

We know that 10% of all bottles filled will contain less than 16 ounces of spring water. Therefore, we let $p = 0.10$ which will be the probability that a bottle selected at random will contain less than 16 ounces. Now the inspector will select 20 filled bottles. Since the production line will be shut down if at least one bottle contains less than 16 ounces, we need to use the binomial table for N = 20 , p = 0.10 and P{X ≥ 1} where X represents the number of bottles with less than 16 ounces.

From the table we have P{X ≥ 1} = 0.8784, the probability that at least one bottle will contain less than 16 ounces. Next, we find the probability that for five different inspections, the production line will be shut down three times. This can happen if during three of the five inspections, the inspector finds a bottle containing less than 16 ounces.

This is a binomial problem where N = 5, and X is the random variable representing the number of times the production line is shut down. For the probability of shutting down a production line, this can only happen if at least one bottle of the twenty has less than 16 ounces. Therefore p = 0.8784 which is computed above.

Finally we need to compute $P\{X = 3\} = \binom{5}{3}(0.8784)^3\ (0.1216)^2\ = 0.1002$, the probability that

three out of five inspections will cause the production line to be shut down.

Unsolved Problems With Answers

21.2 - Problem 1: A recent statistical report of the Association of Fire Chiefs shows that over the last twenty years 5% of all fires reported by the public have been false alarms. Recently 20 fires were reported. Find the probability that

(a). at least 2 of these 20 reports were false alarms.

(b). exactly 5 were false alarms.

(c). less than 5 are false alarms.

Answers:

➤ **(a).** 0.2642

➤ **(b).** 0.0023

➤ **(c).** 0.9974

⇑ *Refer back to* **21.2 - Example 1 & 21.2 - Solved Problem 1**.

21.2 - Problem 2: Mrs. Smith teaches a class in Modern American Literature. Her past records show that over the last five years, 25% of students taking this class earn a final grade of B or better. Using this result, if twenty students are randomly selected find the probability that

(a). less than 12 of these students will receive a grade of B or better.

(b). less than half earn a grade less than a B.

(c). more than 12 students earn a grade less than a B.

(d). between 13 and 16 students earn a grade of B or better.

Answers:

➤ **(a)**. 0.9991

➤ **(b)**. 0.0039

➤ **(c)**. 0.8982

➤ **(d)**. 0.0002

⇑ *Refer back to* **21.2 - Example 2 & 21.2 - Solved Problem 2**.

21.2 - Problem 3: In Mr. Slager's statistics class, 5 multiple-choice exams are given. Each test consists of 20 questions, where the student has 4 possible choices for each question. To get a grade of c or better on a test, at least 10 questions have to be answered correctly. A student decides to randomly select answers on each of the five tests. Out of the 5 tests, find the probability that the student will get a c or better on only one test.

Answer:

0.066

⇑ *Refer back to* **21.2 - Example 3 & 21.2 - Solved Problem 3**.

Supplementary Problems

1. From the Binomial table in this lesson, construct a binomial table for $P\{X \le x\}$ for $N = 20$, $p = 0.5, 0.10, 0.15, ..., 0.50$ and $x = 0, 1, 2, ..., 20$.

2. From the binomial table in this lesson, construct a binomial table for $P\{X = x\}$ for $N = 20$, $p = 0.05, 0.10, 0.15,..., 0.50$ and $x = 0, 1, 2, 4, 5$.

3. The Department of Defense stated in a report that 65% of all Army personnel are over 70" tall and 55% of all Navy personnel are over 70" tall. A sample of 20 Army personal and 20 Navy personnel are randomly selected. Using the Binomial tables, find the probability that

a. exactly 10 Army personnel are over 70" tall.

b. at most 11 Navy personnel are over 70" tall.

c. at least 9 Army personnel are 70" tall or under.

d. at most 9 Navy and 8 Army personnel are over 70" tall.

e. at most 9 Navy or 8 Army personnel are over 70" tall.

4. Assume a binomial experiment consists of 15 independent trials. Let X be the random variable for the number of successes and Y be the random variable for the number of failures. If $P\{X \le 7\} = 0.33$. Find $P\{Y \le 7\}$.

5. Assume a binomial experiment consists of 20 independent trials. Define X and Y as being two independent binomial random variables if $P[\{X = x\} \cap \{Y = y\}] = P\{X = x\}P\{Y = y\}$.

For these random variables, assume the probability of success for each trial is $p_x = 0.50$ and $p_y = 0.35$ respectively. Define $P\{X = x; Y = y\} = P[\{X = x\} \cap \{Y = y\}]$.

Using the binomial tables, find $P\{X = 12; Y = 10\}$.

6. Assume we need to use the binomial $P\{X \ge x\}$ for values of p not on the table. For this problem we can use the following linear interpellation formula

$$y = y_1 + (\frac{y_2 - y_1)}{p_2 - p_2})(p - p_1).$$

where y_1 and y_2 are the binomial probability values associated with p_1 and p_2 respectively. The value y is the binomial probability value interpellated for the value p. Using this formula and the binomial table, find the best estimate for $P\{X \ge 10\}$ for $p = 0.47$.

7. Assume X is a binomial distribution for n = 20. Find $P\{X \ge 5 | X \le 10\}$ for $p = 0.45$.

There are two versions of the Poisson distribution: one finds the probability of an event over a

---

A random variable X has a Poisson time distribution if it takes on values k = 0, 1, 2,... with probability

$$P\{X = k\} = \frac{(\mu t)^k}{k!}e^{-\mu t}, \quad \text{where}$$

1. X is the random variable that counts the number of occurrences of an event over time t, such as the number of phone calls over 24 hours, the number of automobile accidents over 30 days, etc.

2. t is the time where the number of events occur.

3. $P\{X = k\}$ is the probability that the event occurs k times over the time period t.

4. μ is the average number of times the event occurs over a unit of time.

5. $\mu_t = \mu t$ is the average number of times the event occurs over the time period t.

6. $e \approx 2.718$

---

given time and the other in a given space.

22.1 - The Poisson Time Distribution

To apply the Poisson time distribution, two rules must apply:

1. The occurrence of an event during any sub-interval of time is independent of the occurrence of an event over a non-overlapping sub-interval of time.

2. If the time t is small, the chance of two or more events occurring during this time is essentially zero.

22.1 - Example 1: A time study at a large supermarket showed that on average, 2 customers are served at an express lane over a one-minute period. During a 1-minute period, find the probability that three customers are served at the express lane.

Solution:

Step 1: $\mu_t = \mu t = 2$, the average number of customers served at an express lane over a-one minute period where $t = 1$.

Step 2: $k = 3$, the number of customers occurring over a one-minute period.

Step 3: $P\{X = 3\} = \dfrac{(\mu t)^k}{k!}e^{-\mu t} \approx \dfrac{2^3}{3!}2.718^{-2} \approx 0.180$

22.1 - Example 2: A health vitamin manufacturer receives on average 150 orders over a 24-hour period. Find the probability that they receive exactly 5 orders over two hours.

Solution:

Step 1: $k = 5$, since we are interested in the probability that they receive exactly 5 orders.

Step 2: The time period we are concerned with is $t = 2$ hours. Therefore, we need to compute μ.

From the statement of the problem, we know that $24\mu = 150$. Therefore,

$\mu = 150/24 = 6.25$ average number of orders per hour.

Since $t = 2$, we have $t\mu = 2(6.25) = 12.50$ average number of orders over 2 hours.

Step 3: $P\{X = 5\} = \dfrac{(\mu t)^k}{k!}e^{-\mu t} \approx \dfrac{(12.5)^5}{5!}2.718^{-12.5} \approx 0.0095,$

the probability that 5 orders occur over two hours.

22.1 - Example 3: The Department of Motor Vehicles studied the number of auto accidents that occurred in 1993 at the corner of Main and Broadway located in the downtown of a large California city. The study showed that over 30 days, on average, 3 accidents occurred. Find the probability that over 10 days

(a). at least 2 accidents occurred.

(b). 3 accidents did not occur.

Solutions:

➤ **(a).**
Step 1: Compute μ.

Since $\mu_t = 30\mu = 3$, then $\mu = 0.10$ average auto accidents per day.

Step 2: Compute the average number of accidents over 10 days gives

$\mu_t = \mu t = 10\mu = 10(0.10) = 1$, the average number of auto accidents over 10 days.

Step 3: Compute the probability that at least 2 accidents occurred by using the formula

$P\{X \geq 2\} = 1 - P\{X \leq 1\} = 1 - [P\{X = 0\} + P\{X = 1\}]$.

Therefore, we need to use the Poisson formula for $k = 0$ and 1.

For $k = 0$, and $t = 10$ days, we have

$$P\{X = 0\} = \frac{(\mu t)^k}{k!} e^{-\mu t} \approx \frac{(1)^0}{0!} 2.718^{-1} \approx 0.368 \ .$$

For $k = 1$, and $t = 10$ days, we have

$$P\{X = 1\} = \frac{(\mu t)^k}{k!} e^{-\mu t} \approx \frac{(1)^1}{1!} 2.718^{-1} \approx 0.368 \ .$$

Therefore,

$P\{X \geq 2\} = 1 - P\{X \leq 1\} = 1 - [P\{X = 0\} + P\{X = 1\}] = 1 - 0.368 - 0.368 = 0.264$,

the probability that at least 2 accidents occur over 10 days.

➤ **(b).**
The probability that 3 accidents do not occur over a 10-day period can be written

$P\{X \neq 3\} = 1 - P\{X = 3\}$.

From (a), we know that $\mu_t = \mu t = 10\mu = 10(0.10) = 1$, the average number of auto accidents over 10 days.

Since we need to compute $P\{X = 3\}$, use $k = 3$.

$$P\{X = 3\} = \frac{(\mu t)^k}{k!} e^{-\mu t} \approx \frac{(1)^3}{3!} 2.718^{-1} \approx 0.06132$$

$P\{X \neq 3\} = 1 - P\{X = 3\} = 1 - 0.06132 \approx 0.939$

22.1 - Example 4: A machine prints on average 7,200 labels an hour. Statistical studies have shown that 0.25% of these labels are defective. Find the probability that over 3 minute, 2 of the labels are defective.

Solution:

Step 1: The total number of minutes in an hour is t = 60 minutes.

Step 2: The average number of defective labels over t = 60 minutes is

$\mu_t = \mu t = (0.25\%)7200 = (0.0025)7200 = 18$ defective labels.

Step 3: Computing the average number of defective labels per minute μ gives $\mu_t = \mu t = 18$ where t = 60 and

$\mu_t = \mu t = \mu 60 = 18$.

Therefore, $\mu = 18/60 = 0.30$, the average defective per minute.

Step 4: The average number of defective labels over t = 3 minutes is

$\mu_t = 3\mu = 3(0.3) = 0.9$, the average defective labels over 3 minutes.

Step 5: The number is defective labels to check for is k = 2.

Step 6: $P\{X = 2\} = \dfrac{(\mu t)^k}{k!}e^{-\mu t} \approx \dfrac{(0.9)^2}{2!}2.718^{-0.9} \approx 0.165$

Solved Problems

22.1 - Solved Problem 1: A time study at the corner of a major street crossing, showed that over a two hour period an average of 2.5 cars violated the red light. Find the probability that during any 2 hours, that 3 cars will violated the red light.

Solution:

Step 1: time duration: t = 2 hours.

Step 2: The average number of automobiles violating a red line is $\mu_t = \mu t = 2\mu = 2.5$ average number of violations.

Step 3: The number of cars that violate the red light is k = 3.

Step 4: The probability that during any 2 hours, 3 cars will violate the red light:

$P\{X = 3\} = \dfrac{(\mu t)^k}{k!}e^{-\mu t} \approx \dfrac{2.5^3}{3!}2.718^{-2.5} \approx 0.214.$

22.1 - Solved Problem 2: A gasoline station noticed that on average, 13.44 customers purchase super-high octane gasoline over 10 hours. Find the probability that over a 3-hour period, 5 customers purchase high-octane gasoline.

Solution:

Step 1: time duration is $t = 10$ hours.

Step 2: The average number of customers μ_t that purchase high octane gasoline over 10 hours is

$\mu_t = \mu t = 10\mu = 13.44$.

Step 3: Find the average number μ of customers that purchase high-octane gasoline over 1 hour.

Since $10\mu = 13.44$, then $\mu = 13.44/10 = 1.344$.

Step 4: The average number of customers that purchase high octane gasoline over 3-hour period is

$\mu_t = \mu t = (1.344)3 = 4.032$.

Step 5: The probability that 5 customers purchase high-octane gasoline is

$P\{X = 5\} = \dfrac{(\mu t)^k}{k!}e^{-\mu t} \approx \dfrac{4.032^5}{5!}2.718^{-4.032} \approx 0.156$.

22.1 - Solved Problem 1.3: The Department of Motor Vehicles studied the number of auto accidents that occurred in 1993 at the corner of Main and Broadway located in the downtown of a large California city. The study showed that over 30 days, on average, 3 accidents occurred. Find the probability that over 5 days

(a). at most 2 accidents occurred.

(b). between 2 and 4 accidents occur.

Solutions:

Step 1: time duration is $t = 30$ days.

Step 2: The average number of accidents over 30 days is $\mu_t = \mu t = 30\mu = 3$.

Step 3: Since $\mu_t = 3$, the average number $\mu = 3/30 = 0.10$.

Step 4: The average number of accidents over a 5-day period is $\mu_t = \mu t = (0.10)5 = 0.50$.

➤ **(a).**
The probability that at most 2 accidents occur is

$P\{X \le 2\} = P\{X = 0\} + P\{X = 1\} + P\{X = 2\} \approx$

$\dfrac{(0.5)^0}{0!}2.718^{-0.5} + \dfrac{(0.5)^1}{1!}2.718^{-0.5} + \dfrac{(0.5)^2}{2!}2.718^{-0.5} \approx 0.99$.

➤ **(b).**

$P\{2 \leq X \leq 4\} = P\{X = 2\} + P\{X = 3\} + P\{X = 4\} \approx$

$$\frac{(0.5)^2}{2!}2.718^{-0.5} + \frac{(0.5)^3}{3!}2.718^{-0.5} + \frac{(0.5)^4}{4!}2.718^{-0.5} \approx 0.09$$

22.1 - Solved Problem 4: A machine fills 1,000 bottles with 16 oz of spring water over a 24-hour period. A study shows that the machine over-fills 2% of the bottles. Find the probability that over 6 hours, 5 of the bottles are over-filled.

Solution:

Step 1: The total time: t = 24 hours.

Step 2: The average number of over-filled bottles for t = 24 hours is

$\mu_t = \mu t = 2\%(1,000) = (0.020)1000 = 20,$

the average number of over-fills.

Step 3: Computing the average number of over-filled bottles for 1 hour is

$\mu_t = \mu t = 20$ where, t = 24.

Therefore, $\mu = 20/24 = 5/6 \approx 0.83$ average number of over-fills.

Step 4: The average number of bottles over-filled over 6 hours is $\mu_t = 6\mu = 6(0.833) \approx 5$ average.

Step 5: The number of over-filled bottles is k = 5.

Step 6: The probability of 5 over-filled bottles over 6 hours is

$$P\{X = 5\} = \frac{(\mu t)^k}{k!}e^{-\mu t} \approx \frac{5^5}{5!}2.718^{-5} \approx 0.18 \; .$$

Unsolved Problems with Answers

22.1 - Problem 1: Over a 2-hour period, an average of 3 students at a local university violate curfew restrictions. Find the probability that over a 2-hour period, exactly one student violates curfew.

Answer:

0.15

⇑ *Refer back to* **22.1 - Example 1** & **22.1 - Solved Problem 1**.

22.1 - Problem 2: A gasoline station noticed that on average, 3 customers request an oil change over a 2 hour period. Find the probability that over a 3 hour period, 5 customers request an oil change.

Answer:

0.17

⇑ *Refer back to* **22.1 - Example 2 & 22.1 - Solved Problem 2**.

22.1 - Problem 3: A recent study shows that over 1 year, the average number of major fires in a small Oregon city is 6. Find the probability that over 2 months, the probability that

(a). at least 1 fire occurs.

(b). at most 1 fire occurs.

Answers:

➤ **(a)**. 0.63

➤ **(b)**. 0.72

⇑ *Refer back to* **22.1 - Example 3 & 22.1 - Solved Problem 3**.

22.1 - Problem 4: Over a 7-day period, a national mail order firm sends out on average 2,000 orders. Of this number, 3% are returned to the company.

Find the probability that on 1 day, 5 of the orders are returned.

Answer:
0.073

⇑ *Refer back to* **22.1 - Example 4 & 22.1 - Solved Problem 4**.

22.2 - The Poisson Spatial Distribution

A random variable X has a Poisson spatial distribution if it takes on values k = 0, 1, 2,... with probability

$$P\{X = k\} = \frac{(\mu t)^k}{k!} e^{-\mu t}, \quad \text{where}$$

1. X is the random variable that counts the number of occurrences over a spatial region or over a group of units, such as the number of students in a classroom, or the number of spelling errors in a book.

2. t is the number of units in the group.

3. P{X = k} is the probability that the event occurs k times over t.

4. μ is the average number of times the event occurs over a unit of the group.

To apply the Poisson spatial distribution, two rules must apply:

1. The occurrence of an event during any sub-interval of the group is independent of the occurrence of an event over a non-overlapping sub-domain or subgroup.

2. If the group is small, the chance of two or more events occurring over this group is essentially zero.

22.2 - Example 1: A major publishing firm stated in its annual report that the average number of errors in a manuscript is 5 per 1,000 words. Find the probability that in a certain manuscript, only 2 errors occurred over 1,000 words.

Solution:

Step 1: Here a unit is defined as 1 word. The total number of units is t = 1,000.

Step 2: The average number of times an error occurs over 1,000 words is μ = 5.

$$\mu_t = \mu t = 5$$

Step 3: The number of errors is k = 2.

Step 4: The probability that only 2 errors occur over 1,000 words is

$$P\{X = k\} = \frac{(\mu t)^k}{k!} e^{-\mu t} = P\{X = 2\} \approx \frac{5^2}{2!}(2.718^{-5}) \approx 0.084 \ .$$

22.2 - Example 2: A chemical laboratory tested the coverage of a new commercial paint on a stucco wall. For every 100 square feet of coverage, on average, 2 square feet was not adequately covered. Find the probability that for 20 square feet of coverage, 1 square foot of stucco wall was not covered adequately.

Solution:

Step 1: Define a unit is defined as 1 square feet of coverage. Therefore, t = 20.

Step 2: The average number per unit that is not adequately covered is $\mu = 2/100 = 0.02$.

Step 3: The average number of square feet that is not adequately covered over 20 square feet is

$\mu_t = t\mu = (20)(0.02) = 0.40$.

Step 4: The probability that 1 square foot is not adequately covered is

$$P\{X = k\} = \frac{(\mu t)^k}{k!}e^{-\mu t} = P\{X = 1\} \approx \frac{0.4^1}{1!}2.718^{-0.4} = 0.27 \; .$$

22.2 - Example 3: A machine fills 16 oz of spring water in bottles. The machine fill 1,000 bottles over a one-hour period. Each hour, 100 bottles are randomly sampled and checked for the amount filled in each bottle. The following decision rule is used to decide if the filling process is properly working:

If at least 2 bottle from the sample have 16.1 oz or more, then the process is stopped; otherwise allow the process to continue.

(a). Find the probability that the process will be stopped when 1,000 bottles have an average overfill of 1% of all bottles.

(b). Find the probability that the process will not be stopped when 1,000 bottles have an average overfill of 6% of all bottles.

Solutions:

➤ **(a).**
Step 1: Define a unit as a single bottle.

The average overfill for each 1,000 bottles is $\mu_t = \mu t = 1\%(1,000) = (0.01)(1000) = 10$ and

$\mu = 10/1000 = 0.01$.

Step 2: The average overfill for each 100 bottles is $\mu_t = 100\mu = 100(0.01) = 1$.

Step 3: The probability of stopping the process (at least two bottle are overfilled) is

$P\{X \geq 2\} = 1 - P\{X = 0\} - P\{X = 1\}$

$$\approx 1 - \frac{1^0}{0!}2.718^{-1} - \frac{1^1}{1!}2.718^{-1} = 1 - 0.368 - 0.368 = 0.264 \ .$$

➤ **(b).**

Step 1: The average overfill for each 1,000 bottles is

$\mu_t = \mu t = 6\%(1,000) = (0.06)(1000) = 6$ and $\mu = 60/1000 = 0.06$.

Step 2: The average overfill for each 100 bottles is $\mu_t = 100\mu = 100(0.06) = 6$.

Step 3: The probability of not stopping the process (less than 2 bottles are overfilled) is

$$P\{X \leq 1\} = P\{X = 0\} + P\{X = 1\} \approx \frac{6^0}{0!}2.718^{-6} + \frac{6^1}{1!}2.718^{-6} = 0.0025 + 0.0149 = 0.0174 \ .$$

Solved Problems

22.2 - Solved Problem 1: A machine packs 100 light bulbs per box. The average number of bulbs broken is 3 per box. Two boxes are selected at random. Find the probability that no bulbs are broken.

Solution:

Step 1: Define a unit as a bulb.

The total number of units is $t = 200$.

Step 2: $\mu = 3$, the average number of broken bulbs per box.

The average number of bulbs broken in the two boxes is $\mu 200 = 6$.

Step 3: The number of broken bulbs is $k = 0$.

Step 4: The probability that 0 bulbs are broken in the two boxes is

$$P\{X = k\} = \frac{(\mu t)^k}{k!}e^{-\mu t} = P\{X = 0\} \approx \frac{6^0}{0!}2.718^{-6} \approx 0.0025 \ .$$

22.2 - Solved Problem 2: A research study by the United States Department of Agriculture showed that in a certain mid-western state, for every 20 acres farming , 1 acre is not suitable to grow corn. Find the probability that for 5 acres of farming, 2 acres are not suitable for growing corn.

Solution:

Step 1: Define the unit as 1 acre.

Here we have t = 5.

Step 2: μ = 1/20 = 0.05, the average per acre not suitable to grow corn.

Step 3: μ_t = μt = 0.05(5) = - 0.25, the average number acres that is not suitable for growing corn over 5 acres.

Step 4: The probability that 2 acres out of 5 are not suitable for growing corn is

$$P\{X = k\} = \frac{(\mu t)^k}{k!}e^{-\mu t} = P\{X = 2\} \approx \frac{0.25^2}{2!}2.718^{-0.25} \approx 0.0243 \ .$$

22.2 - Solved Problem 3: A manuscript of 300 pages is recently received by a publishing firm. A random sample of 25 pages is done to check for typing errors.

The following decision rule is used to decide if the manuscript should be returned to the author:

If at least 2 pages of the sample have an error, then the manuscript is returned to the author; otherwise it is not returned.

(a). Find the probability that the manuscript will be returned when only 2% of all pages of the manuscript have errors.

(b). Find the probability that the manuscript will not be returned when 5% of all pages of the manuscript have errors.

Solutions:

➤ **(a).**
Step 1: The average number of pages with errors for 300 pages is

μt = 2%(300) = (0.02)(300) = 6 and μ = 6/300 = 0.02 .

Step 2: The average number of pages with errors for the sample is μ_t = 25μ = 25(0.02) = 0.5 .

Step 3: The probability that the manuscript will be returned is

$P\{X \geq 2\} = 1 - P\{X = 0\} - P\{X = 1\} =$

$$1 - \frac{0.5^0}{0!}2.718^{-0.5} - \frac{0.5^1}{1!}2.718^{-0.5} \approx 1 - 0.607 - 0.303 = 0.09 \ .$$

➤ **(b).**

Step 1: The average number of pages with errors for 300 pages is

$$\mu_t = \mu t = 5\%(300) = (0.05)(300) = 15 \text{ and } \mu = 15/300 = 0.05.$$

Step 2: The average number of pages with errors for the sample is $\mu_t = 25\mu = 25(0.05) = 1.25$.

Step 3: The probability that the manuscript will not be returned is

$$P\{X \le 1\} = P\{X = 0\} + P\{X = 1\}$$

$$\approx \frac{1.25^0}{0!}2.718^{-1.25} + \frac{1.25^1}{1!}2.718^{-1.25} \approx 0.287 + 0.358 = 0.645.$$

Unsolved Problems with Answers

22.2 - Problem 1: In a new computer science textbook, an instructor found that the average number of errors per page is 0.25. Over 10 pages, find the probability that 2 errors exist.

Answer:

0.26

⇑ *Refer back to* **22.2 - Example 1 & 22.2 - Solved Problem 1**.

22.2 - Problem 2: A local telephone company reported that for every 100,000 feet of coaxial cable installed, 10 feet are defective. If 10,000 feet of coaxial cable is inspected, find the probability that 2 feet are defective.

Answer:

0.07

⇑ *Refer back to* **22.2 - Example 2 & 22.2 - Solved Problem 2**.

22.2 - Problem 3: A computer manufacturer has 1,000 clients. It sends to 10 clients, a new type computer on a ten-day trial basis. If clients like the computer, then the company will manufacture the new computer. The following decision rule is used:

If at least 2 of these clients like the computer then manufacture the computer; otherwise discontinue the computer.

(a). Find the probability that the company will not continue the computer when 35% of all its clients would like the computer.

(b). Find the probability that the company will manufacture the computer when only 20% of all its clients would like the computer.

Answers:

➤ (a). 0.136

➤ (b). 0.594

⇑ *Refer back to* **22.2 - Example 3 & 22.2 - Solved Problem 3**.

Supplementary Problems

1. A machine packs 100 light bulbs per box. The average number of bulbs broken is 3 per box. Five boxes are randomly selected. Find the probability that three of the boxes have exactly one broken bulb each.

2. An auto manufacturer receives 10,000 transmissions from a company on the East Coast and 15,000 from a company on the West Coast each month. Each of the shipments come in boxes of 100 transmissions. Studies have shown that on average, 500 each month of all transmissions from the East Coast company are defective and on average 450 each month are defective from the West Coast. Assume a box of transmissions is check for defects.

a. Find the probability that the box contains exactly 1 defective transmission.

b. Given that the box contains exactly 1 defective transmission, find the probability that it came from the East Coast.

3. Assume X is a random variable with a Poisson distribution where $\mu = 4.5$. For $t = 1$, find the value k where $P\{X = k\}$ is the maximum.

4. A 500 page book has 500 errors.

a. Find the probability that a random selected page contains at least 3 errors.

5. A certain disease occurs in 1 percent of the general population. How large must a random sample be if the probability of its containing at least 1 person with this disease to be 0.95 or more?

6. Mr. Jones sells cordless phones. How many potential customers would he have to contact, on average, to assure that the probability of a least one sale to be 0.99 or more.

7. A full house in a hand of 5 card poker is 3 of a kind and 2 of a kind. For example, a full house is three kings and two aces.

a. If 10 independent hands are dealt, find the probability that at least 1 hand has a full house.

b. How many hands would have to be dealt to get at least 1 full hand with a probability greater then 0.5 .

8. At a certain university, 1,000 students are majoring in English.

a. If we divided the 1,000 students into 100 equal random groups, find the probability that exactly 10% of this group has only 2 students with the same birthday and the other 8 students different birthdays.
(See supplementary problem 16, Lesson 17).

b. If we divided the 1,000 students into 200 equal random groups, find the probability that exactly 10% of this group has only 2 students with the same birthday and the other 3 students different birthdays.
(See supplementary problem 16, Lesson 17).

9. A publisher of a new history book estimates that there are on average λ spelling errors per page. Assume the book has n pages. Find the probability that at least 1 page has more then k spelling errors.

10. The Dinks Security Co. makes a study over a 72 hour period of the number of false alarms they receive per hour. The following table is the result of this study:

k:Number of false alarms.	0	1	2	3	4	5
H_k: Number of hours where k false alarms occurred .	29	2	1	4	1	0

a. Find the average μ of the number of false alarms per hour.

b. Assume the distribution of the number of false alarms per hour, is approximately Poisson. Using the Poisson distribution, estimate H_k for k = 0,1,2,3,4,5.

11. Concerned about the spread of a non-contagious disease in a large American city, a state government divided the city into 500 small areas of 0.5 square miles each and in each area checks for the number of infections. The following table is the result of this study:

k: Number of infections	0	1	2	3	4	6	7	8	9
A_k: Number of areas where k infections exists.	200	112	80	51	27	15	4	6	5

a. Find the average μ of the number of infections per area.

b. Assume the distribution of the number of infections is approximately Poisson. Using the

Poisson distribution, estimate A_k for k = 0,1,2,3,4,5.

12. According to a report from a agency of the U.S. Justice Dept., a large city has 3 murders per 100,000 population each year. Assuming the distribution of the number of murders has a Poisson distribution, find the probability that for among 200,000 residents there will be for a given year

a. 8 murders.

b. between 4 and 8 murders.

c. less than 3 murders.

d. more than 3 murders.

13. A local police department recently reported that the average number of 911 calls received per minute is 2.5 calls. For a arbitrary given minute, find the probability that there will be

a. 3 calls.

b. less than 5 calls.

c. more than 6 calls.

14. Assume X is a random variable with the distribution $P(X = k) = 2^k e^{-2}/k!$. Find

a. $P(X \geq 2 | X \leq 4)$.

b. $P(X \leq 4 | X \geq 2)$.

15. Assume X and Y are independent Poisson distributions where $P(X = k) = \lambda^k e^{-\lambda}/k!$ and

a. $P(Y = k) = \mu^k e^{-\mu}/k!$. Show Z = X + Y is a Poisson distribution where

$$P(Z = m) = (\lambda + \mu)^m e^{-(\lambda + \mu)}/m!$$

b. Show $P(X = k | X + Y = n)$ has a binomial distribution.

16. A large automobile assembly plant has on its assembly line two robots that detect defects in each automobile. A quality control engineer reported the following data about the detection of defects by the two robots: :

  1. The average number of defects detected by robot A is 0.23 defects per automobile.
  2. The average number of defects detected by robot B is 0.28 defects per automobile.
  3. Robot A only checks for defects in the automobiles' engines.
  4. Robot B only checks for defects on the remaining parts of the automobiles.

5. The two robots operate independently of each other

6. The distribution of defects detected by each robot is Poisson.

Assume on a give day, 575 automobiles were assembled.

a. Complete the following table for the automobiles assembled on that day. (Hint: See Problem 15.)

Number of k defects	0	1	2	3	4	5
Number of autos with k						

b. An automobile is randomly selected and found to have 3 defects. Find the probability that the number of defects detected by robot A is 1.

c. An automobile is randomly selected and found to have 1 defect in its engine. Find the probability that the total number of defects in the automobile is 3.

The exponential distribution.   A random variable X has a exponential distribution if

$P(X \leq x) = 1 - e^{-\lambda x}$ for $\lambda > 0$, $x \geq 0$ and $P(X < x) = 0$ for $x < 0$.

17. Show how the exponential distribution is related to the Poisson distribution. Also give an interpretation of the meaning of the exponential distribution.

18. If X has the exponential distribution, show that

a. $P(X > u + v | X > u) = P(X > v)$ for all u, v > 0.

b. A random variable which satisfies a. is said to have the 'lack of memory property'. Explain the meaning of this phase (See lesson 14, problem 29).

19.Show that if X is a positive continuous random variable with the lack-of-memory property, then X has the exponential distribution.

By using Table B, the Poisson distribution table, much of the computation can be avoided. Table B is limited to $\mu = 0.5$, 1, 1. 5, 2, ..., 5. and x = 0, 1, 2, ..., 16.

The table gives the values for $P\{X \leq x\}$.

23.1 - Using The Poisson Time Distribution Table.

23.1 - Example 1: A time study at a large supermarket showed that on average 2 customers are served at an express line over a one-minute period. During a 1-minute period, find the probability that

(a). at most 3 customers are served at the express line.

(b). at least 5 customers are served at the express line.

(c). between 3 and 6 customers are served at the express line.

(d). 5 customers are served at the express line.

Solutions:

To use the table, we must first find μ.

Step 1: $\mu t = 2$ = average number of customers served at an express line over a one-minute period.

Step 2: Since the time period is one minute, t = 1.

Step 3: Since $\mu t = 2$ and t = 1 then $\mu = 2$.

➤ **(a).**
The probability that at most 3 customers are served at the express line is $P\{X \leq 3\}$.

Step 4: From the table, check the column $\mu = 2.0$.

Step 5: From the table, check the row x = 3.

Step 6: From the table, the intersection of the row and column gives $P\{X \le 3\} \approx 0.8571$.

➤ **(b).**
The probability that at least 5 customers are served at the express line is $P\{X \ge 5\}$. However, the table only gives values $P\{X \le x\}$. Therefore, $P\{X \ge 5\} = 1 - P\{X \le 4\}$.

Step 1: From the table, check the column $\mu = 2$.

Step 2: From the table, check the row x = 4.

Step 3: From the table, the intersection of the row and column gives $P\{X \le 4\} \approx 0.9473$.

Step 4: $P\{X \ge 5\} = 1 - P\{X \le 4\} \approx 1-0.9473 = 0.0527$

➤ **(c).**
The probability that between 3 and 6 customers are served at the express line can be written
$P\{3 \le X \le 6\} = P\{X \le 6\} - P\{X \le 2\}$.

Step 1: From the table, check the column $\mu = 2$.

Step 2: From the table, check the row x = 6.

Step 3: From the table, the intersection of the row and column gives $P\{X \le 6\} \approx 0.9955$.

Step 4: From the table, check the row x = 2.

Step 5: From the table, the intersection of the row and column gives $P\{X \le 2\} \approx 0.6767$.

Step 6: $P\{3 \le X \le 6\} = P\{X \le 6\} - P\{X \le 2\} \approx 0.9955 - 0.6767 = 0.3188$

➤ **(d).** The probability that 5 customers are served at the express line is written

$P\{X = 5\} = P\{X \le 5\} - P\{X \le 4\}$.

From the table we find

$P\{X \le 5\} \approx 0.9834$

$P\{X \le 4\} \approx 0.9473$

Therefore,

$P\{X = 5\} = P\{X \le 5\} - P\{X \le 4\} \approx 0.9834 - 0.9473 = 0.0361$

23.1 - Example 2: A health vitamin manufacturer receives on average 12 orders over a 6-hour period. Find the probability that they receive exactly 5 orders over two hours.

Solution:

$k = 5$, since we are interested in the probability that they receive exactly 5 orders. The time period we are concerned is $t = 2$ hours. Therefore we need to compute μ.

Step 1: Over 6 hours, the average is 12 orders. Therefore,

$6\mu = 12$ and $\mu = 12/6 = 2$. the average number of orders per hour.

Step 2: Since $t = 2$, we have $t\mu = 2(2) = 4$ average number of orders over 2 hours.

Step 3: To use the table, we redefine $\mu = 4$.

Step 4: $P\{X = 5\} = P\{X \le 5\} - P\{X \le 4\}$, the probability that 5 orders occur over two hours.

Step 5: Going to the column for $\mu = 4$ from the table, we have

$P\{X \le 5\} \approx 0.7851$,

$P\{X \le 4\} \approx 0.6288$.

Step 6: $P\{X = 5\} = P\{X \le 5\} - P\{X \le 4\} \approx 0.7851 - 0.1563 = 0.1563$

23.1 - Example 3: A medical drug company has recently developed a new drug to lower blood pressure. Before releasing the drug, they will test the drug on 150 patients with high blood pressure. Their main concern is the possible number of side effects. They decide to use the following decision rule:

If 6 or more patients experience side effects, they will not release the drug to the public. If less than 6 experience side effects, then the drug will be released to the public.

(a). Assume the drug will cause 2% adverse side effects. Find the probability the drug will not be released to the public.

(b). Assume the drug will cause 3% adverse side effects. Find the probability the drug will be released to the public.

Solutions:

➤ **(a).**
Step 1: The drug will cause 2% adverse side effects to the public: $\mu = 150(0.02) = 3$.

Step 2: To find the probability the drug will not be released to the public is determined by the decision rule:

If 6 or more patients experience side effects, do not release the drug to the public: $P\{X \ge 6\}$.

Step 3: Using the Poisson distribution table, we have for $\mu = 3$,

$$P\{X \geq 6\} = 1 - P\{X \leq 5\} \approx 1 - 0.9161 = 0.0839 \,.$$

➤ **(b).**
Step 1: The drug will cause 3% adverse side effects to the public: $\mu = 150(0.03) = 4.5$.

Step 2: To find the probability the drug will be released to the public is determined by the decision rule:

If 5 or less patients experience side effects, release the drug to the public P{X ≤ 5}.

Step 3: Using the Poisson distribution table, we have for $\mu = 4.5$, $P\{X \leq 5\} \approx 0.7029$.

Solved Problems

23.1 - Solved Problem 1: A time study at the corner of a major street crossing showed that over a two-hour period an average of 2.5 cars violated the red light. Find the probability that for a two-hour period:
(a). less than 6 cars violated the red light.

(b). more than 2 cars violated the red light.

(c). less than 3 or more than 6 cars violated the red light.

(d). 1 car violated the red light.

Solutions:

Step 1: time duration: t = 2 hours.

Step 2: The average number of automobiles violating a red line:

$\mu t = 2\mu = 2.5$ average number of violations.

Step 3: The number of cars that violate the red light To use the table we need to redefine $\mu = 2.5$ average number of violations.

➤ **(a).**
the probability that less 6 cars violated the red light can be written:

$$P\{X < 6\} = P\{X \leq 5\} \approx 0.9580 \,.$$

➤ **(b).**
the probability that more than 2 cars violated the red light can be written:

$$P\{X > 2\} = 1 - P\{X \leq 2\} \approx 1 - 0.5438 = 0.4562 \,.$$

➤ **(c).**

the event that less than 3 or more than 6 cars violated the red light can be written

$$\{X < 3\} \cup \{X > 6\} = \{X \leq 2\} \cup \{X \geq 7\}.$$

Therefore,

$$P[\{X < 3\} \cup \{X > 6\}] = P[\{X \leq 2\} \cup \{X \geq 7\}] = P\{X \leq 2\} + P\{X \geq 7\}$$

$$P\{X \leq 2\} \approx 0.5438$$

$$P\{X \geq 7\} = 1 - P\{X \leq 6\} \approx 1 - 0.9858 = 0.142$$

$$P[\{X < 3\} \cup \{X > 6\}] = P\{X \leq 2\} + P\{X \geq 7\} \approx 0.5438 + 0.0142 = 0.558$$

➤ **(d).**

The probability that 1 car violated the red light is

$$P\{X = 1\} = P\{X \leq 1\} - P\{X \leq 0\} \approx 0.2873 - 0.0821 = 0.2052.$$

23.1 - Solved Problem 2: A gasoline station noticed that on average 1.5 customers purchase super-high octane gasoline over 1 minute. Find the probability that over a two minutes, 5 customers purchase super high octane gasoline.

Solution:

Step 1: time duration: For t = 1 minute, $1\mu = \mu = 1.5$ average customers.

Step 2: The average number of customers μt that purchase super high octane gasoline over t = 2 minutes is
$\mu t = 2\mu = 2(1.5) = 3$ average number of customers.

Step 3: Redefine $\mu = 3$.

Step 4: The probability that 5 customers purchase high octane:

$$P\{X = 5\} = P\{X \leq 5\} - P\{X \leq 4\} \approx 0.9161 - 0.8153 = 0.1008.$$

23.1 - Solved Problem 3: A publisher receives a new manuscript from an author. One major concern, is the number of errors in the manuscript. The publisher randomly checks 100 pages of the manuscript and uses the following rule:

If 5 or more pages of the 100 pages have errors, the manuscript is returned to the author for corrections. If less than 5 pages have errors, then the manuscript is not returned.

(a). Find the probability that the manuscript is returned if 3% of all pages of the manuscript has errors.

(b). Find the probability that the manuscript is not returned if 5% of all pages of the manuscript has errors.

Solutions:

➤ **(a).**
Step 1: Assume the manuscript has 3% errors: $\mu = 100(0.03) = 3$.

Step 2: From the Poisson table B, $P\{X \geq 5\} = 1 - P(X \leq 4) \approx 1 - 0.8153 = 0.1847$.

➤ **(b).**
Step 1: Assume the manuscript has 5% errors: $\mu = 100(0.05) = 5$.

Step 2: From the Poisson table B, $P\{X \leq 4\} \approx 0.4405$.

Unsolved Problems with Answers

23.1 - Problem 1: Over a 2-hour period, an average of 3 students at a local university violate curfew restrictions. Find the probability that over a 2-hour period, that

(a). at least 3 students violates curfew.

(b). at most 6 students violates curfew.

(c). between 2 and 7 students violates curfew.

Answers:

➤ **(a).** 0.5768

➤ **(b).** 0.9665

➤ **(c).** 0.7890

⇑ *Refer back to* **23.1 - Example 1 & 23.1 - Solved Problem 1**.

23.1 - Problem 2: A gasoline station noticed that on average 3 customers request an oil change over a 2-hour period. Find the probability that over a 3-hour period, 5 customers request an oil change.

Answer:

0.1708

⇑ *Refer back to* **23.1 - Example 2 & 23.1 - Solved Problem 2**.

23.1 - Problem 3: A large retail store sells about 10,000 television sets a year. Each month they receive a shipment of 1,000 television sets from a wholesaler. The store is concerned about the number of defective television sets. They use the following decision rule:

Sample 25 sets from the shipment. If 4 or more T.V. sets are defective, then return the entire shipment. If less than 4 are defective, then the entire shipment is not returned.

(a). Find the probability that the entire shipment will not be returned given that 10% of all the sets in the shipment are defective.

(b). Find the probability that the entire shipment will be returned given that 2% of all sets in the shipment are defective.

Answers:

➤ (a). 0.7576
➤ (b). 0.0018
⇑ *Refer back to* **23.1 - Example 3 & 23.1 - Solved Problem 3**.

Supplementary Problems

1. Construct a partial Poisson table for $P\{X \geq x\}$ for $\mu = 1, 2, 3, 4, 5$ and $x = 0, 1, 2, 3, 4, 5$.

2. A major publishing firm stated in its annual report that the average number of errors in a manuscript is 5 per 1,000 words. Find the probability that in a certain manuscript, at least 4 errors occurred over 1,000 words.

3. A chemical laboratory tested the coverage of a new commercial paint on a stucco wall. For every 100 square feet of coverage, on average, 2 square feet was not adequately covered. Find the probability that for 200 square feet of coverage, between 2 and 5 square feet of stucco wall was not covered adequately.

4. A machine fills 16 oz of spring water in bottles. The machine fills 1,000 bottles over a one hour period. Each hour, 100 bottles are randomly sampled and checked for the amount filled in each bottle. The following decision rule is used to decide if the filling process is properly working:

If at least 5 bottles from the sample have 16.1 oz or more than the process is stopped; otherwise allow the process to continue.

a. Find the probability that the process will be stopped when 1,000 bottles have an average overfill of 1.5% of all bottles.

b. Find the probability that the process will not be stopped when 1,000 bottles have an average overfill of 5% of all bottles.

5. A machine produces lots of 100 computer chips. To check the quality of each lot, a sample of 50 chips are taken from a lot of 100 chips. If at least 5 of the fifty chips are defective, the entire lot is inspected; otherwise no more inspection is carried out. Assume at a particular time the machine is producing 4% defective chips. Find the average number of chips per lot inspected.

Probability Theory
Lesson 24
The Poisson
Approximation To The
Binomial Distribution

392

Let X be a binomial distribution defined on a binomial sample space. If the sample size N is large (N ≥ 50) and the probability of success p is small (p ≤ 5/N) then we can use the Poisson distribution to approximate the binomial probability distribution:

$$P\{X = k\} \approx \frac{\mu^K}{k!}e^{-\mu} \text{ , where } \mu = pN$$

24.1 - Applications

24.1 - Example 1: A large university estimates that 2% of its students will declare a major in mathematics. A sample of 100 students is taken. Find the probability that only 2 of these students will declare a major in mathematics using

(a). the binomial distribution.

(b). the Poisson approximation to the binomial distribution.

Solutions:

➤ (a)
Step 1: Since this is a Binomial distribution, we have:

N = 100, the sample size,

p = 0.02, the probability that a student will declare a math major,

q = 1 - p = 0.98, the probability that a student will not declare a math major,

k = 2, the number of students that will declare a major in mathematics.

Step 2: Assume X is the binomial distribution:

$$P\{X = 2\} = \begin{pmatrix} N \\ k \end{pmatrix} p^k q^{N-k} = \begin{pmatrix} 100 \\ 2 \end{pmatrix} (0.02)^2 (0.98)^{98} \approx 0.2734.$$

➤ **(b).**

Step 1: Since we are approximating the binomial distribution with the Poisson distribution:

$\mu = Np = 100(0.02) = 2.$

Step 2: $P\{X = 2\} = \dfrac{\mu^k}{k!} e^{-\mu} \approx \dfrac{2^2}{2!}(2.718)^{-2} = 0.2707$

24.1 - Example 2: A machine produces microchips. It is estimated that 3% are defective. A sample of 150 chips is selected. Use the Poisson distribution to find the probability that at least 10 chips are defective.

Solution:

Step 1: This is a binomial distribution X where $\{X = k\}$ is the event that there are k defective chips.

$p = 0.03$, probability that a chip is defective

Step 2: The sample size is $N = 150$.

Step 3: The probability that at least 10 chips are defective is $P\{X \geq 10\}$.

Step 4: Since we are approximating the Binomial distribution with the Poisson distribution,

$\mu = Np = 150(0.03) = 4.5$ average.

Step 5: Using the Poisson table in Lesson 14 for $\mu = 4.5$, we have

$P\{X \geq 10\} = 1 - P\{X \leq 9\} \approx 1 - 0.9829 = 0.0171.$

Solved Problems

24.1 - Solved Problem 1: A science class has 100 students. Find the probability that 2 students are born on January 1 using:

(a). the binomial distribution.

(b). the Poisson approximation to the binomial distribution.

Solution:

➤ **(a).**
Step 1: Since this is a Binomial distribution, we have:

N = 100, the sample size,

$p = \dfrac{1}{365}$, the probability that a student is born on January 1,

$q = 1 - p = \dfrac{364}{365}$, the probability that a student is not born on January 1,

k = 2, the number of students that are born on January 1.

Step 2: Assume X is the Binomial distribution:

$$P\{X = 2\} = \begin{pmatrix} N \\ k \end{pmatrix} p^k \, q^{N-k} = \begin{pmatrix} 100 \\ 2 \end{pmatrix} (\frac{1}{365})^2 \, (\frac{364}{365})^{98} \approx 0.029 \ .$$

➤ **(b).**
Step 1: Since we are approximating the Binomial distribution with the Poisson distribution:

$$\mu = Np = (100)\frac{1}{365} = \frac{100}{365} = 0.274 \ .$$

Step 2: $P\{X = 2\} = \dfrac{\mu^k}{k!} \, e^{-\mu} \approx \dfrac{.274^2}{2!}(2.718)^{-.274} = 0.028$

24.1 - Solved Problem 2: At a large university, it is estimated that only 5% of the graduate students in mathematics receive their masters degree. A sample of 100 graduate students in mathematics is taken. Use the Poisson distribution to find the probability that at least 5 of these students have received their masters degree.

Solution:

Step 1: This is a Binomial distribution X where $\{X = k\}$ the event that there are k students that receive their masters degree.

p = 0.05, probability that a student receives a masters degree

Step 2: The sample size is N = 100 .

Step 3: The probability that at least 5 students receive a masters degree is $P\{X \geq 5\}$.

Step 4: Since we are approximating the Binomial distribution with the Poisson distribution,

$\mu = Np = 100(0.05) = 5$ average.

Step 5: Using the Poisson table in Lesson 14 for $\mu = 5$, we have

$P\{X \geq 5\} = 1 - P\{X \leq 4\} \approx 1 - 0.4405 = 0.5595$.

Unsolved Problems with Answers

24.1 - Problem 1: The Fun Toy Company has 200 retail stores. Last year 5% of these stores had over a million dollars in sales. If a sample of 4 stores is selected at random , find the probability that 2 stores had a million dollars in sales using

(a). the binomial distribution.

(b). the Poisson approximation to the Binomial distribution.

Answers:

➤ **(a)**. 0.014

➤ **(b)**. 0.016

⇑ *Refer back to* **24.1 - Example 1 & 24.1 - Solved Problem 1**.

24.1 - Problem 2: A recent study showed that 1% of males over the age of 18 suffer from paranoia. A sample of 300 males over 18 years old is taken. Use the Poisson distribution to find the probability that at least 3 of these males suffer from paranoia.

Answer:

0.5768

⇑ *Refer back to* **24.1 - Example 2 & 24.1 - Solved Problem 2**.

Supplementary Problems

For the following problems, use the Poisson approximation to the Binomial distribution.

1. A study shows that the chance that a pilot will die in an airplane accident is 0.001. Find the probability that out of 2,000 pilots, more than 2 pilots will die.

2. Mr. Smith claims that only 3% of his students fail English. Find the probability that out of a sample of 100 of his students, none will fail.

3. A sample of 200 college students was taken. If only 2% of college students will continue their studies overseas, find the probability that from this sample:

a. at least 10 students will study overseas.

b. At most 7 students will study overseas.

c. Exactly 5 students will study overseas.

4. Assume a binomial experiment where N = 100, p = 0.1 .

a. Compute the probability that $P\{X = 50\}$, using the binomial formula.

b. Compute the probability that $P\{X = 50\}$, using the Poisson formula.

5. A quality control engineer stated in a report that 2% of all floppy disk drives manufactured by his company are defective. If a box of 100 drives are shipped to a customer, the probability of at least 2 defective drives using
a. the binomial formula.
b. the Poisson formula.
c. What is the minimum number of disk drives a box should contain to assure, with a probability of at least 80%, that there are at least 100 non-defective drives amongst them.

6. A computer generates random digits (0-9). What is the minimum number of digits needed to be generated to assure with 90% probability that at least one 5 will appear?

7. a. Show $\dbinom{n}{0}\dbinom{m}{r} + \dbinom{n}{1}\dbinom{m}{r-1} + \; ... \; + \dbinom{n}{r}\dbinom{m}{0} = \dbinom{n+m}{r}.$

(Hint: The sum of the Hypergeometric distribution equals 1).

b. Show $\dbinom{N}{0}^2 + \dbinom{N}{1}^2 + \dbinom{N}{2}^2 + \; ... \; + \dbinom{N}{N}^2 = \dbinom{2N}{N}.$

8. Jack and Jill each have a fair coin. They play the following game: Each tosses their coin N times and the person at the end with the most heads wins. Find a formula for computing the probability that neither wins using

a. the binomial formula.

b. the Poisson formula.

9. Assume $P(X = k) = \dfrac{\mu^K}{k!}e^{-\mu}$ where $\mu = pN, 0 < p < 1$, and μ is a constant.

Present an argument that shows $\sigma^2 = \mu.$

Probability Theory
Lesson 25
The Normal Distribution

The normal distribution is the most important continuous distribution. Frequently studied as the bell shaped curve, this distribution is important in studying the distribution of sample means.

The normal distribution $P\{X \le x\}$ equals the shaded area under the bell-shaped curve: The mean is μ and the variance is σ^2. The positive number σ is called the standard deviation of the normal distribution. Note that 50% of the area under the curve lies to the left and right of μ.

For the normal distribution , the following rules hold:

1. The total area under the curve is 1.

2. The total area under the right side of the curve is 0.5 .

3. The total area under the left side of the curve is 0.5 .

4. The left and right sides of the curve are symmetric.

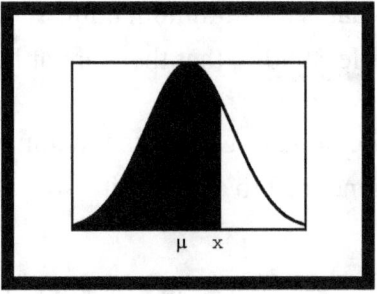

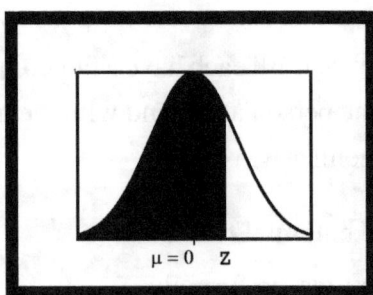

25.1-The Standard Normal Distribution.

A random variable Z has a standard normal distribution $P\{Z \le z\}$ equal to the area shaded under the curve of a normal distribution with $\mu = 0$ and variance $\sigma^2 = 1$ ($\sigma = 1$).

The area of the shaded area from 0 to z is given in the standard normal table C for specific values of z.

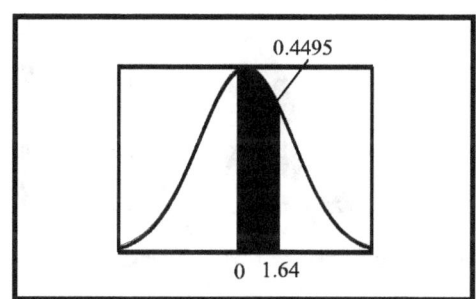

Given z, find the area shaded under the curve.

25.1 - Example 1 - From the normal distribution Table C, find the shaded area for the figures.

a

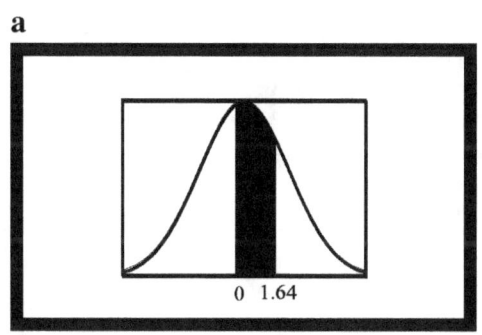

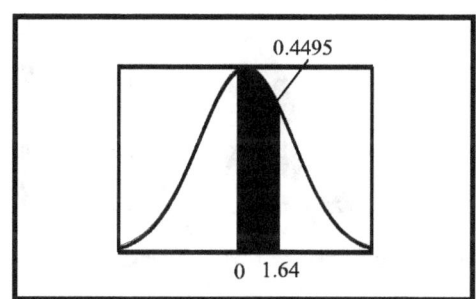

Solution:

Step 1: Go to the top of table C. Since 1.64 = 1.6 + .04, select the column marked 0.04 .

Step 2: Move down the column marked z to the row 1.6 .

Step 3: The intersection of this row and column is the number 0.4495 .

b

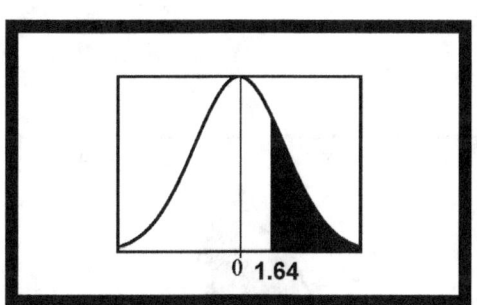

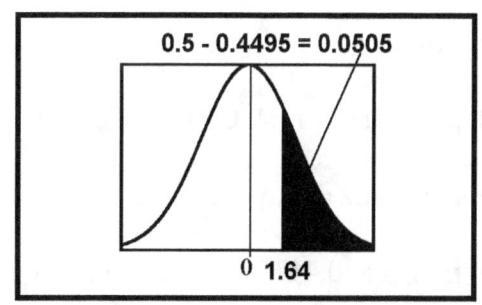

Solution:

Step 1: The area from the table for z = 1.64 is 0.4495

Step 2: Subtract 0.4495 from 0.5 .

Step 3: The shaded area is 0.5 - 0.4495 = 0.0505 .

c

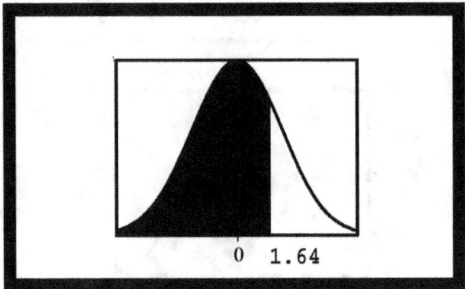

Solution:

Step 1: The shaded area for z = 1.64 is 0.4495 .

Step 2: Add 0.5 to 0.4495 .

Step 3: The shaded area is 0.9495 .

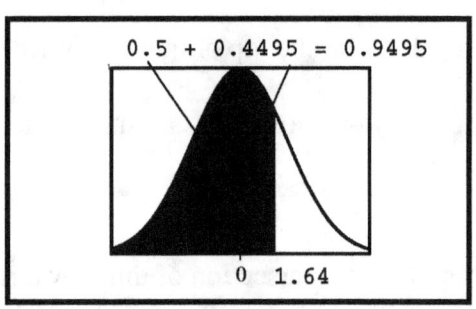

d

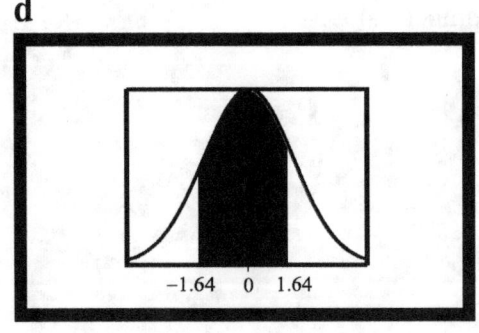

Solution:

Step 1: For z = 1.64, the area is 0.4495 .

Step 2: For z = -1.64, the area is 0.4495 .

Step 3: Add 0.4495 + 0.4495 = 0.8990 .

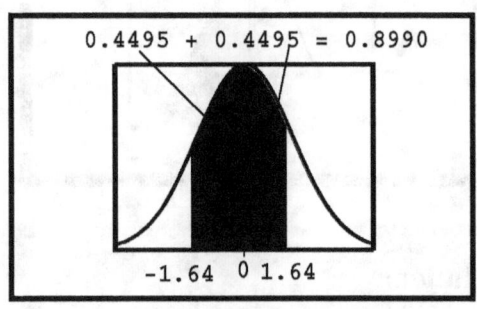

e

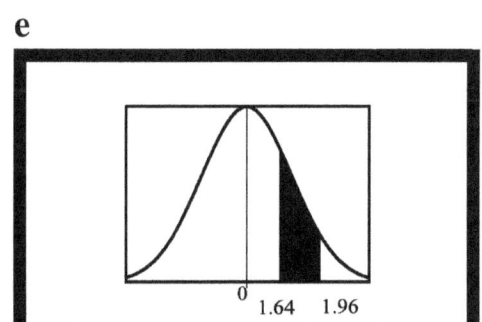

Solution:

Step 1: Look up the area in the table for z = 1.96 .

Step 2: The area is 0.475 .

Step 3: Look up the area in the table for z = 1.64 .

Step 4: The area is 0.4495 .

Step 5: Subtracting these two areas gives the shaded area under the curve which is 0.0255 .

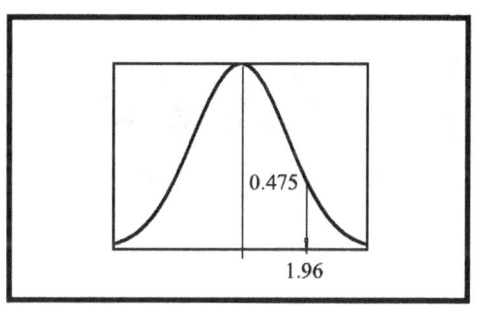

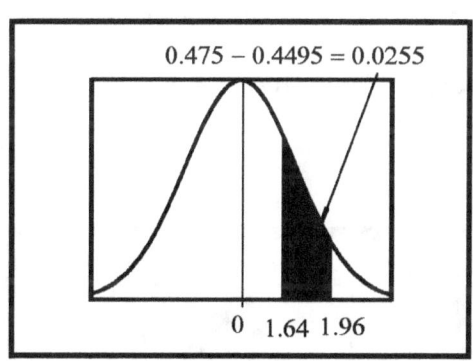

Given the area shaded under the curve, find z.

For the next two examples, the area is given and the value z is to be found from table C.

25.1 - Example 2 - For the shaded area in the figures, find z.

a

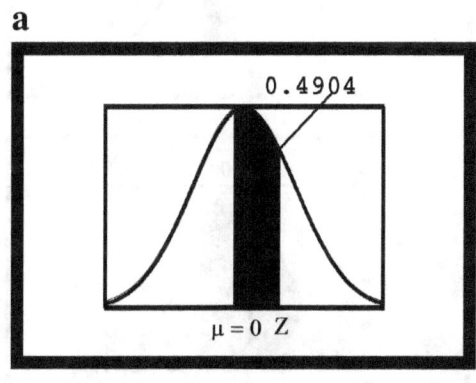

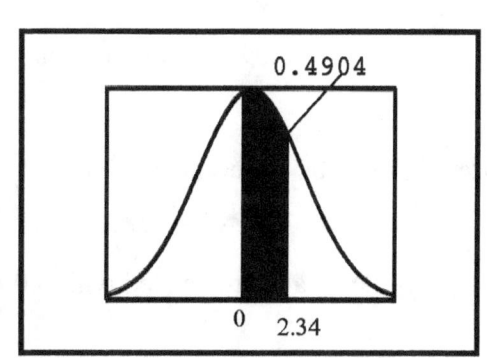

Solutions:

Step 1: Search the area portion of the table for the value 0.4904 .

Step 2: For this area, we find the z value is z = 2.34 .

b

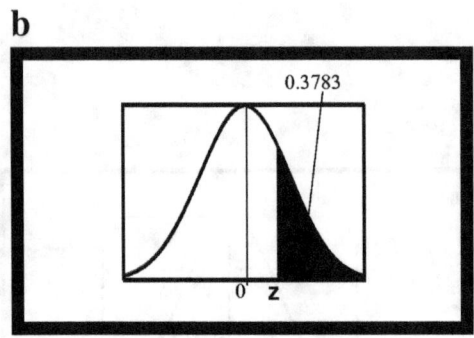
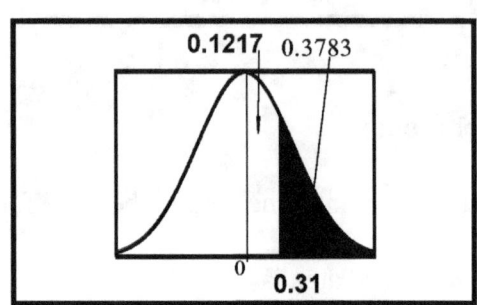

Solution:

Step 1: Compute the area from 0 to z: 0.5 - 0.3783 = 0.1217 .

Step 2: Looking up in the area part of table C, the value 0.1217 .

Step 3: The corresponding z value is z = 0.31 .

c

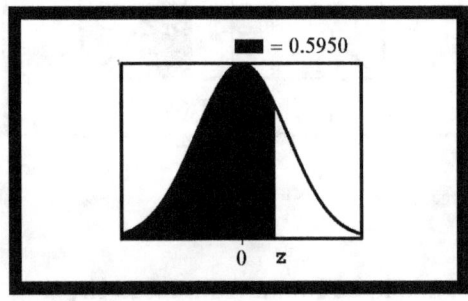

Solution:

Step 1: Compute the shaded area: 0.5950 - 0.5 = 0.0950 .

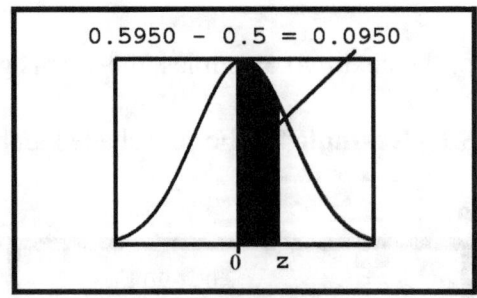

Step 2: The area 0.0950 is **not** in the area portion of the table.

Step 3: Take the closest area to 0.0950 which is 0.0948 .

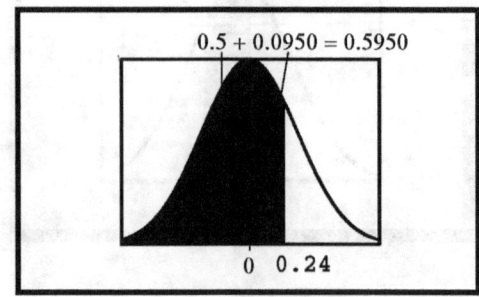

Step 4: The z value associated with 0.0948 is z = 0.24 .

Solved Problems

25.1 - Solved Problem 1: Find the shaded area for the following figures:

➤ **(a).**

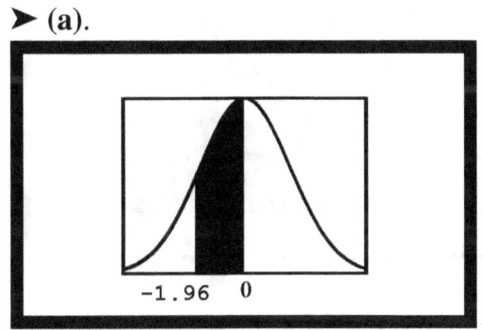

Solution:

Using the symmetry of the normal distribution, look up z = 1.96. We find the area is 0.4750 .

➤ **(b).**

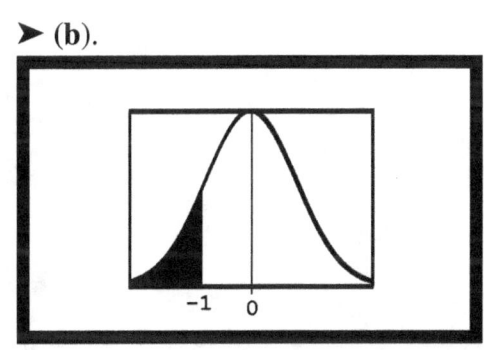

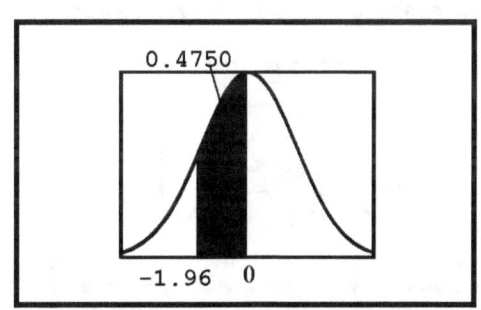

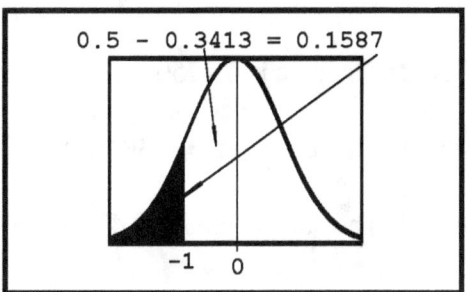

Solution:

Step 1: From symmetry of the normal distribution, let z = 1.

Step 2: From the table, we find an area equal to 0.3413 .

Step 3: Since the total left area is 0.5, then the shaded area is .5 - 0.3413 = 0.1587 .

➤ **(c).**

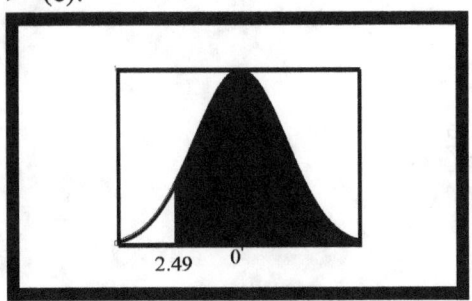

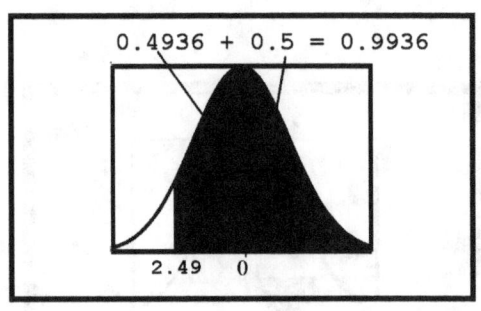

Solution:

Step 1: Look up in the table the area for z = 2.49 .

Step 2: Since the right side of the curve has an area equal to 0.5, the total shaded area is 0.4936 + 0.5 = 0.9936 .

➤ **(d).**

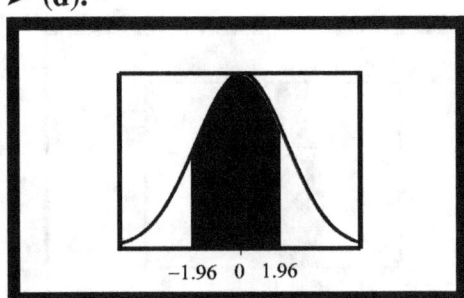

Solution:

Step 1: Look up the area for z = 1.96 .

Step 2: The area is 0.4750 .

Step 3: The area for z = -1.96 is 0.4750 .

Step 4: The shaded area is 0.4750 + 0.4750 = 0.95 .

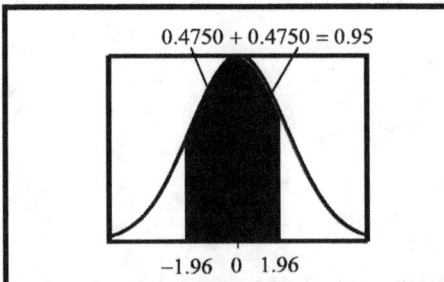

➤ **(e).**

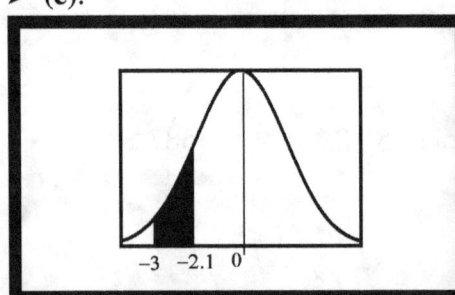

Solution:

Step 1: From the table, the area from 0 to -3 is 0.4987

.

Step 2: From the table, the area from 0 to -2.1 is 0.4821 .

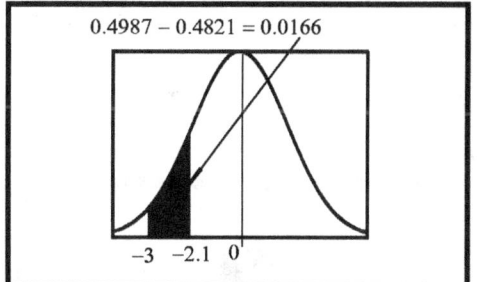

Step 3: The difference, 0.0166, is the shaded area.

25.1 - Solved Problem 2: For the shaded area in the figure, find z.

➤ **(a).**

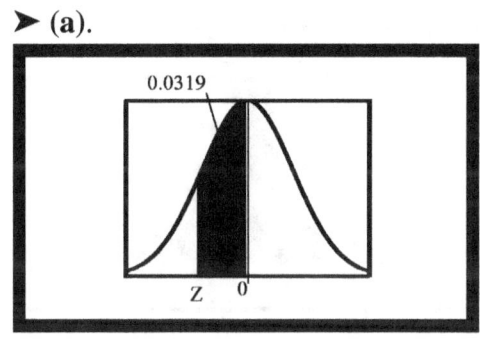

Solution:

Step 1: Search the area portion of the table for the value 0.0319 .

Step 2: For this area, we find the z value is z = -0.08 .

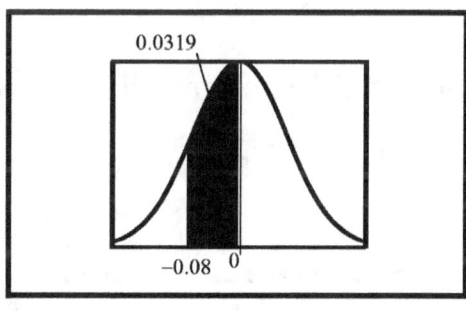

➤ **(b).**

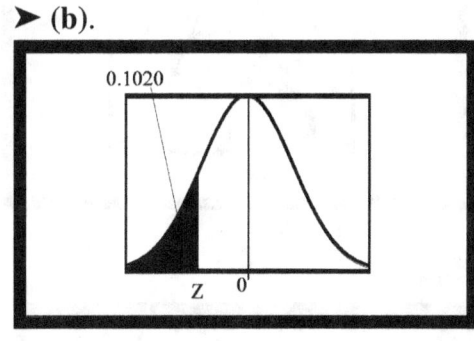

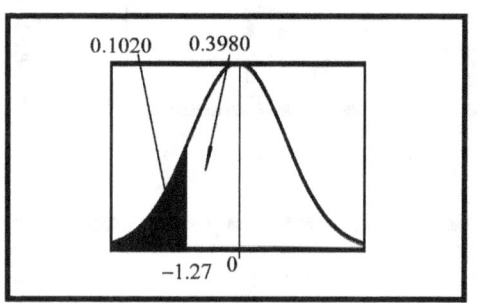

Solution:

Step 1: Compute the area from 0 to z: 0.5 - 0.1020 = 0.3980 .

Step 2: Look up the area part of the table for the value 0.3980 .

Step 3: The corresponding z value is z = -1.27 .

➤ **(c).**

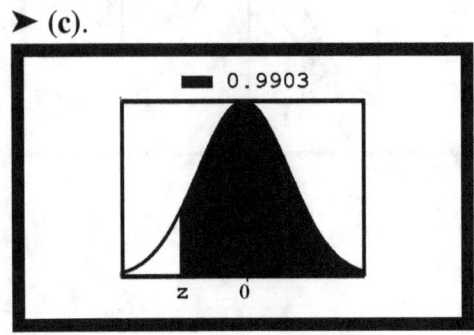

Solution:

Step 1: Compute the area 0.9903 - 0.5 = 0.4903 .

Step 2: The value 0.4903 does not exist in the area portion of the table.

Step 3: The closest value is 0.4904 .

Step 4: The correspond z value is -2.34 .

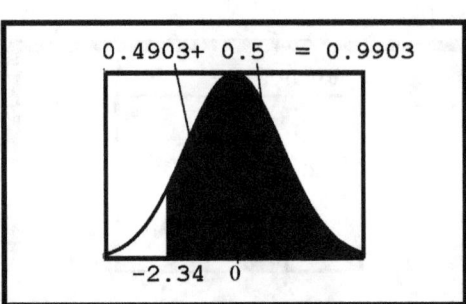

Unsolved Problems with Answers

25.1 - Problem 1: Find the shaded area for the figures below.

➤ **(a).**

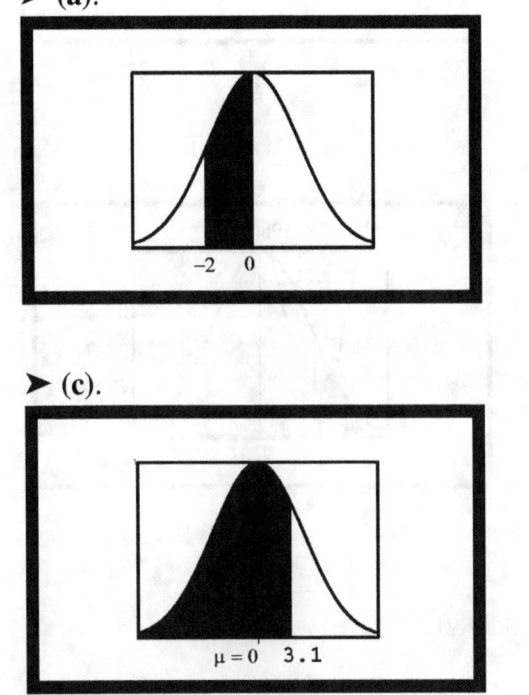

➤ **(b).**

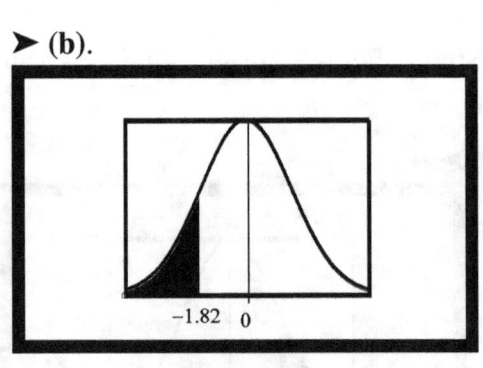

➤ **(c).**

➤ **(d).**

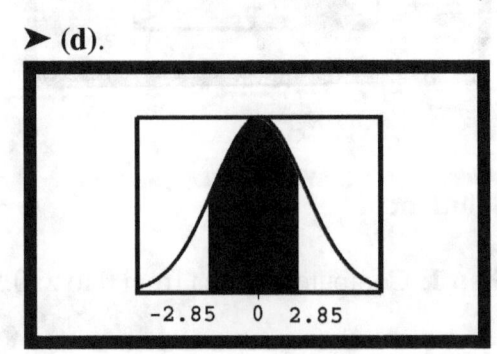

➤ **(e).**

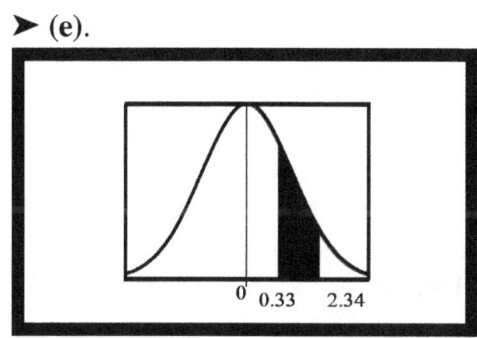

Answers:

➤ **(a).** 0.4772

➤ **(b).** 0.0344

➤ **(c).** 0.9990

➤ **(d).** 0.9956

➤ **(e).** 0.3611

⇑ *Refer back to* **25.1 - Example 1 & 25.1 - Solved Problem 1.**

25.1 - Problem 2: For the shaded area in the figures below, find z.

➤ **(a).**

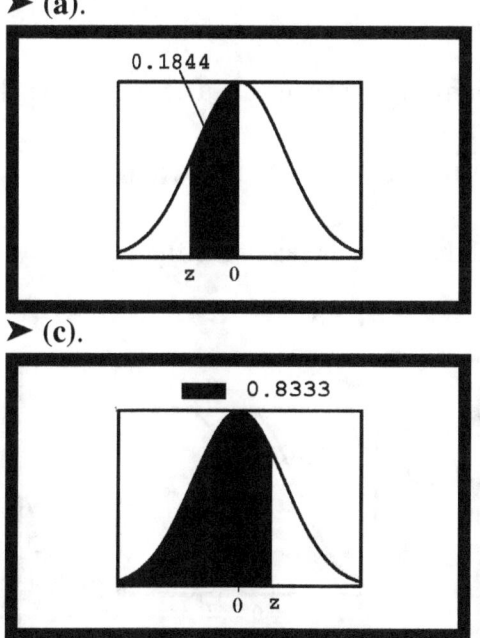

➤ **(b).**

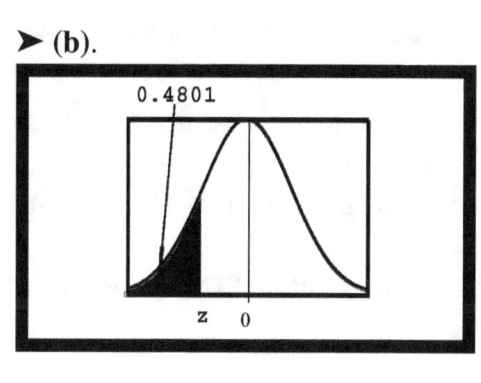

➤ **(c).**

Answers:

➤ **(a).** z = -0.47

➤ **(b).** z = -0.05

➤ **(c).** z = 0.97

⇑ *Refer back to* **25.1 - Example 2 & 25.1 - Solved Problem 2.**

25.2 - The Normal Distribution

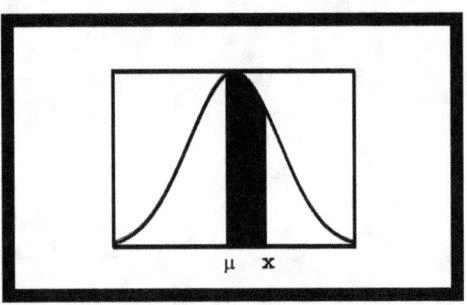

To find the area of the normal distribution, the distribution must be converted to a Standard normal distribution by using the formula:

$$P\{X \le x\} = P\{Z \le z\} \text{ where } z = \frac{x - \mu}{\sigma}$$

The value z measures the number of standard deviations from the mean μ (See Lesson 3, Descriptive Statistics).

The position of the z value must match the position of the x value.

25.2 - Example 1: For $\mu = 2.5$, $\sigma = 0.5$, and $x = 3.5$, find the shaded area in the figure.

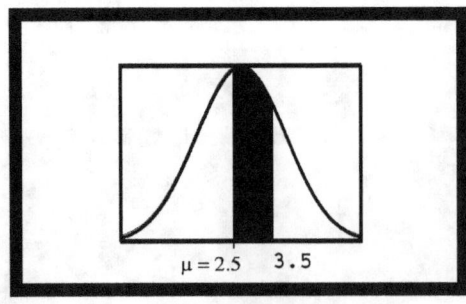

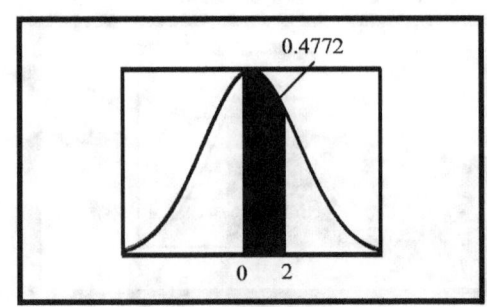

Solution:

Step 1: $z = \dfrac{3.5 - 2.5}{0.5} = 2$

Step 2: Looking up in the table $z = 2$, the area is 0.4772 .

25.2 - Example 2: For $\mu = 10$, $\sigma = 5$, and $x = 3$, find the shaded area in the figure.

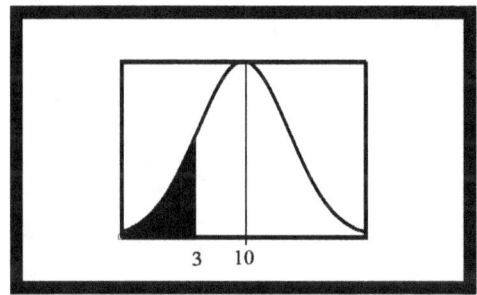

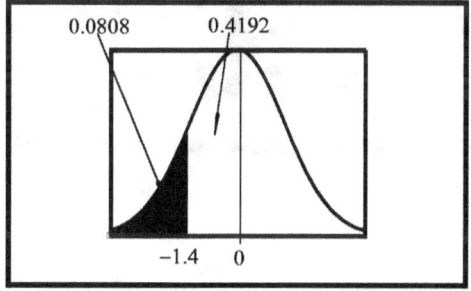

Solution:

Step 1: $z = \dfrac{3 - 10}{5} = -1.4$

Step 2: For $z = -1.4$, the area from the table is 0.4192
.

Step 3: The shaded area is $0.5 - 0.4192 = 0.0808$.

25.2 - Example 3: For $\mu = 120$, $\sigma = 10$, $x_1 = 125$ and $x_2 = 135$, find the shaded area in the figure.

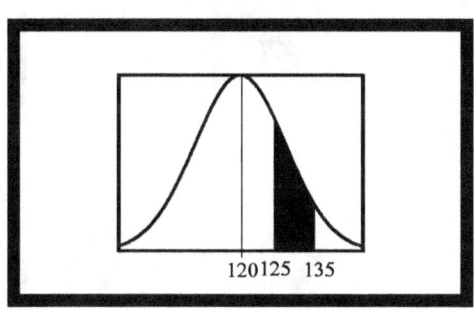

Solution:

Step 1: $z = \dfrac{125 - 120}{10} = 0.5$

Step 2: $z = \dfrac{135 - 120}{10} = 1.5$

Step 3: For z = 0.5, the area is 0.1915 .

Step 4: For z = 1.5, the area is 0.4332 .

Step 5: The shaded area is 0.4332 - 0.1915 = 0.2417 .

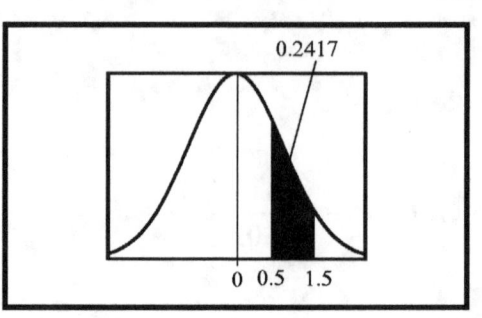

25.2 - Example 4: For μ = 120, σ = 10, x = 110 and x = 135, find the shaded area in the figure.

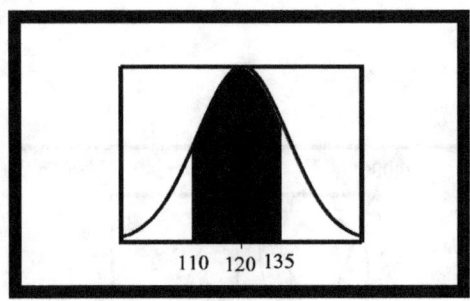

Solution:

Step 1: $z = \dfrac{135 - 120}{10} = 1.5$

Step 2: $z = \dfrac{110 - 120}{10} = -1$

Step 3: The area associated with z = 1.5 is 0.4332 .

Step 4: The area associated with z = -1 is 0.3413 .

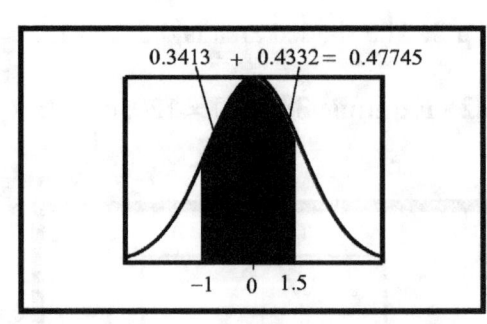

Step 5: The shaded area is the sum: 0.3413 + 0.4332 = 0.7745 .

25.2 - Solved Problems

25.2 - Solved Problem 1: For μ = 12.5, σ = 1.50 and x = 11.5, find the shaded area in the figure.

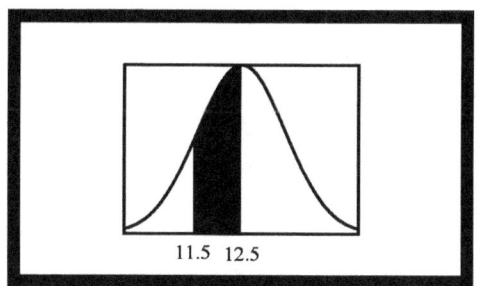

Solution:

Step 1: $z = \dfrac{11.5 - 12.5}{1.5} = -0.67$

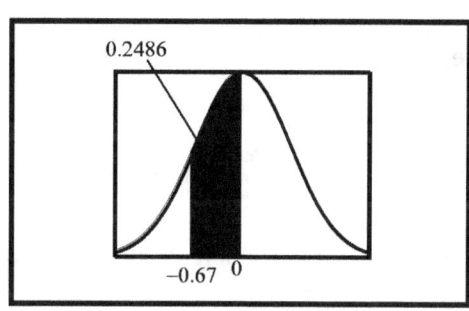

Step 2: For z = -0.67, the table gives an area equal to 0.2486 .

25.2 - Solved Problem 2: For $\mu = 100$, $\sigma = 50$, and x = 130, find the shaded area in the figure.

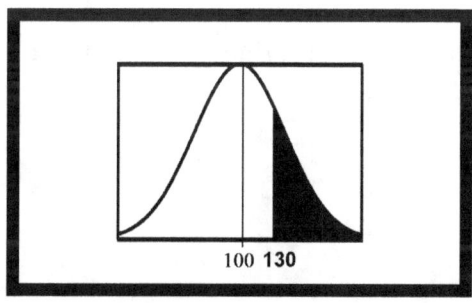

Solution:

Step 1: $z = \dfrac{130 - 100}{50} = 0.6$

Step 2: From the table for z = 0.6, the area is 0.2257 .

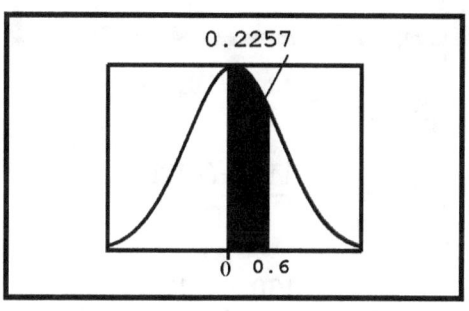

Step 3: The shaded area is 0.5 - 0.2257 = 0.2743 .

25.2 - Problem 3: For $\mu = 5$, $\sigma = 2$, $x_1 = 4$ and $x_2 = 3$, find the shaded area in the figure.

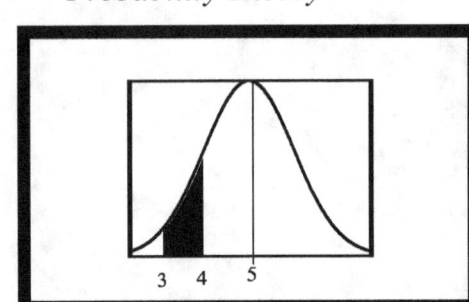

Solution:

Step 1: $z = \dfrac{3 - 5}{2} = -1$

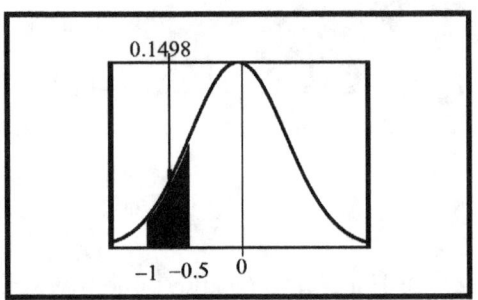

Step 2: $z = \dfrac{4 - 5}{2} = -0.5$

Step 3: The area associated with -1 is 0.3413 .

Step 4: The area associated with -0.5 is 0.1915 .

Step 5: The shaded area is 0.3413 - 0.1915 = 0.1498 .

25.2 - Solved Problem 4: For $\mu = -120$, $\sigma = 10$, $x_1 = -130$ and $x_2 = -118$, find the shaded area in the figure.

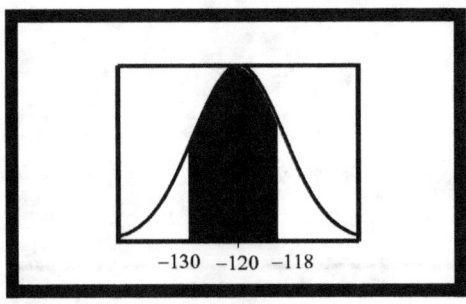

Solution:

Step 1: $z = \dfrac{-118 - (-120)}{10} = 0.2$

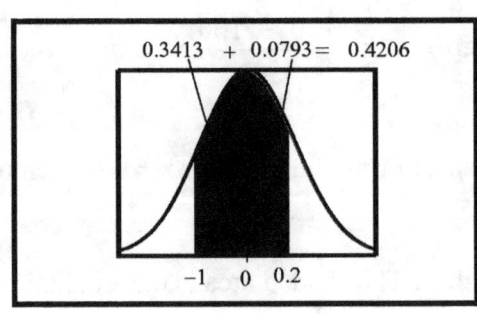

Step 2: $z = \dfrac{-130 - (-120)}{10} = -1$

Step 3: The area associated with z = 0.2 is 0.0793 .

Step 4: The area associated with z = 1 is 0.3413 .

Step 5: The total area is 0.0793 + 0.0.3413 = 0.4206 .

Unsolved problems with Answers

25.2 - Problem 1: For $\mu = 0.5$, $\sigma = 0.15$, and $x = 1$, find the shaded area in the figure.

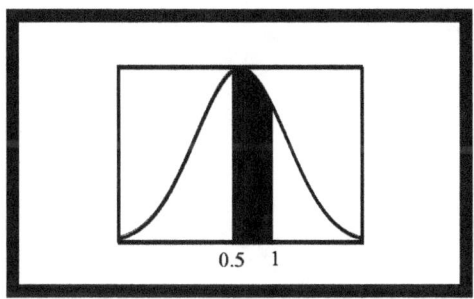

Answer:

0.4996

⇑ *Refer back to* **25.2 - Example 1 & 25.2 - Solved Problem 1.**

25.2 - Problem 2: For $\mu = 1$, $\sigma = 5$, and $x = 13$, find the shaded area in the figure.

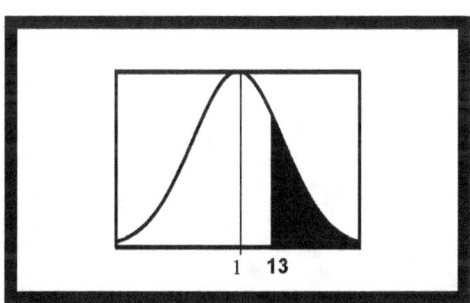

Answer:

0.0082

⇑ *Refer back to* **25.2 - Example 2 & 25.2 - Solved Problem 2.**

25.2 - Problem 3: For $\mu = 1000$, $\sigma = 500$, and $x_1 = 1300$ and $x_2 = 1400$, find the shaded area in the figure.

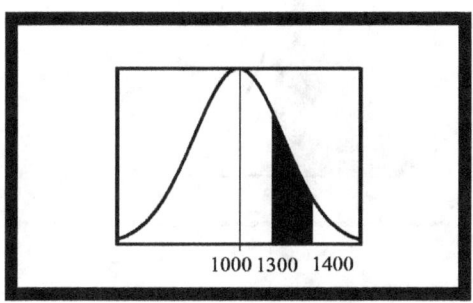

Answer: 0.0624

⇑ *Refer back to* **25.2 - Example 3 & 25.2 - Solved Problem 3.**

25.2 - Problem 4: For $\mu = -12$, $\sigma = 1$, $x_1 = -13$ and $x_2 = -11$, find the shaded area in the figure.

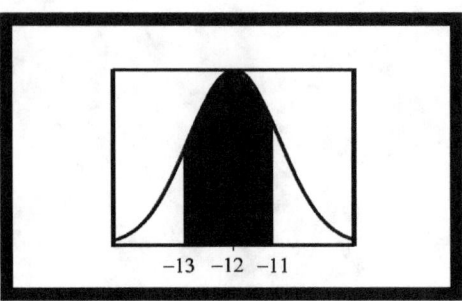

Answer: 0.6826

⇑ *Refer back to* **25.2 - Example 4 & 25.2 - Solved Problem 4.**

25.3 - Important Formulas.

1. $Z = \dfrac{X - \mu}{\sigma}$

2. $X = \mu + Z\sigma$

3. $\mu = X - Z\sigma$

4. $\sigma = \dfrac{X - \mu}{Z}$

25.3 - Example 1: Assume $\sigma = 2$ and $P\{X \le 13\} = 0.7054$ Find μ.

Solution:

Step 1: We use formula 3: $\mu = x - z\sigma$.

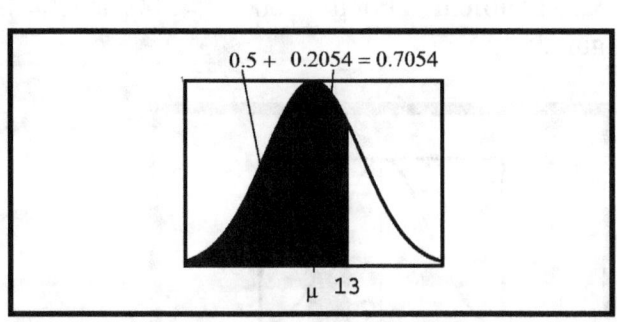

Step 2: $x = 13$

Step 3: From the table, for the area 0.2054,

$z = 0.54$.

Step 4: $\mu = x - z\sigma = 13 - 0.54(2) = 13 - 1.08$
$= 11.92$

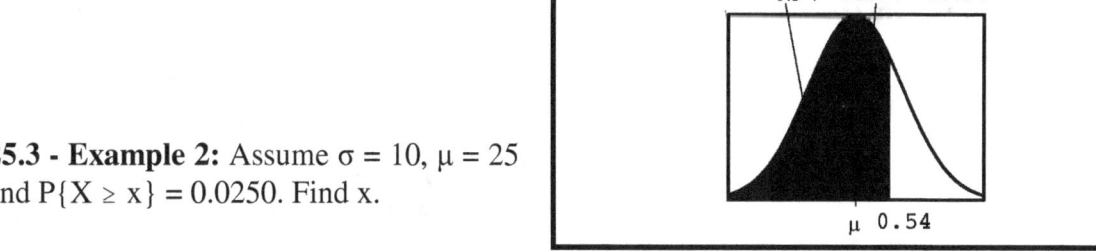

25.3 - Example 2: Assume $\sigma = 10$, $\mu = 25$ and $P\{X \geq x\} = 0.0250$. Find x.

Solution:

Step 1: Use formula 2: $x = \mu + z\sigma = 25 + z(10)$.

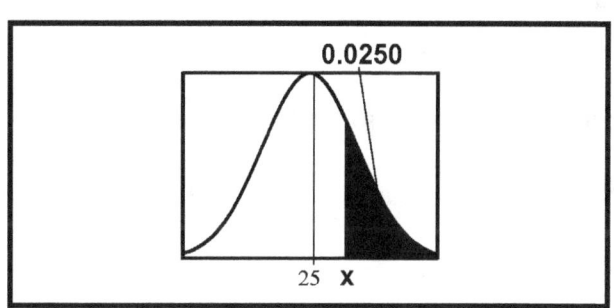

Step 2: To find z, use the area from the table $0.5 - 0.0250 = 0.4750$.

Step 3: From the table, $z = 1.96$

Step 4: Using the equation in Step 1 and $z = 1.96$ gives

$x = \mu + z\sigma = 25 + z(10) = 25 + 1.96(10) =$

$25 + 19.6 = 44.60$.

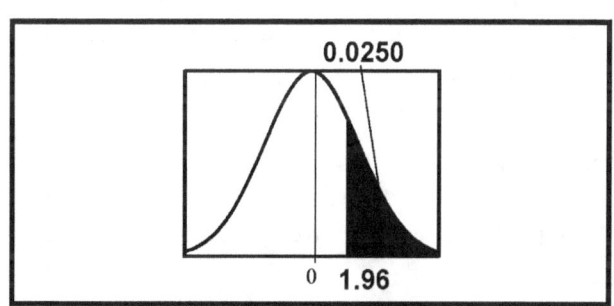

25.3 - Example 3: Assume $\mu = 4.75$ and

$P\{X < 7.51\} = 0.9808$. Find σ.

Solution:

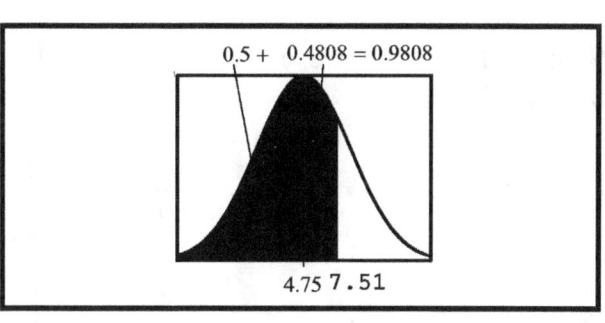

Step 1: Use formula 4: $\sigma = \dfrac{x - \mu}{z}$.

Step 2: Using the area portion of the table for 0.4808, we find $z = 2.07$.

Step 3:

$\sigma = \dfrac{x - \mu}{z} = \dfrac{7.51 - 4.75}{2.07} = \dfrac{2.76}{2.07} \approx 1.33$

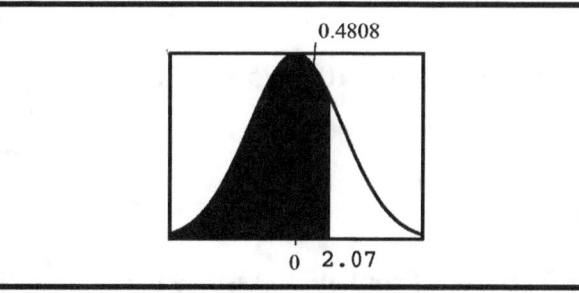

Solved Problems

25.3 - Solved Problem 1: Assume $\sigma = 10$ and
$P\{X \le 13\} = 0.0336$. Find μ.

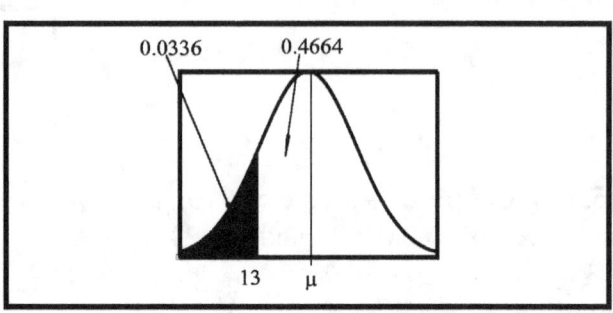

Solution:

Step 1: We use formula 3: $\mu = x - z\sigma$.

Step 2: $x = 13$

Step 3: Since 13 is to the left of μ, z will be a negative number.

Step 4: Using the area portion of the table for 0.4664 we find $z = -1.83$.

Step 5: $\mu = x - z\sigma = 13 - (-1.83)(10) =$
$13 + 18.3 = 31.3$

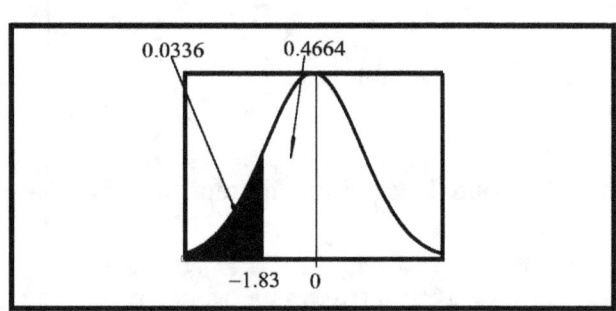

25.3 - Solved Problem 2: Assume $\sigma = 10$,
$\mu = 25$ and $P\{X \ge x\} = 0.9131$. Find x.

Solution:

Step 1: Use formula 2: $x = \mu + z\sigma =$
$25 + z(10)$

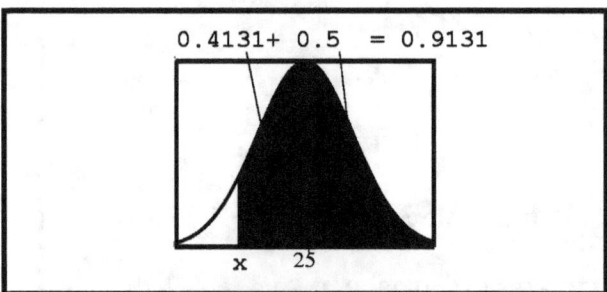

Step 2: To find z, we use $P\{X \ge x\} =$
0.9131 .

Step 3: In the area portion of the table, 0.4131 has a $z = -1.36$.

Step 4: Using the equation in Step 1 and $z =$
-1.36 gives

$x = \mu + z\sigma = 25 + z(10) = 25 + -1.36(10) =$
$25 + -13.6 = 11.4$.

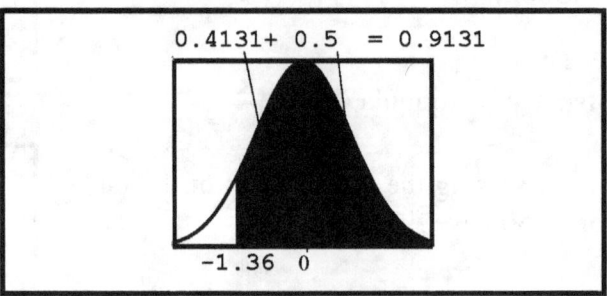

25.3 - Solved Problem 3: Assume $\mu = -5$ and $P\{X < 1\} = 0.9808$. Find σ.

Solution:

Step 1: Use formula 4: $\sigma = \dfrac{x - \mu}{z}$.

Step 2: $P\{X < 1\} = P\{X \le 1\} - P\{x = 1\}$

Step 3: Using the area portion of the table for 0.4808, we find $z = 2.07$.

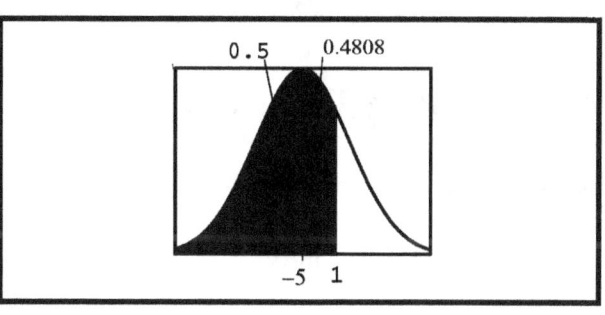

Step 4: $\sigma = \dfrac{x-\mu}{z} = \dfrac{1-(-5)}{2.07} = \dfrac{6}{2.07} \approx 2.90$

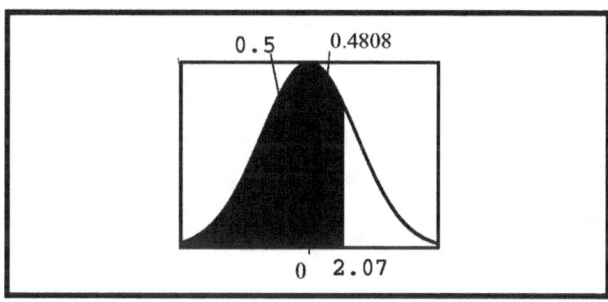

Unsolved Problems with Answers

25.3 - Problem 1: Assume $\sigma = 10$ and $P\{X \le 13\} = 0.5$. Find μ.

Answer:

13

⇑ *Refer back to* **25.3 - Example 1 & 25.3 - Solved Problem 1.**

25.3 - Problem 2: Assume $\sigma = 1$, $\mu = 2.5$ and $P\{X \ge x\} = 0.4129$. Find x.

Answer:

2.72

⇑ *Refer back to* **25.3 - Example 2 & 25.3 - Solved Problem 2.**

25.3 - Problem 3: Assume $\mu = 5$ and $P\{X < -7\} = 0.1736$. Find σ.

Answer:

$\sigma = 12.77$

⇑ *Refer back to* **25.3 - Example 3 & 25.3 - Solved Problem 3.**

Supplementary Problems

Assume Z and X are normally distributed

1. If $P\{Z < z\} = 0.003$, find z.

2. If $\mu = 2$, $\sigma = 5$, and $P\{X < x\} = 0.003$, find x.

3. Assume two distributions where $\mu = 10$ when $P\{X < x\} = 0.05$ and $\mu = 2$ when $P\{X > x\} = 0.01$
.
Find x and σ.

4. Assume $P\{\mu - 2 < X < \mu + 2\} = 0.2206$. Find σ.

5. Find:

a. $P\{\mu - 2\sigma \le X \le \mu + 2\sigma\}$

b. $P\{\mu - 3\sigma \le X \le \mu + 3\sigma\}$

6. Assume X has a standard deviation $\sigma = a$ and Y has a standard deviation 2a. Which is larger,

$P\{\mu - a \le X \le a + \mu\}$ or $P\{\mu - a \le Y \le a + \mu\}$?

7. If $P\{X \le -5\} = 0.01$ and $P\{X \ge 7\} = 0.05$. Find μ and σ.

8. From the equation $z = \dfrac{x - \mu}{\sigma}$, algebraically derive:

a. $x = \mu + z\sigma$

b. $\mu = x - z\sigma$

c. $\sigma = \dfrac{x - \mu}{z}$.

For the following problems, shade the appropriate area under the normal distribution and find the area.

9. $P\{Z \le -2.77\}$

10. $P\{Z > 0.13\}$

11. $P\{-2.79 \le Z \le 3.33\}$

12. $P\{2.15 \le Z \le 3.20\}$

For the following problems, shade the appropriate area under the normal distribution and find z.

13. $P\{Z \le z\} = 0.9671$

14. $P\{Z \le z\} = 0.1492$

For the following problems, shade the appropriate area under the normal distribution and find the area.

15. $P\{Z \le -2\}$

16. $P\{1.13 < Z\}$

17. $P\{-3.11 \le Z \le 3.11\}$

18. Assume $\mu = 5$, $\sigma = 2$. Find $P\{X \le 10\}$.

19. Assume $\mu = 15$, $\sigma = 3$. Find $P\{10 \le X \le 19\}$.

20. Assume $\mu = 15$, $\sigma = 3$. Find $P\{X \ge 18\} + P\{X \le 12\}$.

21. If $P\{X \ge 10\} = 0.4$ and $P\{X \le 5\} = 0.3$, find μ and σ.

For the remaining problems assume Z is a random variable with the standard distribution and X are random variables with means standard distributions μ, σ.

22. Find $P(Z \ge -1 \mid Z \le 1)$.

23. Find x given $P(X \le x \mid X > 2) = 0.70$, $\mu = 4$, $\sigma = 2$.

24. Find μ, σ, given $P(X \ge 5 \mid X \le 10) = 0.40$, $P(X \le 10) = 0.85$.

25. Assume X has a normal distribution. Show that $P(X = x) = 0$.

For all examples and problems in this lesson, we assume a normal distribution. We also will need the following formulas from the previous lesson:

$$1 \quad Z = \frac{X - \mu}{\sigma}$$

$$2. \quad X = \mu + Z\sigma$$

$$3. \quad \mu = X - Z\sigma$$

$$4. \quad \sigma = \frac{X - \mu}{Z}$$

26.1- Real Life Applications

26.1 - Example 1: Recently a computer trade journal published a study on the number of hours computer programmers worked each day, in the year 1999. The article stated that the average number of hours worked each day is $\mu = 10$ hours with a standard deviation of $\sigma = 3$ hours. Assume a computer programmer is selected at random from the study. Find the probability that the number of hours he/she **1** worked each day in 1999 was

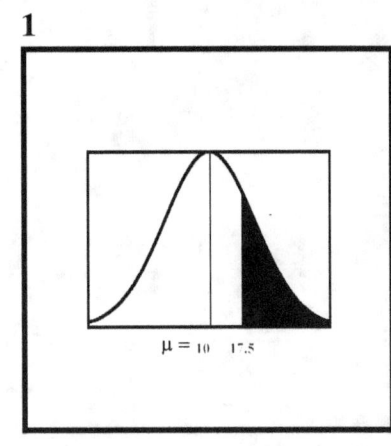

(a). greater than 17.5 hours.

(b). between 12.5 and 15 hours.

(c). less than 16.5 hours.

Solutions:

For each problem we need the conversion formula:

$$Z = \frac{X - \mu}{\sigma} \; .$$

➤(a).
Step 1: $P\{X \geq 17.5\} = 0.5 - P\{10 \leq X \leq 17.5\}$
fig. 1
Step 2: Since x = 17.5,

$$z = \frac{x - \mu}{\sigma} = \frac{017.5 - 10}{3} = 2.5$$

Step 3: $P\{10 \leq X \leq 17.5\} = P\{0 \leq Z \leq 2.5\}$
fig. 2

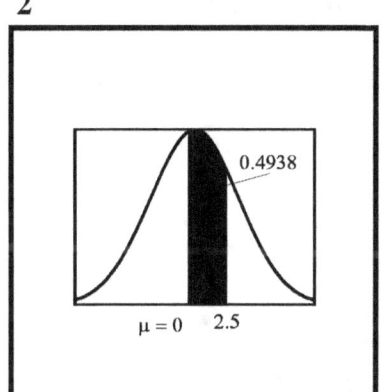

Step 4: From the normal distribution table for z = 2.5

$P\{10 \leq X \leq 17.5\} = P\{0 \leq Z \leq 2.5\} = 0.4938$

Step 5: $P\{X \geq 17.5\} = 0.5 - P\{10 \leq X \leq 17.5\} =$

$0.5 - 0.4938 = 0.0062$.

➤(b).
fig. 3

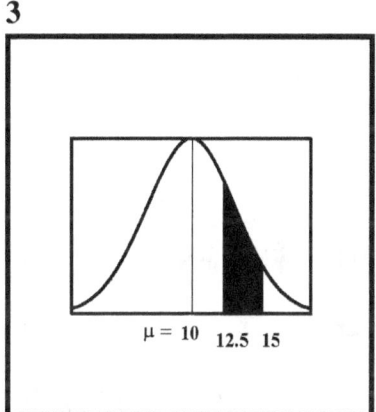

Step 1: $P\{12.5 \leq X \leq 15\} =$

$P\{10 \leq X \leq 15\} - P\{10 \leq X \leq 12.5\}$

Step 2: For x = 15

$$z = \frac{x - \mu}{\sigma} = \frac{15 - 10}{3} = 1.67$$

fig. 4

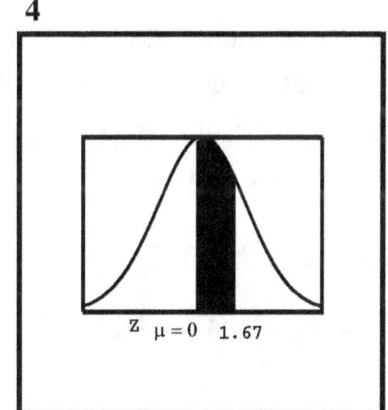

Step 3: $P\{10 \leq X \leq 15\} =$

$P\{0 \leq Z \leq 1.67\} = 0.4525$, from the table.

Step 4: For x = 12.50

$$z = \frac{x - \mu}{\sigma} = \frac{12.5 - 10}{3} \approx 0.83$$

fig. 5

Step 5: From the table, $P\{10 \leq X \leq 12.5\} =$

$P\{0 \leq Z \leq 0.833\} = 0.2967$

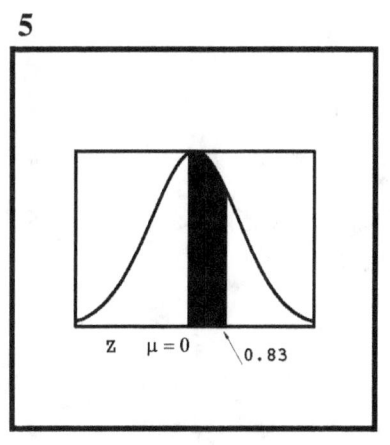

Step 6: $P\{12.5 \leq X \leq 15\} =$

$P\{10 \leq X \leq 15\} - P\{10 \leq X \leq 12.5\} =$

$P\{0 \leq Z \leq 1.67\} - P\{0 \leq Z \leq 0.833\} = 0.4525 - 0.2967 = 0.1558$

fig. 6

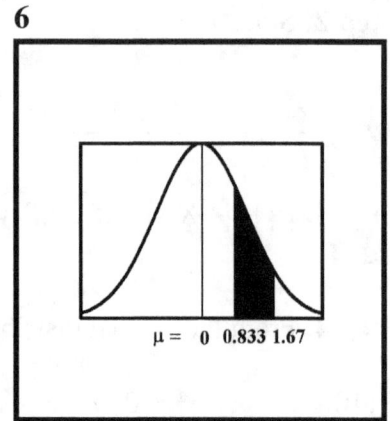

➤(c).
Step 1: $P\{X < 16.5\} = 0.5 + P\{10 \leq X \leq 16.5\}$

Step 2: For x = 16.5,

$$z = \frac{x - \mu}{\sigma} = \frac{16.5 - 10}{3} \approx 2.17$$

fig. 7

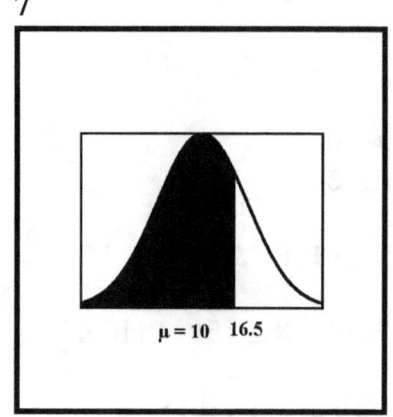

Step 3: From the table, $P\{0 \leq Z \leq 2.17\} = 0.4850$

Step 4: $P\{X < 16.5\} = 0.5 + P\{10 \leq X \leq 16.5\} =$

$0.5 + P\{0 \leq Z \leq 2.17\} = 0.5 + 0.4850 = 0.9850$

26.1 - Example 2: Ms. Jones wrote a computer program that computes random numbers with a mean $\mu = 45.60$ and a standard deviation $\sigma = 5.25$. Find the probability that a random number generated is

(a). greater than 35.75 .

(b). between 50 and 55.

(c). less than 55.25 .

(d). between 42.11 and 47.51.

Solutions:

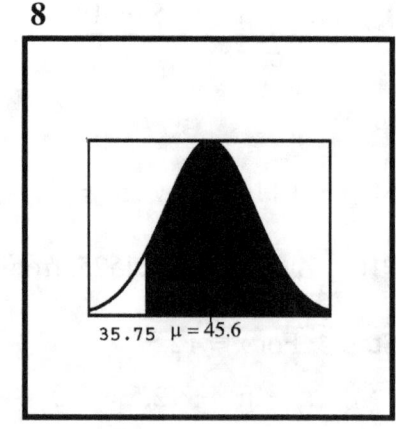

➤(a).
Step 1: $P\{35.75 < X\} = P\{35.75 \leq X \leq 45.60\} + 0.5$
fig. 8

Step 2: For x = 35.75,

$$z = \frac{x - \mu}{\sigma} = \frac{35.75 - 45.60}{5.25} = -1.876$$

fig. 9

Step 3: $P\{35.75 \leq X \leq 45.60\} = P\{-1.876 \leq Z \leq 0\} =$

$P\{0 \leq Z \leq 1.876\} = 0.4699$

Step 4: $P\{35.75 < X\} = P\{35.75 \leq X \leq 45.60\} + 0.5 =$
$0.5 + 0.4699 = 0.9699$

➤**(b).**
Step 1: $P\{50 \leq X \leq 55\} =$
$P\{45.60 \leq X \leq 55\} - P\{45.60 \leq X \leq 50\}$

Step 2: For x = 55,

$$z = \frac{x - \mu}{\sigma} = \frac{55 - 45.60}{5.25} = 1.79$$

fig 10

Step 3: For x = 50,

$$z = \frac{x - \mu}{\sigma} = \frac{50 - 45.60}{5.25} = 0.84$$

fig. 11

Step 4: $P\{50 \leq X \leq 55\} =$
$P\{45.60 \leq X \leq 55\} - P\{45.60 \leq X \leq 50\} =$

$P\{0 \leq Z \leq 1.79\} - P\{0 \leq Z \leq 0.84\} = 0.4633 - 0.2995 = 0.1638$

➤**(c).**
Step 1: $P\{X < 55.25\} = P\{45.6 \leq X \leq 55.25\} + 0.5$
fig. 12

Step 2: For x = 55.25,

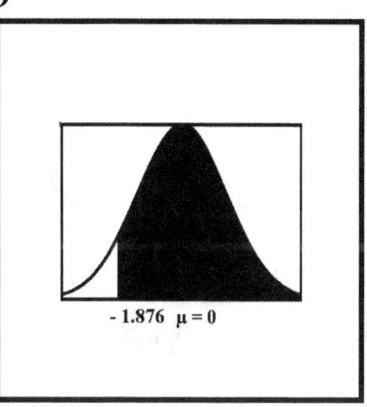

9

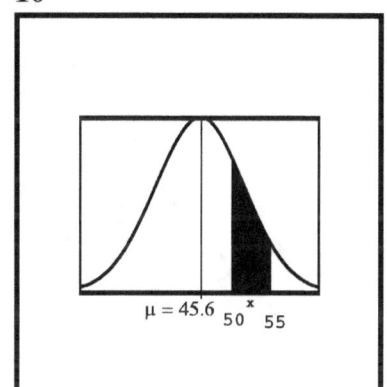

10

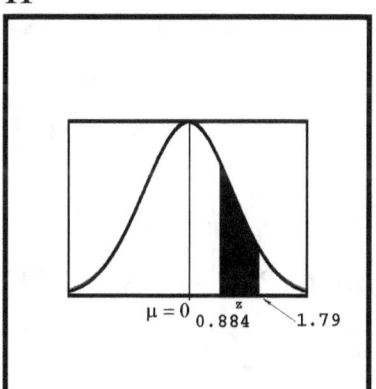

11

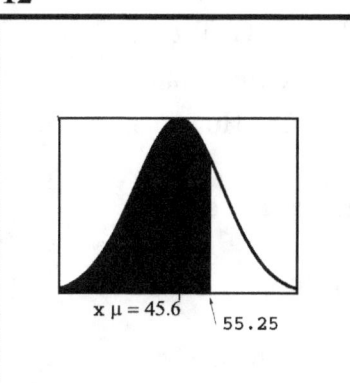

12

$$z = \frac{x - \mu}{\sigma} = \frac{55.25 - 45.60}{5.25} = 1.84$$

fig. 13

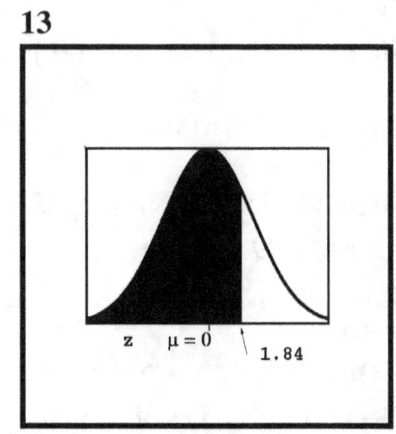

13

Step 3: $P\{45.6 \le X \le 55.25\} = p\{0 \le Z \le 1.84\} = 0.4671$

Step 4: $P\{X < 55.25\} = P\{45.6 \le X \le 55.25\} + 0.5 =$

$0.4671 + 0.5 = 0.9671$

➤**(d).**
Step 1: $P\{42.11 \le X \le 47.51\} =$

$P\{45.60 \le X \le 47.51\} + P\{42.11 \le X \le 45.60\}$

Step 2: For x = 47.51,

$$z = \frac{x - \mu}{\sigma} = \frac{47.51 - 45.60}{5.25} = 0.36$$

fig. 14

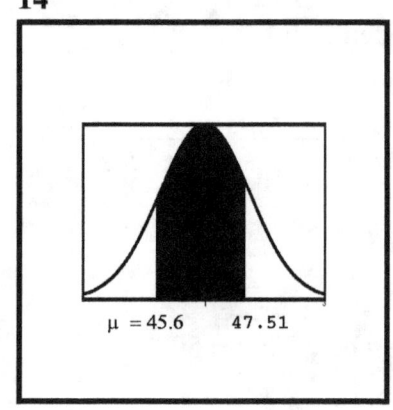

14

Step 3: $P\{45.60 \le X \le 47.51\} = P\{0 \le Z \le 0.36\} = 0.1406$

Step 4: For x = 42.11,

$$z = \frac{x - \mu}{\sigma} = \frac{42.11 - 45.60}{5.25} = -0.66$$

Step 5: $P\{42.11 \le X \le 45.60\} = P\{-0.66 \le Z \le 0 \} =$

$P\{0 \le Z \le 0.66 \} = 0.2454$

Step 6: $P\{42.11 \le X \le 47.51\} = P\{46.60 \le X \le 47.51\} + P\{42.11 \le X \le 45.60\} =$

$P\{-0.66 \le Z \le 0\} + P\{0 \le Z \le 0.36\} = 0.2454 + 0.1406 = 0.3860$

26.1 - Example 3: Professor Kline has 125 students in his Western Civilization class. At the end of the semester, the average final grade was $\mu = 68.85$ and a standard deviation $\sigma = 7.25$. How many students received a final score

(a). better than 90?

(b). between 70 and 85?

(c). less than 60.50?

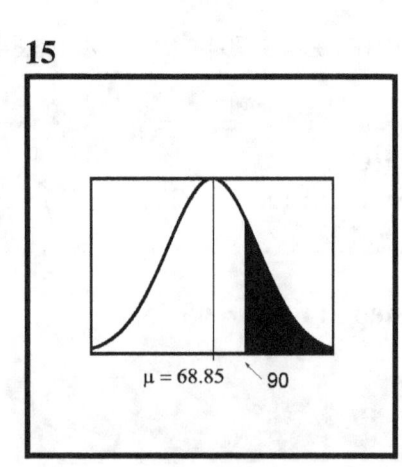

15

(d). between 60 and 70?

16

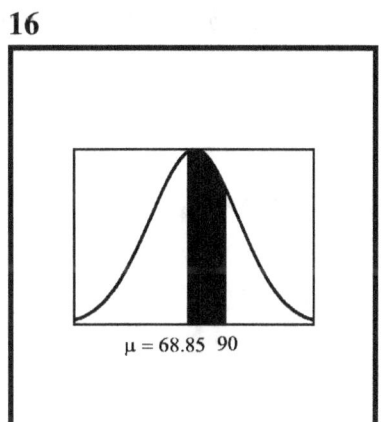

μ = 68.85 90

Solutions:

➤(a).

Step 1: $P\{X > 90\} = P\{X \geq 90\} = 0.5 - P\{68.85 \leq X \leq 90\}$
fig. 15

Step 2: For x = 90,

$$z = \frac{x - \mu}{\sigma} = \frac{90 - 68.85}{7.25} = 2.92$$

Step 3: $P\{68.85 \leq X \leq 90\} = P\{0 \leq Z \leq 2.92\} = 0.4982$
fig 16

17

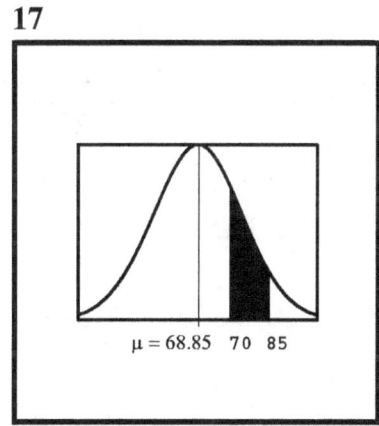

μ = 68.85 70 85

Step 4: $P\{X > 90\} = P\{X \geq 90\} = 0.5 - P\{68.85 \leq X \leq 90\} = 0.5 - 0.4982 = 0.0018$

Step 5: $125(0.0018) \approx 0$ students

➤(b).
Step 1: $P\{70 \leq X \leq 85\} =$
$P\{68.85 \leq X \leq 85\} - P\{68.85 \leq X \leq 70\}$

fig 17

Step 2: For x = 85,

$$z = \frac{x - \mu}{\sigma} = \frac{85 - 68.85}{7.25} = 2.23$$

18

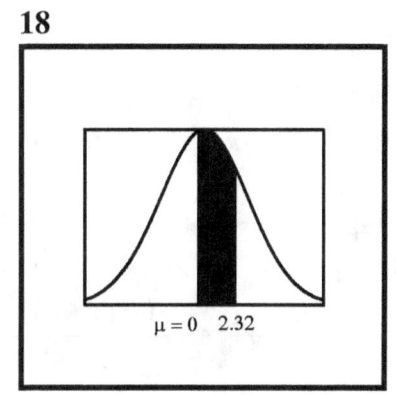

μ = 0 2.32

Step 3: $P\{68.85 \leq X \leq 85\} = P\{0 \leq Z \leq 2.23\} = 0.4871$

Step 4: For x = 70,

fig. 18

$$z = \frac{x - \mu}{\sigma} = \frac{70 - 68.85}{7.25} = 0.16$$

Step 5: $P\{8.85 \le X \le 70\} = P\{0 \le Z \le 0.16\} = 0.0636$

fig. 19

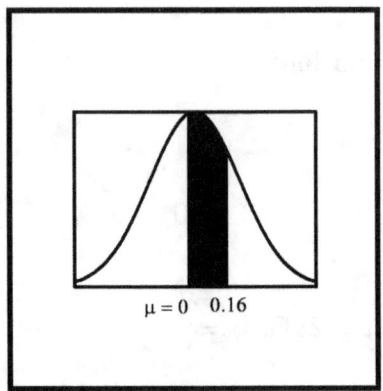

19

Step 6: $P\{70 \le X \le 85\} =$
$P\{68.25 \le X \le 85\} - P\{68.25 \le X \le 70\} =$

$0.4871 - 0.0636 = 0.4235$

fig. 20

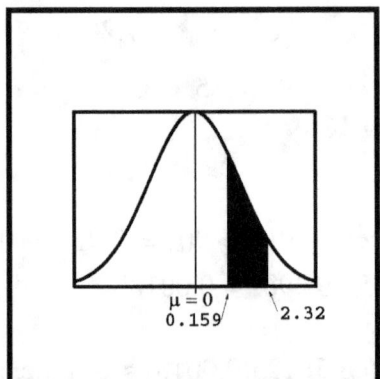

20

Step 7: $125(0.4235) \approx 53$ students

➤(c).
Step 1: $P\{X \le 60.50\} = 0.5 - P\{60.50 \le X \le 68.8\}$
fig. 21

Step 2: For x = 60.50,

$$z = \frac{x - \mu}{\sigma} = \frac{60.50 - 68.85}{7.25} = -1.15$$

Step 3: $P\{60.50 \le X \le 68.25\} = P\{-1.15 \le Z \le 0\} =$
$P\{0 \le Z \le 1.15\}$

Step 4: $P\{0 \le Z \le 1.15\} = 0.3749$

fig. 22

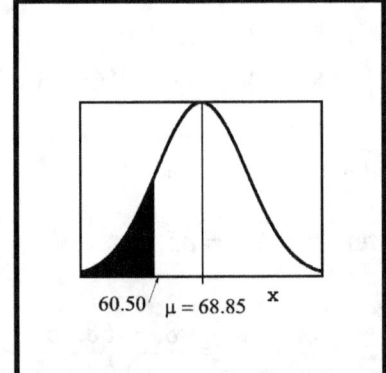

21

Step 5: $P\{X \le 60.50\} = 0.5 - P\{60.50 \le X \le 68.25\} =$

$0.5 - 0.3749 = 0.1251$

Step 6: $125(0.1251) \approx 16$ students

➤(d).
Step 1: $P\{60 \le X \le 70\} =$
$P\{60 \le X \le 68.85\} + P\{68.85 \le X \le 70\}$

Step 2: For x = 60,
fig. 23

$$z = \frac{x - \mu}{\sigma} = \frac{60 - 68.85}{7.25} = -1.22 = -1.22$$

fig. 24
Step 3: $P\{60 \le X \le 68.85\} = P\{-1.22 \le Z \le 0\} =$

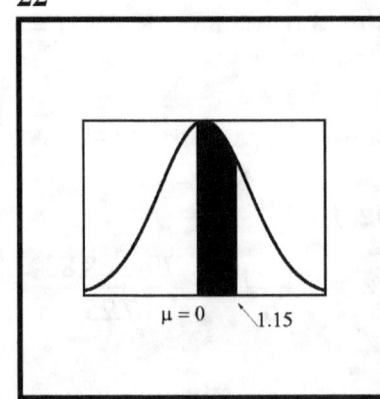

22

$P\{0 \le Z \le 1.22\} = 0.3888.$

Step 4: For x, = 70
fig. 25

$$z = \frac{x - \mu}{\sigma} = \frac{70 - 68.85}{7.25} = 1.66$$

Step 5: $P\{68.85 \le X \le 70\} = P\{0 \le Z \le 0.16\} = 0.0636$

Step 6: $P\{60 \le X \le 70\} =$
$P\{60 \le X \le 68.85\} + P\{68.85 \le X \le 70\} =$
$0.3888 + 0.0636 = 0.4524$

Step 6: $125(0.4524) \approx 57$ students **fig. 26**

23

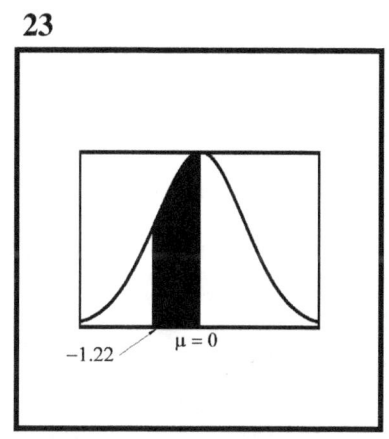

24

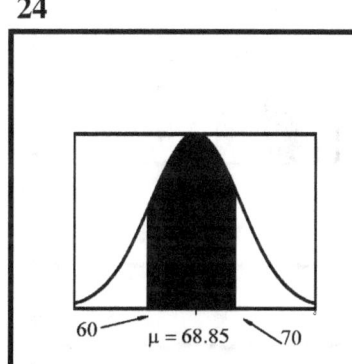

25

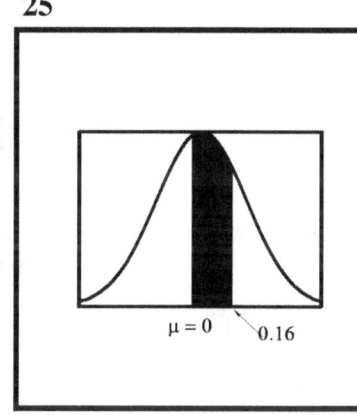

26

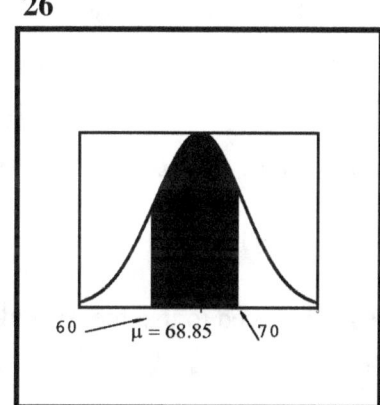

26.1 - Example 4: A newly designed machine fills liquid into bottles. The machine is so designed that the user can set the mean μ and standard deviation σ to any desired values. Assume the machine is purchased by the Bubble Bottling Company to fill 16 oz of soda into bottles.

(a). State regulations require that at most 2% of the bottles can contain less than 16 oz. Assuming the standard deviation is set so that $\sigma = 0.02$ oz, find the value of μ that the machine should be set to.

(b). The Company is also concerned about overfilling the bottles. They want the mean and standard deviation set at values so that only 2% of the bottles contain less than 16 oz and at most 5% of the bottles can contain more than 16.5 oz. Find μ, σ.

(c). The Company finally decides to set the mean at $\mu = 16.25$ and the standard deviation at a value to assure that 96% of all bottles filled contain between 16 oz and 16.5 ounces of soda. Find σ.

27

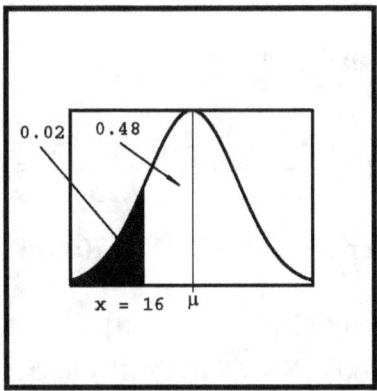

Solutions:

➤**(a).**
We need to use the formula μ = x - zσ.
fig. 27

Step 1: P{X ≤ 16} = 0.02

Step 2: P{16 ≤ X ≤ μ} = 0.5 - P{X ≤ 16} = 0.5 - 0.02 = 0.48

28

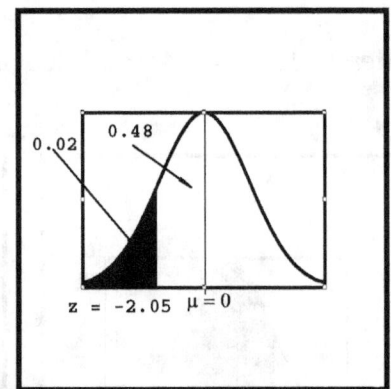

Step 3: P{z ≤ Z ≤ 0} = P{16 ≤ X ≤ μ} = 0.48

Step 4: From the area portion of the table, we find for area = 0.48 that z = -2.05 .
fig. 28

Step 5: Since x = 16,

μ = x - zσ = 16 - (-2.05)(0.02) = 16+ 0.041 = 16.041 oz.

➤**(b).** We need to use the formula μ = x - zσ twice.
fig. 29

29

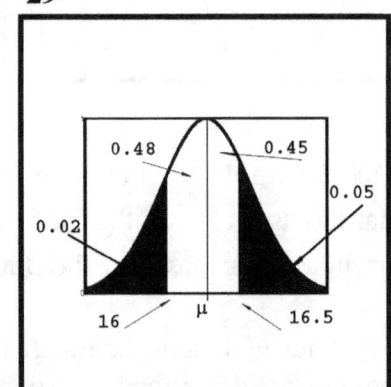

Step 1: P{X ≤ 16} = 0.02

Step 2: P{16 ≤ X ≤ μ} = 0.5 -P{X ≤ 16} = 0.5 - 0.02 = 0.48

Step 3: P{z ≤ Z ≤ 0} = P{16 ≤ X ≤ μ} = 0.48

Step 4: From the area portion of the table, we find for area = 0.48 that z = -2.05 .

Step 5: Since x = 16, μ = x - zσ = 16 - (-2.05σ) = 16 + 2.05σ

30

Step 6: P{X ≥ 16.5} = 0.05
fig. 30

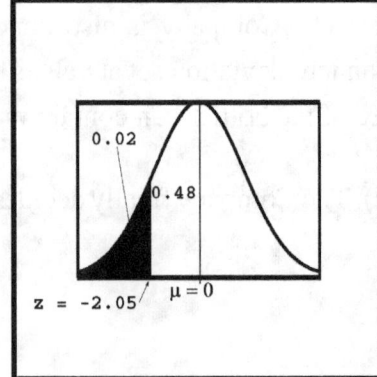

Step 7: P{μ ≤ X ≤ 16.5} = 0.5 - P{16.5 ≤ X} = 0.5 - 0.05 = 0.45

Step 8: P{0 ≤ Z ≤ z} = P{μ ≤ X ≤ 16.5} = 0.45

Step 9: From the area portion of the table, we find for area = 0.45 that z = 1.65 .
fig. 31

Step 10: Since x = 16.5, μ = x - zσ = 16.5 - 1.65σ

Step 11: We have two equations to solve for μ and σ:

μ = 16 + 2.05σ

μ = 16.5 - 1.65σ.

31

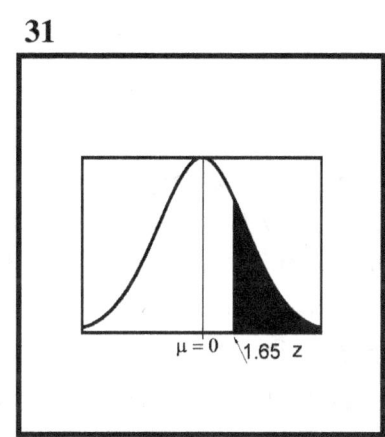

Step 12: Since both means μ are the same, we set the two equations equal and solve for σ:

16 + 2.05σ = 16.5 - 1.65σ.

Step 13: Using elementary algebra,

2.05σ + 1.65σ = 16.5 - 16

3.70σ = 0.5

σ = $\dfrac{0.5}{3.70}$ $\approx$ 0.135

32

Step 14: μ = 16 + 2.05σ = 16 + 2.05(0.135) $\approx$ 16.28

➤(c). We need the formula σ = $\dfrac{x - \mu}{z}$.
fig. 32

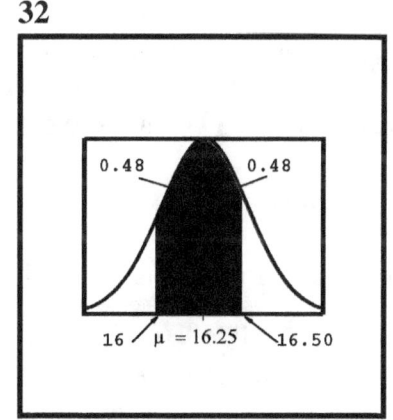

Step 1: For the x value we use x = 16.50 .

Step 2: The value z_0 associated with x = 16.50 can be found since

P{16.25 $\leq$ x $\leq$ 16.50} = P{0 $\leq$ z $\leq$ z_0} = 0.96/2 = 0.48 .

Step 3: Looking in the area portion of the normal distribution table, we find z_0 = 2.05 .

Step 4: σ = $\dfrac{x - \mu}{z}$ = $\dfrac{16.50 - 16.25}{2.05}$ $\approx$ 0.12

26.1 - Example 5: The grades in Mr. Smith's statistics class is normally distributed with an average grade of 65.5 and a standard deviation of 5. Based on these scores, he wants to give 15% of his class a final grade of A. Find the lowest score a student needs to earn an A.

Solution:

Use the formula $x = \mu + z\sigma$.

Step 1: Since $\mu = 65.5$ and $\sigma = 5$, $x = \mu + z\sigma = 65.5 + z5$

Step 2: From the figure, we find z by looking up the value 0.35, in the normal distribution table.

Step 3: From this table $z = 1.04$. Therefore,

$x = \mu + z\sigma = 65.5 + z5 = 65.5 + 1.04(5) = 70.70$

33

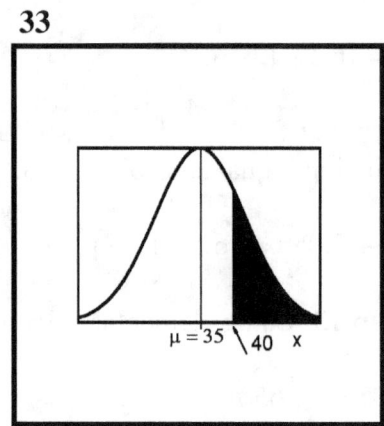

Solved Problems

26.1 - Solved Problem 1: The average time it takes an employee of the Bubble Corp. to get to work in the morning is $\mu = 35$ minutes with a standard deviation of $\sigma = 3.25$ minutes. An employee is selected at random. Find the probability that on the following morning, it takes him or her

(a). greater than 40 minutes to get to work.

(b). between 40 and 45 minutes to get to work.

34

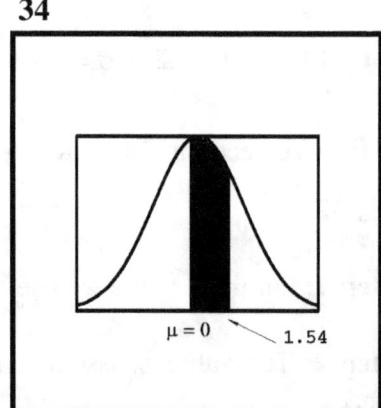

(c). less than 39 minutes to get to work.

Solutions:

For each of problems we need the conversion formula:

$$z = \frac{x - \mu}{\sigma}.$$

➤**(a).**
Step 1: $P\{X > 40\} = 0.5 - P\{35 \le X \le 40\}$
fig. 33

35

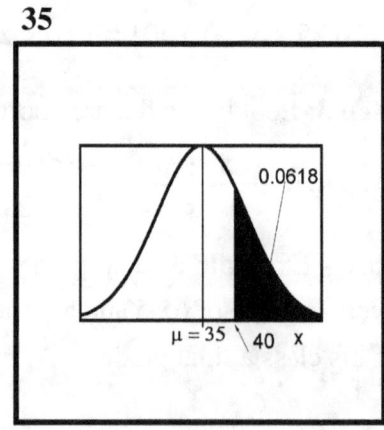

Step 2: Since $x = 40$,

$$z = \frac{x - \mu}{\sigma} = \frac{40 - 35}{3.25} = 1.54$$
fig. 34

Step 3: $P\{35 \le X \le 40\} = P\{0 \le Z \le 1.54\}$

Step 4: From the normal distribution table for $z = 1.54$,

$P\{35 \le X \le 40\} = P\{0 \le Z \le 1.54\} = 0.4382$.
fig. 35

Step 5: $P\{X > 40\} = 0.5 - P\{35 \le X \le 40\} = 0.5 - 0.4382 = 0.0618$

➤**(b).**
Step 1: $P\{40 \le X \le 45\} = P\{35 \le X \le 45\} - P\{35 \le X \le 40\}$
fig. 36

Step 2: For x = 40,

$$z = \frac{x - \mu}{\sigma} = \frac{40 - 35}{3.25} = 1.54$$

fig. 37

Step 3: $P\{35 \le X \le 40\} = P\{0 \le Z \le 1.54\} = 0.4382$

Step 4: For x = 45,

$$z = \frac{x - \mu}{\sigma} = \frac{45 - 35}{3.25} = 3.08$$

fig. 38

Step 5: $P\{35 \le X \le 45\} = P\{0 \le Z \le 3.08\} = 0.4990$
fig. 39

Step 6: $P\{40 \le X \le 45\} = P\{35 \le X \le 45\} - P\{35 \le X \le 40\}$ =

$P\{0 \le Z \le 3.08\} - P\{0 \le Z \le 1.54\} = 0.4990 - 0.4382 = 0.0608$

➤**(c).**
Step 1: $P\{X < 39\} = 0.5 + P\{35 \le X \le 39\}$

Step 2: For x = 39,

$$z = \frac{x - \mu}{\sigma} = \frac{39 - 35}{3.25} = 1.23$$

fig. 40
Step 3: $P\{35 \le X \le 39\} = P\{0 \le Z \le 1.23\} = 0.3907$

Step 4: $P\{X < 39\} = 0.5 + P\{35 \le X \le 39\} = 0.5 + P\{0 \le Z \le 1.23\}$ =

$0.5 + 0.3907 = 0.8907$

36

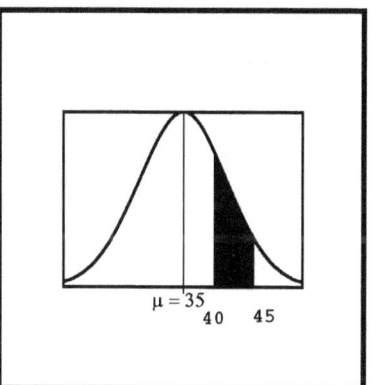

37

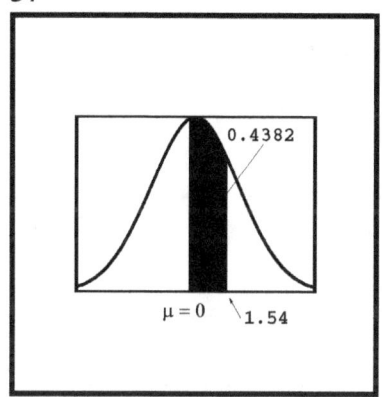

38

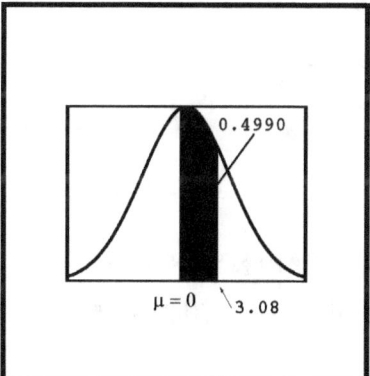

39

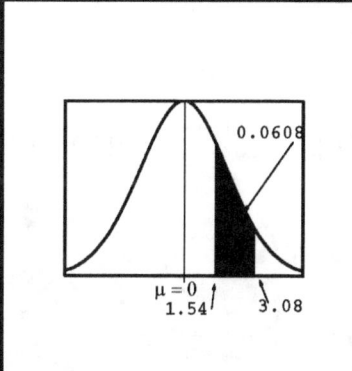

26.1 - Solved Problem 2: The U.S. Army recently reported that the average height of a male combat soldier is $\mu = 70.2''$ with standard deviation $\sigma = 1.5''$. A male combat soldier is selected at random. Find the probability his height is

40

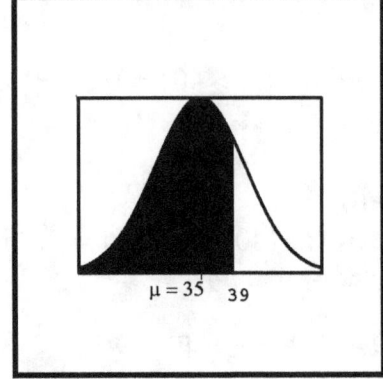

$\mu = 35$ 39

(a). greater than 68.5".

(b). between 72" and 74".

(c). less than 72.25".

(d). between 65" and 72".

41

Solutions:

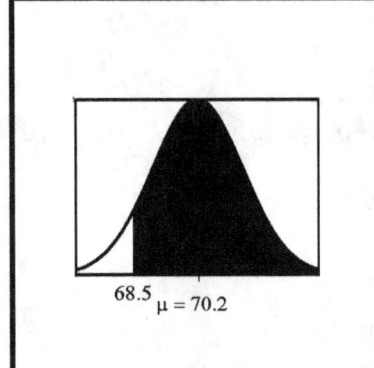

68.5 $\mu = 70.2$

➤(a).

Step 1: $P\{X > 68.5\} = P\{68.5 \le X \le 70.2\} + 0.5$
fig. 41

Step 2: For x = 68.5,

$$z = \frac{x - \mu}{\sigma} = \frac{68.5 - 70.20}{1.5} = -1.13$$

fig. 42

42

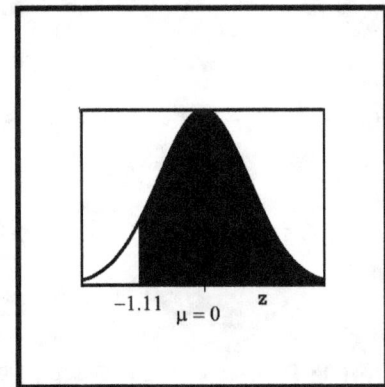

−1.11 $\mu = 0$ **z**

Step 3: $P\{68.5 \le X \le 70.2\} = P\{-1.13 \le Z \le 0\} =$

$P\{0 \le Z \le 1.13\} = 0.3708$

Step 4: $P\{X > 68.5\} =$
$P\{68.5 \le X \le 70.2\} + 0.5 = 0.5 + 0.3708 =$
0.8708
fig. 43

43

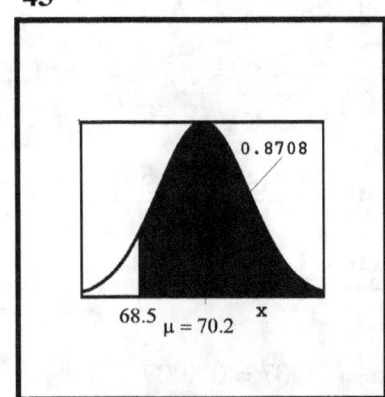

0.8708

68.5 $\mu = 70.2$ **x**

➤(b).
Step 1: $P\{72 \le X \le 74\} =$

$P\{70.2 \le X \le 74\} - P\{70.2 \le X \le 72\}$

Step 2: For x = 74,

$$z = \frac{x - \mu}{\sigma} = \frac{74 - 70.20}{1.5} = 2.53$$

fig. 44

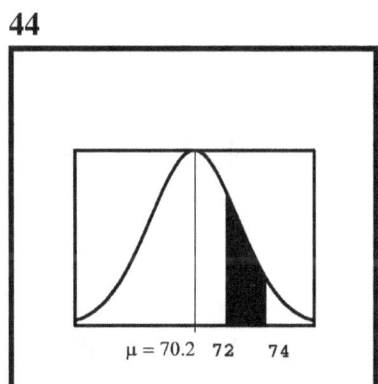

Step 3: For x = 72,

$$z = \frac{x - \mu}{\sigma} = \frac{72 - 70.20}{1.5} = 1.2$$

Step 4: P{72 ≤ X ≤ 74} =
P{70.2 ≤ X ≤ 74} - P{70.2 ≤ X ≤ 72} =

P{0 ≤ Z ≤ 2.53} - P{0 ≤ Z ≤ 1.2} = 0.4943 - 0.3849 = 0.1094

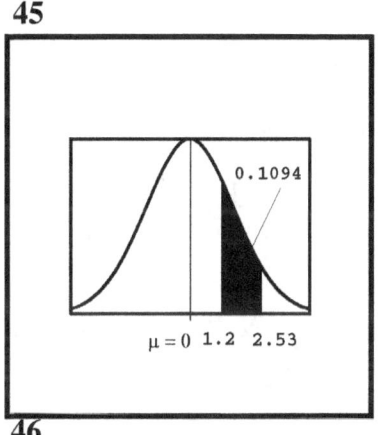

fig. 45

➤(c).

Step 1: P{X < 72.25} = P{70.2 ≤ X ≤ 72.25} + 0.5

fig. 46

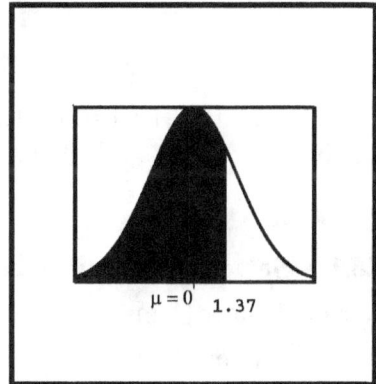

Step 2: For x = 72.25,

$$z = \frac{x - \mu}{\sigma} = \frac{72.25 - 70.20}{1.5} = 1.37$$

fig. 47

Step 3: P{70.2 ≤ X ≤ 72.25} = P{0 ≤ Z ≤ 1.37} = 0.4147

Step 4: P{X < 72.25} = P{70.2 ≤ X ≤ 72.25} + 0.5
= 0.4147 + 0.5 = 0.9147

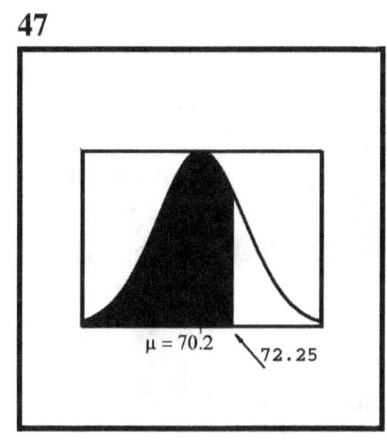

48

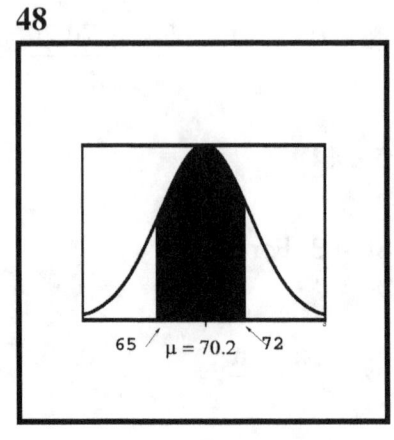

➤(d).

Step 1: $P\{65 \le X \le 72\} = P\{65 \le X \le 70.2\} + P\{70.2 \le X \le 72\}$

Step 2: For x = 72,

$$z = \frac{x - \mu}{\sigma} = \frac{72 - 70.20}{1.5} = 1.12$$

fig. 48

49

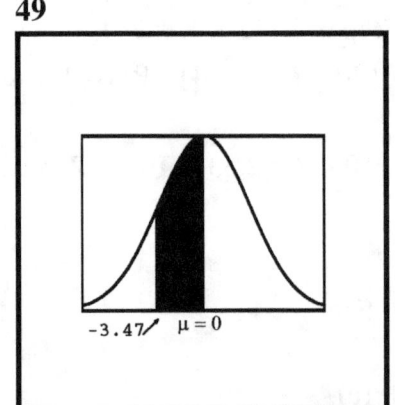

Step 3: $P\{65 \le X \le 72\} = P\{0 \le Z \le 1.2\} = 0.3849$

Step 4: For x = 65,

$$z = \frac{x - \mu}{\sigma} = \frac{65 - 70.20}{1.5} = -3.47$$

fig. 49

Step 5: $P\{65 \le X \le 70.2\} = P\{-3.47 \le Z \le 0\} =$
$P\{0 \le Z \le 3.47\} = 0.5$

50

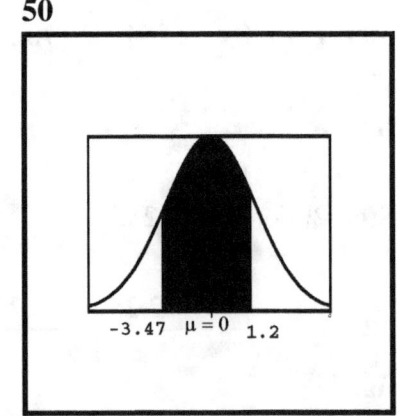

Step 6: $P\{65 \le X \le 72\} = P\{65 \le X \le 70.2\} + P\{70.2 \le X \le 72\}$
=

$P\{-3.2 \le Z \le 0\} + P\{0 \le Z \le 1.2\} = 0.3849 + 0.5 = 0.8849$
fig. 50

26.1 - Solved Problem 3: Last year, a small city collected recycled paper from 10,000 homes. The average weight of this paper per household was $\mu = 18.0$ lbs per month with a standard deviation $\sigma = 2$ lbs. How many households per month recycled

51

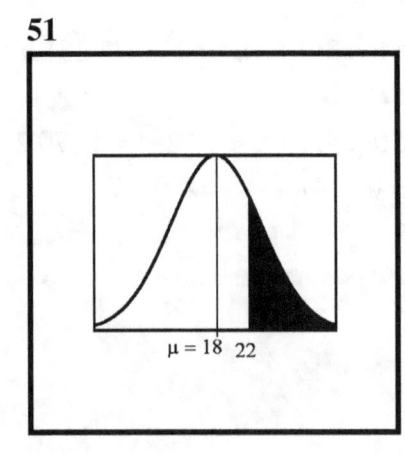

(a). more than 22 lbs?

(b). between 25 and 35 lbs?

(c). less than 16.5 lbs?

(d). between 16 and 20 lbs?

Solutions:

52

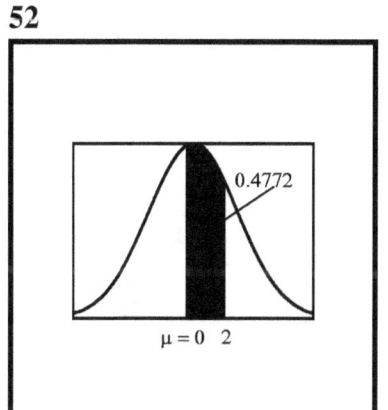

➤ **(a).**
Step 1: $P\{X > 22\} = P\{X \geq 22\} = 0.5 - P\{18 \leq X \leq 22\}$
fig. 51

Step 2: For x = 22,

$$z = \frac{x - \mu}{\sigma} = \frac{22 - 18}{2} = 2$$

Step 3: $P\{18 \leq X \leq 22\} = P\{0 \leq Z \leq 2\} = 0.4772$
fig. 52

53

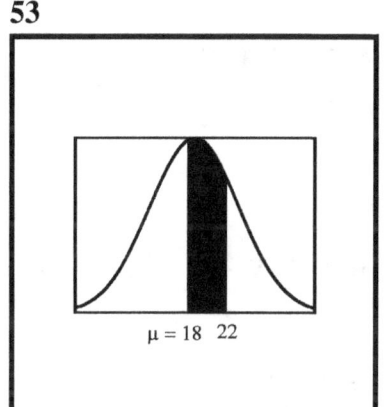

Step 4: $P\{X > 22\} = P\{X \geq 22\} = 0.5 - P\{18 \leq X \leq 22\} =$

$0.5 - 0.4772 = 0.0228$
fig. 53

Step 5: $10{,}000(.0228) \approx 228$ homes

➤**(b).**
Step 1: $P\{25 \leq X \leq 35\} = P\{18 \leq X \leq 35\} - P\{18 \leq X \leq 25\}$

54

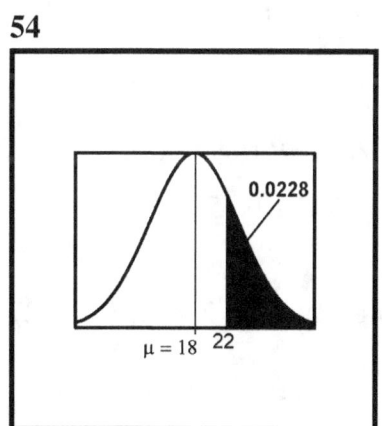

fig. 54

Step 2: For x = 35,

$$z = \frac{x - \mu}{\sigma} = \frac{35 - 18}{2} = 8.5$$

fig. 55

Step 3: $P\{18 \leq X \leq 35\} = P\{0 \leq Z \leq 8.5\} = 0.5$, since we are off the table.

Step 4: For x = 25,

$$z = \frac{x - \mu}{\sigma} = \frac{25 - 18}{2} = 3.5$$

fig. 56
Step 5: $P\{18 \leq X \leq 25\} = P\{0 \leq Z \leq 3.5\} = 0.4998$

Step 6: $P\{25 \leq X \leq 35\} = P\{18 \leq X \leq 35\} - P\{18 \leq X \leq 25\} =$
$0.5 - 0.4998 = 0.0002$

56

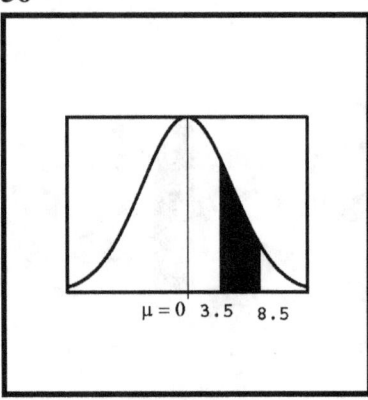

55

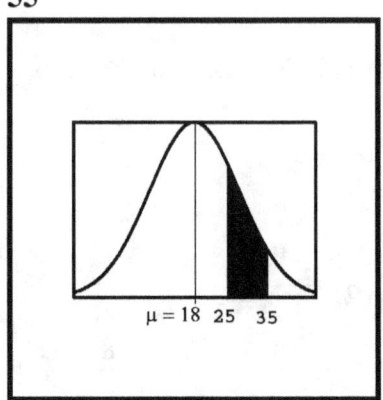

Step 7: $10,000(0.0002) \approx 2$ households

fig. 57

57

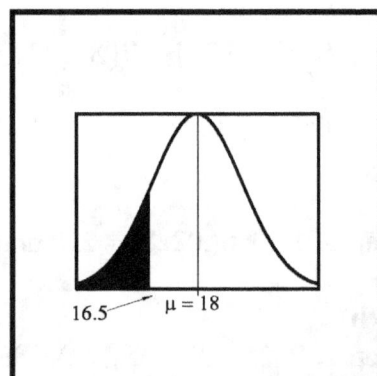

➤(c).

Step 1: $P\{X \le 16.5\} = 0.5 - P\{16.5 \le X \le 18\}$

Step 2: For x = 16.5,

$$z = \frac{x - \mu}{\sigma} = \frac{16.5 - 18}{2} = -0.75$$

fig. 58

58

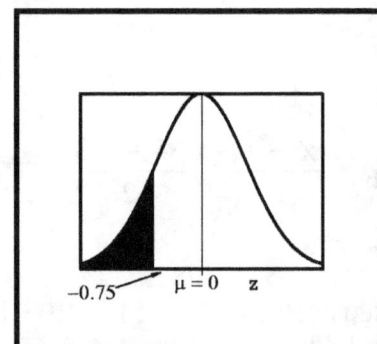

Step 3: $P\{16.5 \le X \le 18\} = P\{-0.75 \le Z \le 0\} =$
$P\{0 \le Z \le 0.75\}$

Step 4: $P\{0 \le Z \le 0.75\} = 0.2734$

fig. 59

59

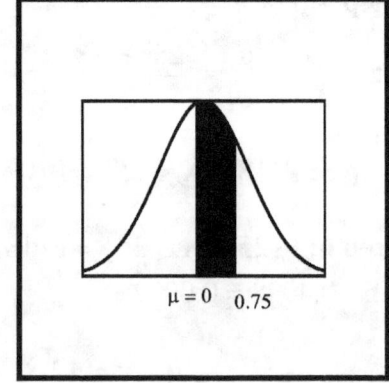

Step 5: $P\{X \le 16.5\} = 0.5 - P\{16.5 \le X \le 18\} = 0.5 - 0.2734$
$= 0.2266$

Step 6: $10,000(0.2266) \approx 2,266$ households

➤(d)..

Step 1: $P\{16 \le X \le 20\} = P\{16 \le X \le 18\} + P\{18 \le X \le 20\}$

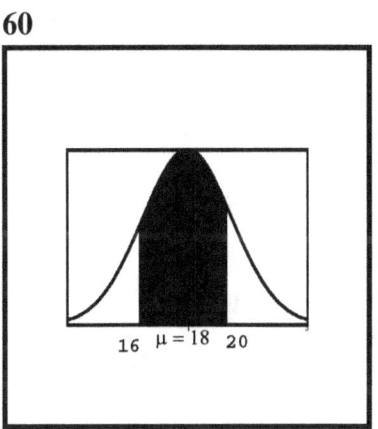

60

Step 2: For x = 16,

$$z = \frac{x - \mu}{\sigma} = \frac{16 - 18}{2} = -1$$

fig. 60

Step 3: $P\{16 \le X \le 18\} = P\{-1 \le Z \le 0\} = P\{0 \le Z \le 1\} = 0.3413$

61

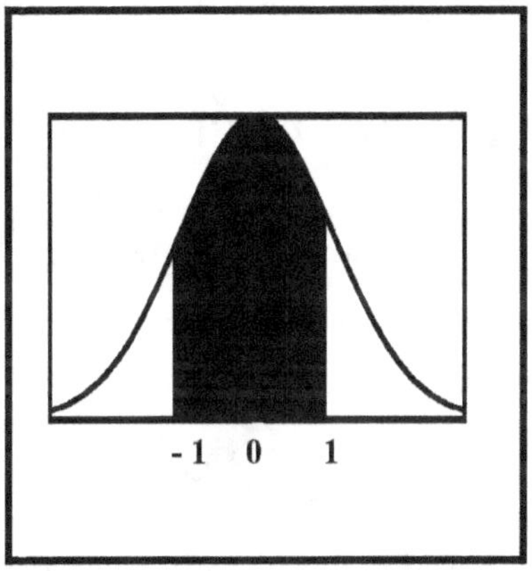

Step 4: For x = 20,

$$z = \frac{x - \mu}{\sigma} = \frac{20 - 18}{2} = 1$$

fig. 61

Step 5: $P\{18 \le X \le 20\} = P\{0 \le Z \le 1\} = 0.3413$

Step 6: $P\{16 \le X \le 20\} = P\{16 \le X \le 18\} + P\{18 \le X \le 20\} =$

$0.3413 + 0.3413 = 0.6826$

Step 7: $10,000(0.6826) = 6,826$

26.1 - Solved Problem 4: Ms. Jones, a new math instructor, wants to grade on a curved based on a normal distribution with points ranging from 0 to 100. In testing, she has complete freedom to set her average μ and standard deviation σ as she wishes.

(a). For her tests, she wants to assure that at most 5% of her students will get below 25 points. Assuming the standard deviation for a certain test is σ = 25 points, Find the average value of the test μ.

(b). She also wants to assure that 15% of her class gets at least a grade of 75 or better and 5% get 25 or less. Find μ, σ.

(c). Assume she decides, in designing her tests, that the average should be 65 points and 95% of all her students should receive a point grade between 55 and 75. Find σ.

62

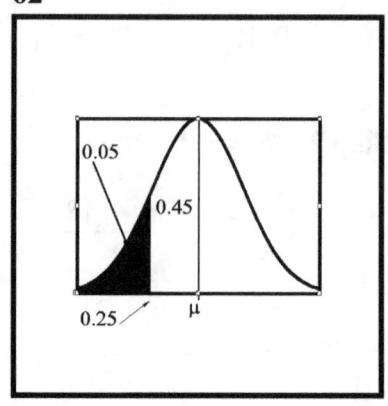

Solutions:

➤(a).
We need to use the formula μ = x - zσ.
fig. 62

Step 1: P{X ≤ 25} = 0.05

Step 2: P{25 ≤ X ≤ μ} = 0.5 -P{X ≤ 25} = 0.5 - 0.05 = 0.45

Step 3: P{z ≤ Z ≤ 0} = P{25 ≤ X ≤ μ} = 0.45

Step 4: From the area portion of the table, we find for area = 0.45 that z = -1.65 .
fig. 63

63

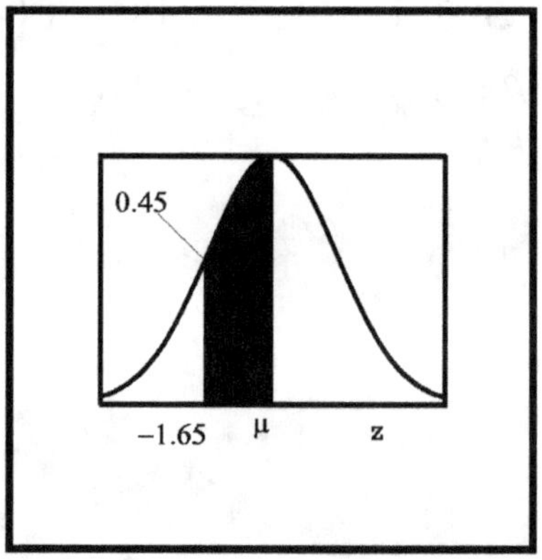

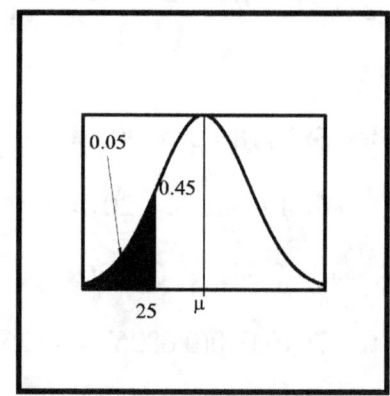

64

Step 5: Since x = 25, μ = x - zσ = 25 -(-1.65)(25) = 25 + 41.25 = 66.25

➤(b).
We need to use the formula μ = x - zσ twice.
fig. 64

Step 1: P{X ≤ 25} = 0.05

Step 2: P{25 ≤ X ≤ μ} = 0.5 - P{X ≤ 25} = 0.5 - 0.05 = 0.45

Step 3: P{z ≤ Z ≤ 0} = P{25 ≤ X ≤ μ} = 0.45

Step 4: From the area portion of the table, we find for area 0.45 that z = -1.65 .
fig. 65

65

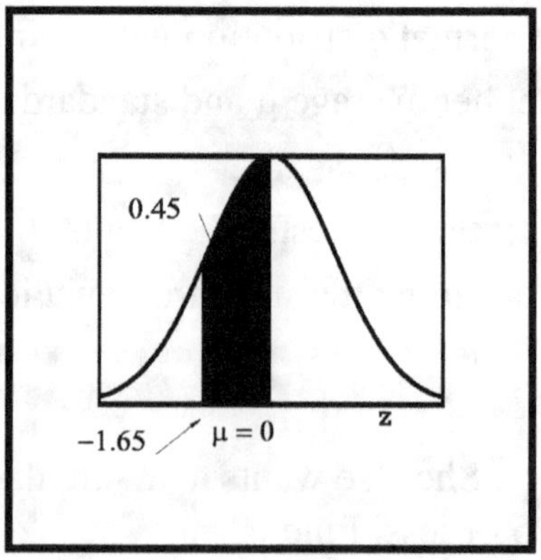

Step 5: Since x = 25, μ = x - zσ = 25 -(-1.65)σ = 25 + 1.65σ

Step 6: P{75 ≤ X} = 0.15

Step 7: P{μ ≤ X ≤ 75} = 0.5 - P{75 ≤ X} = 0.5 - 0.15 = 0.35
fig. 66

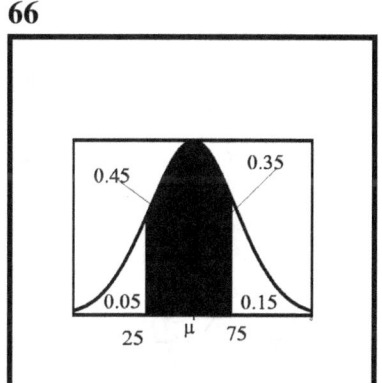

Step 8: P{0 ≤ Z ≤ z} = P{μ ≤ X ≤ 75} = 0.35

Step 9: From the area portion of the table, we find for area = 0.35 that z = 1.03 .
fig. 67

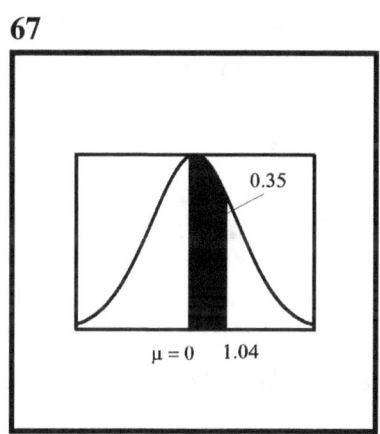

Step 10: Since x = 75, μ = x - zσ = 75 - 1.03σ

Step 11: We have two equations to solve for μ and σ:

μ = 25 + 1.65σ

μ = 75 - 1.03σ

Step 12: Since both means μ are the same, we set the two equations equal and solve for σ:

25 + 1.65σ = 75 - 1.03σ

Step 13: Using elementary algebra

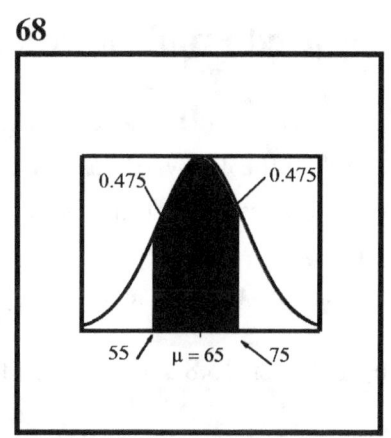

1.65σ + 1.03σ = 75 - 25

2.68σ = 50

σ $= \dfrac{50}{2.68}$ **= 18.66**

μ = 25 + 1.65σ = 25 + (1.65)(18.66) = 25 + 30.79 = 55.79

➤(c).

We need the formula   σ $= \dfrac{x - μ}{z}$.

fig. 68

Step 1: For the x value we use x = 75.

Step 2: The value z_0 associated with x = 75 can be found, since

$P\{65 \leq x \leq 75\} = P\{0 \leq z \leq z_0\} = 0.95/2 = 0.475$.

Step 3: Looking in the area portion of the normal distribution table, we find $z_0 = 1.96$.

Step 4: $\sigma = \dfrac{x - \mu}{z} = \dfrac{75 - 65}{1.96} \approx 5.10$

26.1 - Solved Problem 5: A local supermarket's records indicate that over a one year period, the price of hamburger is normally distributed with an average price of $1.65 and a standard deviation of $ 0.15. What is the highest price of the lowest 10% of all possible prices.

Solution:

Use the formula $x = \mu + z\sigma$.

Step 1: Since $\mu = 1.65$ and $\sigma = 0.15$, $x = \mu + z\sigma = 1.65 + z(0.15)$

Step 2: From the figure., we find z by looking up the value 0.40 in the normal distribution table.

Step 3: From this table z = -1.28 Therefore,

$x = \mu + z\sigma = 65.5 + z5 = 1.65 - 1.28(0.15) \approx \$1.46$

Unsolved Problem with Answers

26.1 - Problem 1: At a local high school's athletic program, the average time it takes a runner to complete the 100 yard dash is 14.6 seconds with a standard deviation of 2 seconds. A runner is selected at random. Find the probability that it will take him/her to run the 100 dash

(a). in more than 15 seconds.

(b). between 15.8 and 16 seconds.

(c). in less than 14 seconds.

Answers:

➤(a). 0.4207

➤(b). 0.0323

➤(c). 0.3821

⇑ *Refer back to* **26.1 - Example 1 & 26.1 - Solved Problem 1.**

26.1 - Problem 2: The U.S. Army recently reported that the average weight of a male combat soldier is μ = 170.75 lbs and standard deviation σ = 15.5 lbs. A male combat soldier is selected at random. Find the probability his weight is

(a). greater than 168.5 lbs

(b). between 172 and 174 lbs

(c). less than 172.25 lbs

(d). between 165 and 172 lbs.

Answers:

➤(a). 0.5596

➤(b). 0.0513

➤(c). 0.5398

➤(d). 0.1762

⇑ *Refer back to* **26.1 - Example 2 & 26.1 - Solved Problem 2.**

26.1 - Problem 3: A New York publishing firm's records over the last 20 years show that the average number of pages in novels it publishes is 354 pages with a standard deviation of 42 pages. Assume over that period, it published 1,321 novels. How many novels have been published with

(a). more than 400 pages?

(b). between 410 and 450 pages?

(c). less than 300 pages?

(d). between 300 and 400 pages?

Answers:

➤(a). 179

➤**(b).**   107

➤**(c).**   130

➤**(d).**   1012

⇑ *Refer back to* **26.1 - Example 3 & 26.1 - Solved Problem 3.**

26.1 - Problem 4: The Weight Reducers of America is promoting a new weight reduction program. In their advertisements, they claim that within 60 days, their new program will cause a weight loss of 20 pounds or more for 90% of the users.

(a). If the standard deviation of weight loss is 2.5 pounds, find the average weight loss one can expect.

(b). If they also claim that there is a 5% chance that a person using this plan will lose more than 27.65 pounds after 60 days, find μ, σ.

(c). After further research, the Company discovers that the average weight loss is 25 pounds and 85% of all users loss between 15 and 35 pounds. Find the standard deviation of weight lose.

Answers:

➤**(a).** $\mu = 23.2$ pounds

➤**(b).** $\mu = 23.35$ pounds,
          $\sigma = 2.62$ pounds

➤**(c).** $\sigma = 6.94$ pounds

⇑ *Refer back to* **26.1 - Example 4 & 26.1 - Solved Problem 4.**

26.1 - Problem 5: The average computer time used by employees of a large corporation is 4.75 hours per day with a standard deviation of 0.5 hours. Find the minimum hours used by the highest 10%.

Answer: 5.39 hours

⇑ *Refer back to* **26.1 - Example 5 & 26.1 - Solved Problem 5.**

Supplementary Problems

1. Suppose that the distribution of the number of items produced by an assembly line during an eight-hour shift can be approximated by a normal distribution with a mean value of 150 and a standard deviation of 10. What is the probability that

a. the number of items produced is at most 130?

b. at least 125 items are produced?

c. between 135 and 160 (inclusive) items are produced?

2. The lifetime of a certain brand of battery is normally distributed with a mean value of 6 hours and a standard deviation of 0.8 hours when it is used in a particular cassette player. Suppose two new batteries are independently selected and put into the player. The player will cease to function as soon as one of the batteries fails. What is the probability that the player functions for at least 7 hours.

3. A machine that cuts corks for wine bottles operates so that the diameter of the corks produced is approximately normally distributed with a mean of 0.3 cm and a standard deviation of 0.01 cm. The specifications call for corks with diameters between 0.29 and 0.31 cm. A cork not meeting these specifications is considered defective.

a. What proportion of corks produced by this machine are defective?

b. To reduce the number of corks rejected, management wants it reduced to 5% the number of corks rejected. What new diameters should be used in considering if a cork is defective?

4. A machine fills containers with a particular product. The standard deviation of filling weights is known from past data to be 0.6 ounces. If only 2% of the containers hold less than 18 ounces, what is the mean filling weight for the machine. Assume the filling weights have a normal distribution.

5. The Environmental Protection Agency has in recent years developed a testing program to monitor vehicle emission levels of several pollutants. Data presented in a recent paper suggests that the normal distribution is a plausible model for the amount of oxides of nitrogen (g/mile) emitted. Suppose that this normal distribution has a mean of 1.6 and a standard deviation of 0.14 . What pollution level C is such that 99% of all such vehicles emit pollution amounts less than C?

6. Suppose that the force acting on a column that provides support for a building is normally distributed with mean 15 Kips and standard deviation 1.25 kips. What is the probability that the

force

a. is at most 17 Kips?

b. is less than 14 Kips?

c. is between 12 and 17 Kips?

d. differs from 15 Kips by more than two standard deviations?

7. An electronic device is used in an ocean-exploration apparatus. The operational life of the device is normally distributed with a mean of 500 hours and a standard deviation of 50 hours.

a. What is the probability that the device would fail prior to 420 hours of operation time?

b. What is the probability that the device would have a operational life between 510 and 530 hours?

8. Arithmetic test scores of 1,000 eleventh graders are normally distributed with a mean of 68 and a standard deviation of 10.

a. The teacher wants to set the score so that only 8% of the students will receive an A. What is the lowest score an A student can receive?

b. If a student is selected at random, what is the chance the student will receive a score between 70

 and 78?

9. The ABC Tire Company produces a radial tire which has a mean lifetime of 50,000 miles and a standard deviation of 5,000 miles. Assume that the lifetime is normally distributed. The manufacturer guarantees to replace free any tire that lasts less than x miles. Determine the value of x so that the manufacturer would have to replace only 2 percent of his tires.

10. The morning commute time between John Wayne airport and L.A. airport is normally distributed with a mean of 45 minutes and a standard deviation of 5 minutes.

Find the probability that

a. it takes at most 40 minutes to commute.

b. it takes between 50 and 60 minutes to commute.

c. 70% of the commute time is at least how many minutes?

11. A local bottling company fills 5,000 16 oz bottles an hour. Assume the fill is normally distributed with a standard deviation of 0.25 oz. What is the average (mean) fill per bottle to assure that at most 75% of the bottles filled have at most 16.1 oz?

12. Suppose that the distribution of the number of items produced by an assembly line during an eight-hour shift can be approximated by a normal distribution with a mean value of 150 and a standard deviation of 10. What is the probability that

a. the number of items produced is at most 130?

b. at least 125 items are produced?

c. between 135 and 160 (inclusive) items are produced?

13. A manufacturer of light bulbs has recently purchased 2 new machines. The life of each light bulb is normally distributed for each machine and $\mu = 2,800$ hours with a standard deviation of 210 hours.

a. Assume 10 light bulbs are randomly selected from machine A. Find the probability that at least 2 bulbs will have a life longer than 3,000 hours.

b. Assume 10 light bulbs are randomly selected from A and 10 are selected from B. Find the probability that at least 2 bulbs from machine A and at least 2 bulbs from machine B will have a life longer than 3,000 hours.

c. Assume 10 light bulbs are randomly selected from A and 10 are selected from B. Find the probability that at least 2 bulbs from machine A or at least 2 bulbs from machine B will have a life longer than 3,000 hours

d. Assume 10 light bulbs are randomly selected from A and 10 are selected from B. Find the probability that a total of 3 light bulbs will have a life longer than 3,000 hours.

14. Recently a computer trade journal published a study on the number of hours computer programmers worked each day, in the year 1999. The article stated that the average number of hours worked each day was $\mu = 10$ hours. The study further showed that 2% of the programmers work more than 17 hours each day. Find the probability that the number of hours he/she worked each day in 1999 was greater than 13 hours.

15 A local supermarket's records indicate that over a one year period, the price of hamburger is normally distributed. From a random sample of hamburger the following information was obtained:

$P(X > \$1.85 | X < \$2.85) = 0.75$ and $P(X < \$2.85) = 0.65.$, Find μ, σ.

16. At a local community college a research study shows that the students' GPA is normally distributed with an average $\mu = 2.6$. Given that 95% of the students have a GPA less than 3.5, what is the minimum GPA required in order for 10% of the students to be on the Dean's List?

Probability Theory
Lesson 27
The Normal Approximation
To
The Binomial Distribution
446

Assume a binomial experiment results in a binomial distribution with a sample size of N and the chance of success on each trial is p. If N is large, the normal distribution X can be used as a good approximation to the binomial distribution where the distribution is expressed as

$$P\{X \leq k + 0.5\} \text{ and } P\{k - 0.5 \leq X\}.$$

The addition and subtraction of 0.5 to k is called the continuity correction and is necessary for accuracy.

1

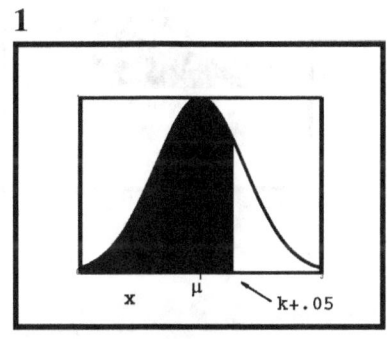

2

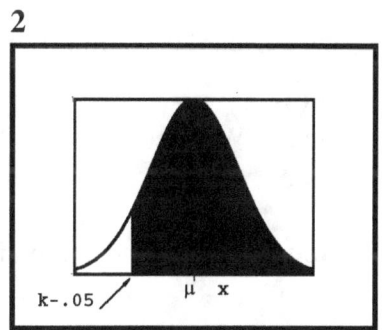

Since we are using the Normal Distribution to approximate the binomial distribution, we use the following formulas for μ and σ:

$$\mu = Np \text{ and}$$
$$\sigma = \sqrt{Npq} \text{ where } q = 1 - p$$

27.1-Real Life Applications

27.1 - Example 1: A well balanced coin is to be tossed 100 times. Find the probability that the number of heads that will occur is (a). at most 60.

(b). less than 55.

(c). at least 45.

(d). more than 41.

(e). is between 45 and 55.

(f). exactly 50.

Solutions:

This is a binomial distribution where $N = 100$ and $p = 0.50$,

the mean: $\mu = Np = 100(.5) = 50$.

and the standard deviation: $\sigma = \sqrt{Npq} = \sqrt{100(0.5)(0.5)} = 5$.

Therefore, the normal distribution has a mean $\mu = 100(.5) = 50$
and standard deviation $\sigma = \sqrt{Npq} \; \sqrt{100(0.5)(0.5)} = 5$.

➤ **(a).**
Step 1: We need to find $P\{X \le 60\}$.
fig. 3

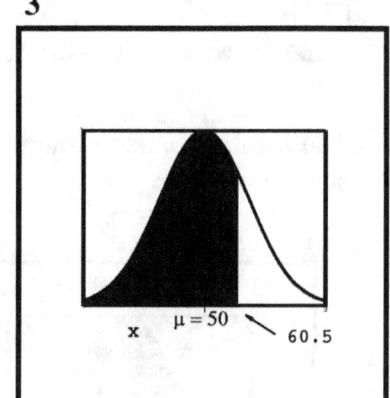

Step 2: Making the continuity correction: $P\{X \le 60\} = P\{X \le 60.5\}$.

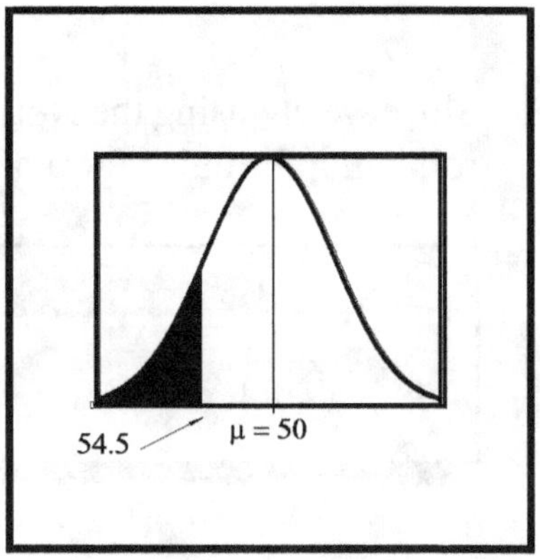

Step 3: $z = \dfrac{x - \mu}{\sigma} = \dfrac{60.5 - 50}{5} = 2.1$

Step 4: From the Standard normal distribution table,

$P\{Z \le 2.1\} = 0.4821 + 0.5 = 0.9821$.

Step 4: $P\{X \le 60.5\} = P\{Z \le 2.1\} = 0.9821$

➤ **(b).**
Step 1: We need to find $P\{X < 55\} = P\{X \le 54\}$.
fig. 4

Step 2: Making the continuity correction: $P\{X \le 54\} = P\{X \le 54.5\}$.

Step 3: $z = \dfrac{x - \mu}{\sigma} = \dfrac{54.5 - 50}{5} = 0.9$

Step 4: From the standard normal distribution table,

$P\{Z \le 0.9\} = 0.3159 + 0.5 = 0.8159$.

Step 5: $P\{X < 55\} = P\{X \le 54\} = P\{X \le 54.5\} = P\{Z \le 0.9\} = 0.8159$

➤ **(c).**
Step 1: We need to find $P\{X \ge 45\}$.
fig. 5

Step 2: Making the continuity correction: $P\{X \ge 44.5\}$.

Step 3: $z = \dfrac{x - \mu}{\sigma} = \dfrac{44.5 - 50}{5} = -1.1$

Step 4: From the standard normal distribution table

$P\{-1.1 \le Z\} = 0.3643 + 0.5 = 0.8643$.

Step 5: $P\{X \ge 45\} = P\{X \ge 44.5\} = P\{Z \ge -1.1\} = 0.8643$

➤ **(d).**
Step 1: We need to find $P\{X > 41\} = P\{X \ge 42\}$.
fig. 6

Step 2: Making the continuity correction: $P\{X \ge 41.5\}$.

Step 3: $z = \dfrac{x - \mu}{\sigma} = \dfrac{41.5 - 50}{5} = -1.7$

Step 4: From the Standard normal distribution table

$P\{Z \ge -1.7\} = 0.4554 + 0.5 = 0.9554$.

Step 5: $P\{41 < X\} = P\{41.5 \le X\} = P\{-1.7 \le Z\} = 0.9554$

➤ **(e).**
Step 1: We need to find $P\{45 \le X \le 55\}$.

Step 2: Making the continuity correction:

5

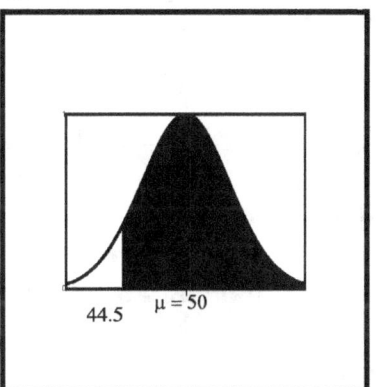

6

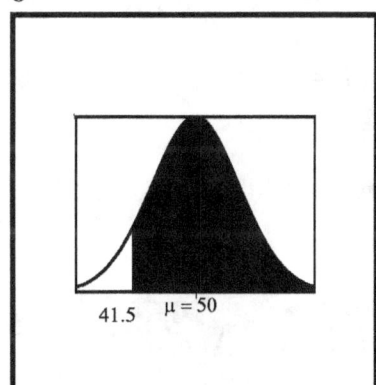

7

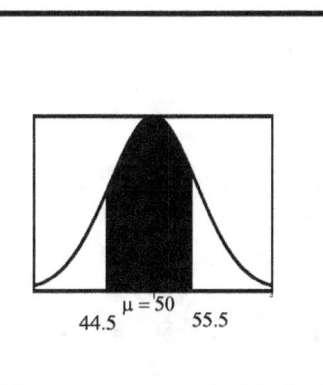

$P\{44.5 \le X \le 55.5\}$.
fig. 7

Step 3: $P\{44.5 \le X \le 55.5\} = P\{50 \le X \le 55.5\} + P\{44.5 \le X \le 50\}$

Step 4: $z = \dfrac{x - \mu}{\sigma} = \dfrac{55.5 - 50}{5} = 1.1$

$P\{50 \le X \le 55.5\} = P\{0 \le Z \le 1.1\} = 0.3643$

Step 5: $z = \dfrac{x - \mu}{\sigma} = \dfrac{44.5 - 50}{5} = -1.1$

$P\{44.5 \le X \le 50\} = P\{-1.1 \le Z \le 0\} = 0.3643$

Step 6: $P\{44.5 \le X \le 55.5\} = P\{50 \le X \le 55.5\} + P\{44.5 \le X \le 50\} = 0.3643 + 0.3643 = 0.7286$

Step 7: $P\{45 \le X \le 55\} = P\{50 \le X \le 55.5\} + P\{44.5 \le X \le 50\} = 0.7286$

➤ **(f).**
Step 1: We need to find $P\{X = 50\}$.

Step 2: Making the continuity correction:

$P\{X = 50\} = P\{49.5 \le X \le 50.5\}$.
fig. 8

Step 3: $P\{49.5 \le X \le 50.5\} = P\{50 \le X \le 50.5\} + P\{49.5 \le X \le 50\}$

Step 4: $z = \dfrac{x - \mu}{\sigma} = \dfrac{50.5 - 50}{5} = 0.1$

$P\{50 \le X \le 50.5\} = P\{0 \le Z \le 0.1\} = 0.0398$

Step 5: $z = \dfrac{x - \mu}{\sigma} = \dfrac{49.5 - 50}{5} = -0.1$

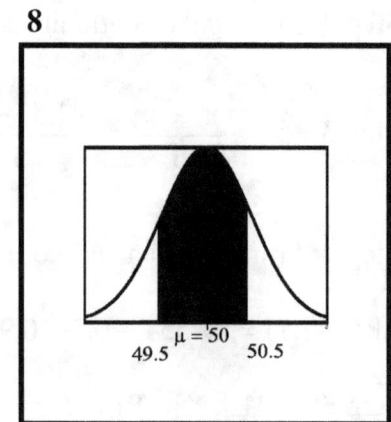

$P\{49.5 \le X \le 50.5\} = P\{-.1 \le Z \le 0\} = 0.0398$

Step 6: $P\{49.5 \le X \le 50.5\} = P\{50 \le X \le 50.5\} + P\{49.5 \le X \le 50\} = 0.0398 + 0.0398 = 0.0796$

Step 7: $P\{X = 50\} = P\{49.5 \le X \le 50\} + P\{50.5 \le X \le 50\} = 0.0796$

Solved Problems

27.1 - Solved Problem 1: College enrollment records show that 35% of all college students enrolled the previous year at a local community college had graduated from high school. A random sample of 200 students are selected. Find the probability that

(a). at most 87 students had graduated from high school the previous year.

(b). less than 80 students had graduated from high school the previous year.

(c). at least 75 students had graduated from high school the previous year.

(d). more than 65 students had graduated from high school the previous year.

(e). between 70 and 75 students had graduated from high school the previous year.

(f). exactly 70 students had graduated, the previous year, from high school.

Solutions:

This is a binomial distribution where N = 200 and p = 0.35 .

The mean: $\mu = Np = 200(.35) = 70$

The standard deviation: $\sigma = \sqrt{Npq} = \sqrt{200(0.35)(0.65)} = 6.75$

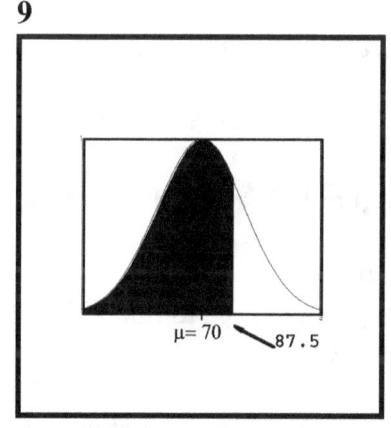

➤(a).
Step 1: We need to find $P\{X \le 87\}$.
fig. 9

Step 2: Making the continuity correction: $P\{X \le 87\} = P\{X \le 87.5\}$.

Step 3: $z = \dfrac{x - \mu}{\sigma} = \dfrac{87.5 - 70}{6.75} = 2.59$

Step 4: From the standard normal distribution table,
$P\{Z \le 2.59\} = 0.4952 + 0.5 = 0.9952$.

Step 5: $P\{X \le 87.5\} = P\{Z \le 2.59\} = 0.9952$

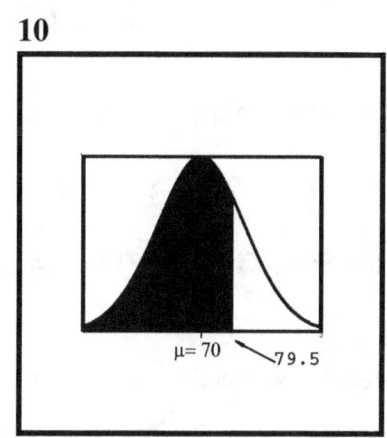

➤(b).
Step 1: We need to find $P\{X < 80\} = P\{X \le 79\}$.
fig. 10

Step 2: Making the continuity correction: $P\{X \leq 79\} = P\{X \leq 79.5\}$.

Step 3: $z = \dfrac{x - \mu}{\sigma} = \dfrac{79.5 - 70}{6.75} = 1.41$

Step 4: From the standard normal distribution table,

$P\{Z \leq 1.41\} = 0.4207 + 0.5 = 0.9207$.

Step 5: $P\{X < 80\} = P\{X \leq 79\} = P\{X \leq 79.5\} = P\{Z \leq 1.41\} = 0.9207$

➤ **(c).**
Step 1: We need to find $P\{X \geq 75\}$.
fig. 11

11

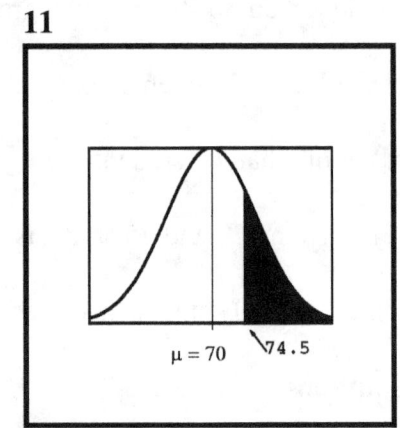

Step 2: Making the continuity correction: $P\{X \geq 74.5\}$.

Step 3: $z = \dfrac{x - \mu}{\sigma} = \dfrac{74.5 - 70}{6.75} = 0.67$

Step 4: From the standard normal distribution table,

$P\{Z \geq 0.67\} = 0.5 - 0.2486 = 0.2514$.

Step 5: $P\{X \geq 75\} = P\{X \geq 74.5\} = P\{Z \geq 0.67\} = 0.2516$

12

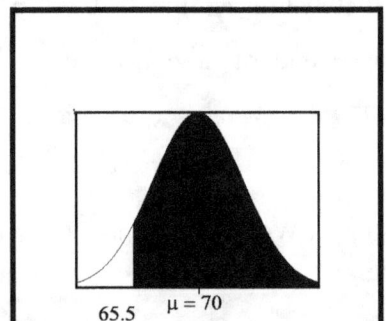

➤ **(d).**
Step 1: We need to find $P\{X > 65\} = P\{X \geq 66\}$
fig. 12

Step 2: Making the continuity correction: $P\{X \geq 65.5\}$.

Step 3: $z = \dfrac{x - \mu}{\sigma} = \dfrac{65.5 - 70}{6.75} = -0.67$

Step 4: From the standard normal distribution table,

$P\{Z \geq -0.67\} = 0.2486 + 0.5 = 0.7486$.

Step 5: $P\{X > 65\} = P\{X \geq 64.5\} = P\{Z \geq -0.67\} = 0.7480$

13

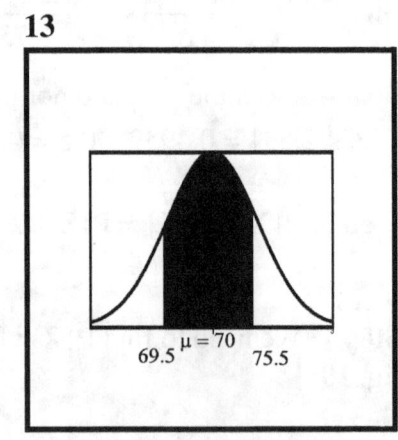

➤ **(e).**
Step 1: We need to find $P\{70 \leq X \leq 75\}$.

Step 2: Making the continuity correction: $P\{69.5 \leq X \leq 75.5\}$.

fig. 13

Step 3: $P\{69.5 \le X \le 75.5\} = P\{69.5 \le X \le 70\} + P\{70 \le X \le 75.5\}$.

Step 4: $z = \dfrac{x - \mu}{\sigma} = \dfrac{69.5 - 70}{6.75} = -.074$

$P\{69.5 \le X \le 70\} = P\{-0.074 \le Z \le 0\} = 0.0279$

Step 5: $z = \dfrac{x - \mu}{\sigma} = \dfrac{75.5 - 70}{6.75} = 0.81$

$P\{70 \le X \le 75.5\} = P\{0 \le Z \le 0.81\} = 0.2910$

Step 6: $P\{69.5 \le X \le 75.5\} = P\{69.5 \le X \le 70\} + P\{70 \le X \le 75\} = 0.0279 + 0.2910 = 0.3189$

Step 7: $P\{70 \le X \le 75\} = 0.3189$

➤ **(f)**.
Step 1: We need to find $P\{X = 70\}$.
fig. 14

Step 2: Making the continuity correction:

$P\{X = 70\} = P\{69.5 \le X \le 70.5\}$.

Step 3: $P\{69.5 \le X \le 70.5\} = P\{69.5 \le X \le 70\} + P\{70 \le X \le 70.5\}$

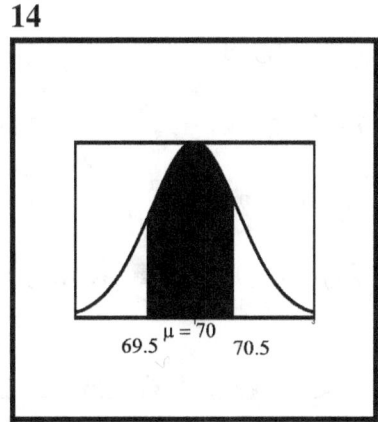

14

Step 4: $z = \dfrac{x - \mu}{\sigma} = \dfrac{70.5 - 70}{6.75} = 0.07$

$P\{70 \le X \le 70.5\} = P\{0 \le Z \le 0.07\} = 0.0279$

Step 5: $z = \dfrac{x - \mu}{\sigma} = \dfrac{69.5 - 70}{6.75} = -0.074$

$P\{69.5 \le X \le 70\} = P\{-0.07 \le Z \le 0\} = 0.0279$

Step 6: $P\{69.5 \le X \le 70.5\} = 0.0279 + 0.0279 = 0.0558$

Step 7: $P\{X = 70\} = P\{69.5 \le X \le 70\} + P\{70 \le X \le 70.5\} = 0.0558$

Unsolved Problems with Answers

27.1 - Problem 1: Mr. Jones purchased a computer program to predict the final outcome of football games. The manufacture's claim that the program has a success rate of 60% in predicting the outcomes of these games. To test this claim Mr. Jones checks the results of the outcome of the predictions on the next 50 games. Find the probability that

(a). at most 35 games were successfully predicted.

(b). less than 33 games were successfully predicted.

(c). at least 35 games were successfully predicted.

(d). more than 35 games were successfully predicted.

(e). between 32 and 35 games were successfully predicted.

(f). exactly 35 games were successfully predicted.

Answers:

➤ **(a).** 0.9441

➤ **(b).** 0.7642

➤ **(c).** 0.0968

➤ **(d).** 0.0548

➤ **(e).** 0.2788

➤ **(f).** 0.042

⇑ *Refer back to* **27.1 - Example 1 & 27.1 - Solved Problem 1.**

27.2 - Binomial Decision Theory

It is possible to test the probability of success of each trial in a binomial experiment. In accepting or rejecting the success claimed, we can compute the probability of making an error in our decision.

27.2 - Example 1: Mr. Jones purchased a computer program to predict the final outcome of football games. The manufacture's claim that the program has a success rate of 60% or more in predicting the outcomes of these games. To test this claim Mr. Jones checks the results of the outcome of the predictions on the next 100 games and uses the following decision rule:

If the outcome of less than 58 games are predicted correctly, then Mr. Jones will return the program for a full refund. However, if 58 or more games are predicted correctly then Mr. Jones will use the program.

(a). Assume the program has a prediction success of $p = 0.62$. Find the probability that Mr. Jones will return the program.

(b). Assume the program has a prediction success of $p = 0.55$. Find the probability that Mr. Jones will not return the program.

(c). Modify the decision rule so that the chance of returning the program is only 0.05 when the success of prediction is $p = 0.62$.

Solutions:

➤ **(a).**
Even though the program's success at predicting the outcome of football games is $p = 0.62$, it is possible, due to random variation, that out of 100 games, less than 58 are successful predicted correctly. We can find this probability by following these steps:

Step 1: Since $p = 0.62$ and $n = 100$,
fig. 15

$\mu = np = 100(.62) = 62.$

$\sigma = \sqrt{npq} = \sqrt{100(0.62)(0.38)} \approx 4.85$

Step 2: The probability of at most 57 successes is $P\{X \le 57\}$ $= P\{X \le 57.5\}$.

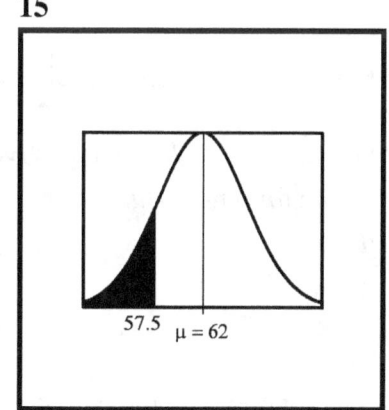

15

57.5 $\mu = 62$

Step 3: $z = \dfrac{57.5 - 62}{4.85} = -0.93$

Step 4: $P\{X \le 57\} = P\{X \le 57.5\} = P\{X \le -0.93\} = 0.5 - 0.3238 = 0.1762$

Step 5: The probability of rejecting the claim and returning the program is $P\{X \le 57\} = 0.1762$

➤ **(b).**
Even though the program's success at predicting the outcome of football games is only $p = 0.55$, it is possible, due to random variation, that out of 100 games, more than 57 games are predicted correctly.

We can find this probability by following these steps:
fig. 16

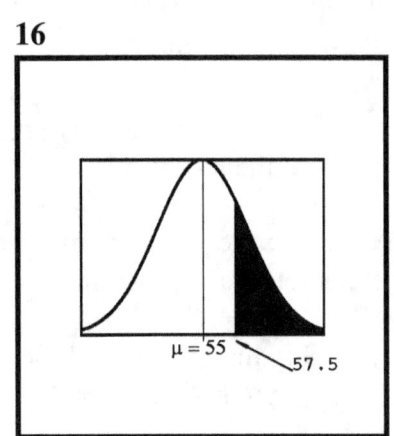

16

Step 1: Since $p = 0.55$ and $n = 100$

$\mu = np = 100(0.55) = 55$

$\sigma = \sqrt{npq} = \sqrt{100(0.55)(0.45)} \approx 4.97$

Step 2: The probability of at least 58 successes $= P\{X \geq 58\}$
$= P\{X \geq 57.5\}$.

Step 3: $z = \dfrac{57.5 - 55}{4.97} = 0.50$

Step 4: $P\{X \geq 58\} = P\{X \geq 57.5\} = P\{X \geq 0.5\} = 0.5 - 0.1915 = 0.3085$

Step 5: The probability of accepting the claim and returning the program is $P\{X \geq 58\} = 0.3085$

➤ **(c).**
Let us restate the decision rule as follows:

If the outcome of less than c^* games are predicted correctly, then Mr. Jones will return the program for a full refund. However, if c^* or more games are predicted correctly, then Mr. Jones will use the program.

We need to find c^* so that the chance of returning the program is only 0.05 when $p = 0.62$, the chance of
predicting correctly the outcomes of games.

To find c^*, we use the formula:

$c^* = \mu + z\sigma.$

Step 1: Find c* where $P\{X \le c^*\} = 0.05$.

fig. 17

Step 2: We use the formula

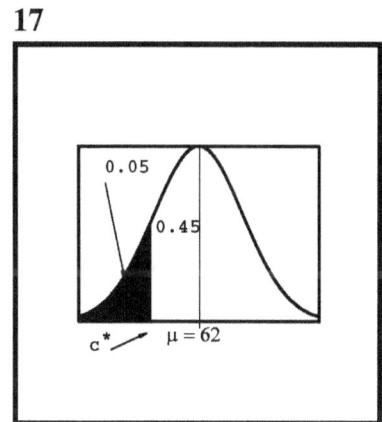

$c^* = \mu + z\sigma$.

Step 3: Since p = 0.62 and n = 100

$\mu = np = 100(0.62) = 62$.

$\sigma = \sqrt{npq} = \sqrt{100(0.62)(0.38)} \approx 4.85$

Step 4: $P\{X \le c^*\} = P\{Z \le z\} = 0.05$ Therefore, $P\{z \le Z \le 0\} = 0.5 - P\{Z \le z\} = 0.45$.

Step 5: From the table, z = -1.64

Step 6: $c^* = \mu + z\sigma = 62 - 1.64(4.85) = 62 - 7.95 \approx 54$ successes

The decision rule now reads:

If the outcome of at most 54 games are predicted correctly, then Mr. Jones will return the program for a full refund. However, if 55 or more games are predicted correctly, then Mr. Jones will use the program.

Solved Problems

27.2 - Solved Problem 1: Mrs. Pillar is running for reelection to Congress. Her opponent is against free trade. She hires a political analyst to take a poll of 200 voters from her district to find out the number in favor of free trade. She decides on the following decision rule for supporting free trade in her campaign:

If at least 105 of the voters say they are in favor of free trade, she will state in her election advertisements that she supports free trade. However if less than 105 say they are in favor of free trade, she will state in her election advertisements that she does not support free trade.

(a). Assume 55% of all voters in the District support free trade. Find the probability that she will not support free trade in her campaign.

(b). Assume 50% of all voters in the District support free trade. Find the probability that she will support free trade in her campaign.

(c). Modify the decision rule so the chance she will support free trade in her campaign is 0.01 even though only 50% of all voters in the district support free trade.

Solutions:

18

➤ (a).
We can find this probability by following these steps:
fig. 18

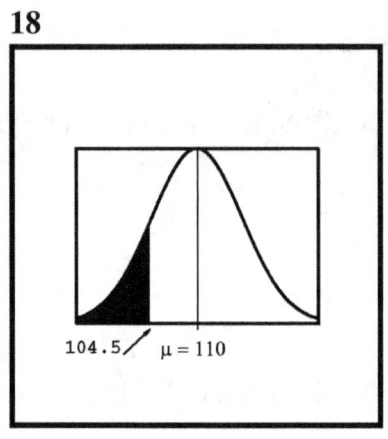

Step 1: Since p = 0.55 and n = 200 .

$\mu = np = 200(.55) = 110$

$\sigma = \sqrt{npq} = \sqrt{200(0.55)(0.45)} \approx 7.04$

Step 2: The probability of less than 105 voters in favor of free trade:

$P\{X \le 104\} = P\{X \le 104.5\}$.

Step 3: $z = \dfrac{104.5 - 110}{7.04} = -0.78$

Step 4: $P\{X \le 104\} = P\{X \le 104.5\} = P\{X \le -.78\} = 0.5 - 0.2823 = 0.2177$

Step 5: The probability of her not supporting free trade is $P\{X \le 104\} = 0.2177$.

➤ (b).
fig. 19

19

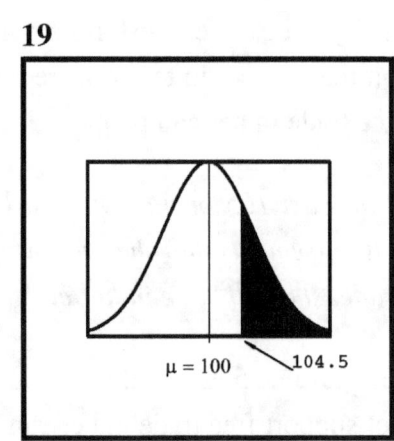

Step 1: Since p = 0.50 and n = 200.

$\mu = np = 200(.5) = 100$

$\sigma = \sqrt{npq} = \sqrt{200(0.5)(0.5)} \approx 7.07$

Step 2: The probability of at least 105 voters in favor of free trade

is $P\{X \ge 105\} = P\{X \ge 104.5\}$.

Step 3: $z = \dfrac{104.5 - 100}{7.07} \approx 0.64$

Step 4: $P\{X \geq 105\} = P\{X \geq 104.5\} = P\{X \geq .64\} = 0.5 - 0.2389 = 0.2611$

Step 5: The probability of her supporting free trade is $P\{X \geq 105\} = 0.2611$.

➤ (c).
Let us restate the decision rule as follows:

If the number of voters is at least c*, then she will support free trade. If the number of voters is less than c*, then she will not support free trade.

We need to find c* so that the chance of her supporting free trade is only 0.01 even though only 50% of all voters support it.

Step 1: Find c* where $P\{c^* \leq X\} = 0.01$.
fig. 20

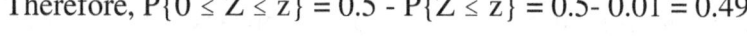

20

Step 2: We use the formula
$c^* = \mu + z\sigma$.

Step 3: Since p = 0.50 and n = 200

$\mu = np = 200(0.50) = 100$.

$\sigma = \sqrt{npq} = \sqrt{200(0.5)(0.5)} \approx 7.07$

Step 5: $P\{c^* \leq X\} = P\{Z \leq z\} = 0.01$

Therefore, $P\{0 \leq Z \leq z\} = 0.5 - P\{Z \leq z\} = 0.5 - 0.01 = 0.49$

Step 6: From the table, z = 2.33

Step 7: $c^* = \mu + z\sigma = 100 + 2.33(7.07) = 100 + 16.4731 \approx 116.47$ voters in favor of free trade.

The decision rule now reads:

If at least 117 of the voters say they are in favor of free trade, she will state in her election advertisements that she supports free trade. However, if less than 116 say they are in favor of free trade, she will state in her election advertisements that she does not support free trade.

Unsolved Problems with Answers

27.2 - Problem 1: A recent study showed that the majority of people that purchase classical music compact disks are adult males. An advertising company was hired by a large classical recording company to check the accuracy of this study. The advertising firm decided to take a sample of 300 adult consumers of classical music compact disks. They use the following decision rule:

If at least 180 of these customers are male, then they conclude the study is correct. If less than 180 of these customers are male, then they conclude the study is faulty and reject it.

(a). Assume 55% of all adults that purchase these compact disks are male. Find the probability they will reject the study.

(b). Assume 48% of all adults that purchase these compact disks are male. Find the probability they will accept the study.

(c). Modify the decision rule so the chance is 0.05 that they will reject the study even though 60% of all adults that purchase these disks are male.

Answers:

➤ **(a).** 0.9535

➤ **(b).** 0

➤ **(c).**
If at least 174 of these customers are male, then they conclude the study is correct. If less than 174 of these customers are male, then they conclude the study is faulty and reject it.

⇑ *Refer back to* **27.2 - Example 1 & 27.2 - Solved Problem 1.**

Supplementary Problems

1. According to a national survey, 35% of all registered voters are Republicans. Two hundred registered voters are sampled. Find the probability that

a. less than 80 are Republicans.

b. less than 122 are not Republicans.

c. more than 110 are not Republicans.

d. between 120 and 140 are not Republicans.

2. The MacroStar Computing Company purchases its micro chips from a large Northern California computing manufacturer. This company claims that 5% of all the chips it produces are defective. Each day, MacroStar receives boxes each containing 500 micro chips. Inspecting each box, MacroStar uses the following decision rule:

A random sample of 50 chips is selected from each box. If four or more chips are defective, the
box is rejected and returned to the manufacturing company.

On a given day, ten boxes are inspected. Find the probability that at most one box is returned.

3. A U.S. Air Force fighter plane fires 75 missiles at different targets. Assuming that the chance
that any missile will hit its target is 0.90 find:

a. at least 50 targets are hit.

b. exactly 70 targets are hit.

c. less than 45 targets are hit.

d. between 70 and 75 targets are hit.

4. The Clear Water Bottling Company each day fills 10,000 bottles with 16 ounces of spring water.
Due to the imperfections of the filling machinery, government regulations require that no more
than 3% of all bottles contain less than 16 ounces. To enforce these rules, government inspectors
will use the following decision rule:

Select at random 50 bottles from the production line. If 2 or more of these bottles contains less
than 16 ounces, then the production line is shut down.

For five different inspections, find the probability that it will be shut down exactly three times.

 5. Mr. Jones purchased a computer program to predict the final outcome of football games. The
manufacture's claim that the program has a success rate of 60% or more in predicting the outcomes
of these games. To test this claim, Mr. Jones checks the results of the outcome of the predictions
on the next N games and uses the following decision rule:

If the outcome of less than 58% are predicted correctly, then Mr. Jones will return the program
for a full refund. However, if 58% or more games are predicted correctly then Mr. Jones will use
the program.

Find the sample size N so that the chance of returning the program is only 0.05 when the success
of prediction is p = 0.62 .

6. A recent study showed that the majority of people that purchase classical music compact disks
are adult males. An advertising company was hired by a large classical recording company to
check the accuracy of this study. The advertising firm decided to take a sample of N adult
consumers of classical music compact disks. They use the following decision rule:

If at least c* of these customers are male, then they conclude the study is correct. If less than c* of these customers are male, then they conclude the study is faulty and reject it.

Find N and c* so that if 55% of all adults that purchase these compact disks are male, the probability they will reject the study is 0.01 and if 48% of all adults that purchase these compact disks are male, the probability they will accept the study is 0.05 .

7. The ABC Tire Company receives each week a large shipment of tire valves. If 5% or more of the valves are defective, the ABC Company will send the entire shipment back. Management decides on the following decision rule:

A sample of 200 valves are randomly selected and tested. If 7% or more valves are found defective, the entire shipment is returned; otherwise the entire shipment is kept.

Find the following probabilities:

a. If a shipment has 4% defective valves, the entire shipment is returned.

b. If a shipment has 6% defective valves, the entire shipment is kept.

8. The Irvine city council is concerned that less than 80% of all households have smoke alarms. If this is so, they wish to pass an ordinance requiring that all households must install such alarms. They hire you to take a random sample of 100 residents. If 85% or less in the survey have smoke alarms, then they will pass the ordinance; otherwise the ordinance will not be passed.

a. Assume that 90% of all households have smoke alarms, what is the chance the city council will pass the ordnance.

b. Assume that 78% of all households have smoke alarms, what is the chance the city council will not pass the ordnance.

9. Mrs. Smith teaches a class in Modern American Literature. Her past records show that over the last five years, 25% of students taking this class earn a final grade of B or better. Using this result, if 150 students are randomly selected, find the probability that

a. less than 50 of these students will receive a grade of B or better.

b. less than half earn a grade less than a B.

c. more than 112 students earn a grade less than a B.

d. between 30 and 46 students earn a grade of B or better.

10. Government regulations require that at most 5% of the bottles can contain less than 16 oz. To monitor the amount of filling per bottle, each hour 100 bottles are selected at random and the

following decision rule is used:

If 4 bottles or more contain less than 16 oz then the filling process is stopped and the machine is adjusted. Otherwise the filling process is allowed to continue.

a. Assume the machine is filling 6% of the bottles with less than 16 oz of soda. What is the probability that the process will not be stopped?

b. Assume the machine is filling 3.5% of the bottles with less than 16 oz of soda. What is the probability that the process will be stopped?

11. Suppose a manufacturer of calculators chooses 200 stamped circuits from the day's production to determine the number of defective circuits in the sample. Suppose that 6% are defective in the manufacturing process.

a. What is the standard deviation of the sample?

b. what is the probability that at least 20 defective are observed?

A decision rule is to be established to stop the production process based on this sample. Assume the decision rule is as follows:

Decision Rule: If at least x calculators from the sample are defective then stop the process; otherwise allow the process to continue.

c. Find the value x so that there is a 5% chance of stopping this process.

12. A quality control engineer stated in a report that 2% of all floppy disk drives manufactured by his company are defective. If a box of 100 drives are shipped to a customer, the probability of at most 2 defective drives using

a. the binomial distribution.

b. the Poisson distribution.

c. the normal approximation to the binomial distribution.

d. the normal approximation to the Poisson distribution, where $\sigma^2 = \mu$. (See Lesson 24, problem 9.)

13. In a small city, 2 movie theaters are showing the same film, *Ghosts of New York*, starting on Saturday, beginning at the same time. Assume 500 people will go to see this film in such a way that their theater selection is a Bernoulli sequential trials with a probability of 1/2 that it will select one of the two theaters.

a. If the Roxy theater has only 275 seats, estimate the probability that they will sell out for the performance.

b. What is the minimum number of seats Roxy has to have to assure with at most 5% chance it will sell out.

The proportion P random variable. Assume X is the binomial random variable where $\mu = pN$, We define the proportion random variable $P = X/N$ where $P = p'$, $0 \le p' \le 1$.

14. Find μ and σ_p^2.

15. *The normal approximation to the proportion distribution.* The standard normal distribution formula for

approximating the binomial distribution is $z = \dfrac{x - pN}{\sqrt{Npq}}$.

Find the standard normal distribution formula for approximating the proportion distribution.

16. A fair coin is tossed 100 times.

a. Find the probability that at least 60% of the tosses resulted in heads using the proportion distribution.

b. Compare the result obtained in a using the normal distribution approximation to the binomial distribution.

Estimating p'. To estimate p' we can use the formula: $p' = p \pm z\,\sigma_p$.

17. Assume a fair coin is tossed 100 times. Estimate with 95% confidence the range of p'.

Confidence interval. We can estimate with a certain confidence the variation of p' from p within a given error

by using $p' - z\sqrt{\dfrac{pq}{N}} \le p \le p' + z\sqrt{\dfrac{pq}{N}}$.

where z is determined by the confidence interval desired and $z\sqrt{\dfrac{pq}{N}}$ is the error.

18. Derive the confidence interval formula.

19. Assume N is fixed. Show that the largest value of $\sigma_P = \sqrt{\dfrac{pq}{N}}$ is $\sigma_P = \dfrac{1}{2\sqrt{N}}$.

20. Given p', find a confidence interval for p using $\sigma_P = \dfrac{1}{2\sqrt{N}}$.

21. *Sampling for p from a large population.* Assume from a large population, we take a survey to estimate the percentage of people of the population that smoke. A random sample of 400 people

is taken. This sample resulted in 23% of the those interviewed are smokers.

a. From these results, find the a 90% confidence interval that will estimate the percentage of people of the population that smoke.

b. If 23% is used to estimate the true percentage of smokers in the population, find the probability that the error created exceeds 2%. (Hint: Error = $|p' - p|$).

c. Find the minimum sample size so that the probability is 0.10 that the error exceeds 2% .

Table A - Cumulative Binomial Distribution: P{X ≥ x} for N = 20

p	0.05	0.10	0.15	0.20	0.25	0.30	0.35	0.40	0.45	0.50
x										
0	1.0	1.0	1.0	1.0	1.0	1.0	1.0	1.0	1.0	1.0
1	0.6415	0.8784	0.9612	0.9885	0.9968	0.9992	0.9998	1.0	1.0	1.0
2	0.2642	0.6083	0.8244	0.9308	0.9757	0.9924	0.9979	0.9995	0.9999	1.0
3	0.755	0.3231	0.5951	0.7939	0.9087	0.9645	0.9879	0.9964	0.9991	0.9998
4	0.0159	0.1331	0.3523	0.5886	0.7748	0.8929	0.9556	0.9840	0.9951	0.9987
5	0.0026	0.0432	0.1702	0.3704	0.5852	0.7625	0.8818	0.9490	0.9811	0.9941
6	0.0003	0.0113	0.0673	0.1958	0.3428	0.5836	0.7546	0.8744	0.9447	0.9793
7	0.0000	0.0024	0.0219	0.0867	0.2142	0.3920	0.5843	0.7500	0.8701	0.9423
8	0.0000	0.0004	0.0059	0.0321	0.1018	0.2277	0.3990	0.5841	0.7480	0.8684
9	0.0000	0.0001	0.0013	0.0100	0.0409	0.1133	0.2376	0.4044	0.5857	0.7483
10	0.0000	0.0000	0.0002	0.0026	0.0139	0.0480	0.1218	0.2447	0.4086	0.5881
11	0.0000	0.0000	0.0000	0.0006	0.0039	0.0171	0.0532	0.1275	0.2493	0.4119
12	0.0000	0.0000	0.0000	0.0001	0.0009	0.0051	0.0196	0.0565	0.1308	0.2517
13	0.0000	0.0000	0.0000	0.0000	0.0002	0.0013	0.0060	0.0210	0.0580	0.1316
14	0.0000	0.0000	0.0000	0.0000	0.0000	0.0003	0.0015	0.0065	0.0214	0.0577
15	0.0000	0.0000	0.0000	0.0000	0.0000	0.0000	0.0003	0.0016	0.0064	0.0207
16	0.0000	0.0000	0.0000	0.0000	0.0000	0.0000	0.0000	0.0003	0.0015	0.0059
17	0.0000	0.0000	0.0000	0.0000	0.0000	0.0000	0.0000	0.0000	0.0003	0.0013
18	0.0000	0.0000	0.0000	0.0000	0.0000	0.0000	0.0000	0.0000	0.0000	0.0002
19	0.0000	0.0000	0.0000	0.0000	0.0000	0.0000	0.0000	0.0000	0.0000	0.0000
20	0.0000	0.0000	0.0000	0.0000	0.0000	0.0000	0.0000	0.0000	0.0000	0.0000

Table A - Cumulative Binomial Distribution: P{X ≥ x} for N = 20

p =	0.55	0.60	0.65	0.70	0.75	0.80	0.85	0.90	0.95
x									
0	1.0	1.0	1.0	1.0	1.0	1.0	1.0	1.0	1.0
1	1.0	1.0	1.0	1.0	1.0	1.0	1.0	1.0	1.0
2	1.0	1.0	1.0	1.0	1.0	1.0	1.0	1.0	1.0
3	1.0	1.0	1.0	1.0	1.0	1.0	1.0	1.0	1.0
4	0.9997	0.9999	1.0	1.0	1.0	1.0	1.0	1.0	1.0
5	0.9985	0.9997	1.0	1.0	1.0	1.0	1.0	1.0	1.0
6	0.9936	0.9984	0.9997	1.0	1.0	1.0	1.0	1.0	1.0
7	0.9786	0.9935	0.9985	0.9997	1.0	1.0	1.0	1.0	1.0
8	0.9420	0.9790	0.9940	0.9987	0.9998	1.0	1.0	1.0	1.0
9	0.8692	0.9435	0.9804	0.9949	0.9991	0.9999	1.0	1.0	1.0
10	0.7507	0.8725	0.9468	0.9829	0.9961	0.9994	1.0	1.0	1.0
11	0.5914	0.7553	0.8782	0.9520	0.9861	0.9974	0.9999	1.0	1.0
12	0.4143	0.5956	0.7624	0.8867	0.9591	0.9900	0.9992	0.9999	1.0
13	0.2520	0.4159	0.6000	0.7723	0.8982	0.9679	0.9962	0.9996	1.0
14	0.1299	0.2500	0.4166	0.6080	0.7858	0.9133	0.9847	0.9976	1.0
15	0.0553	0.1256	0.2454	0.4164	0.6172	0.9042	0.9493	0.9887	0.9999
16	0.0189	0.0510	0.1182	0.2375	0.4148	0.6296	0.8625	0.9568	0.9990
17	0.0049	0.0160	0.0444	0.1071	0.2252	0.4111	0.6959	0.8670	0.9926
18	0.0009	0.0036	0.0121	0.0346	0.0913	0.2061	0.4550	0.6769	0.9561
19	0.0000	0.0005	0.0021	0.0076	0.0243	0.0692	0.2084	0.3917	0.8103
20	0.0000	0.0000	0.0002	0.0008	0.0030	0.0115	0.0490	0.1216	0.4420

Table B - POISSON DISTRIBUTION P{X ≤ x}

μ =	0.5	1.0	1.5	2.0	2.5	3.0	3.5	4.0	4.5	5.0
x										
0	0.6065	0.3679	0.2231	0.1353	0.0821	0.0498	0.0302	0.0183	0.0111	0.0067
1	0.9098	0.7358	0.5578	0.4060	0.2873	0.1991	0.1359	0.0916	0.0611	0.0404
2	0.9856	0.9197	0.8088	0.6767	0.5438	0.4232	0.3208	0.2381	0.1736	0.1247
3	0.9982	0.9810	0.9344	0.8571	0.7576	0.6472	0.5366	0.4335	0.3423	0.2650
4	0.9998	0.9963	0.9814	0.9473	0.8912	0.8153	0.7254	0.6288	0.5321	0.4405
5	1.0000	0.9994	0.9955	0.9834	0.9580	0.9161	0.8576	0.7851	0.7029	0.6160
6		0.9999	0.9991	0.9955	0.9858	0.9665	0.9347	0.8893	0.8311	0.7622
7		1.0000	0.9998	0.9989	0.9958	0.9881	0.9733	0.9489	0.9134	0.8666
8			1.0000	0.9998	0.9989	0.9962	0.9901	0.9786	0.9597	0.9319
9				1.0000	0.9997	0.9989	0.9967	0.9919	0.9829	0.9682
10					0.9999	0.9997	0.9990	0.9972	0.9933	0.9863
11					1.0000	0.9999	0.9997	0.9991	0.9976	0.9945
12						1.0000	0.9999	0.9997	0.9992	0.9980
13							1.0000	0.9999	0.9997	0.9993
14								1.0000	0.9999	0.9998
15								0.9887	1.0000	0.9999
16								0.9568	0.9990	1.0000

Table C - Standard Normal Distribution

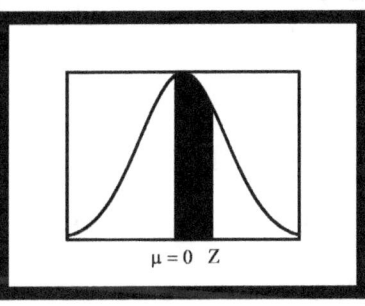

$\mu = 0$ Z

z =	0.00	0.01	0.02	0.03	0.04	0.05	0.06	0.07	0.08	0.09
0	.0000	.0040	.0080	.0120	.0160	.0199	.0239	.0279	.0319	.0359
0.1	.0398	.0438	.0478	.0517	.0557	.0596	.0636	.0675	.0714	.0754
0.2	.0793	.0832	.0871	.0910	.0948	.0987	.1026	.1064	.1103	.1141
0.3	.1179	.1217	.1255	.1293	.1331	.1368	.1406	.1443	.1480	.1517
0.4	.1554	.1591	.1628	.1664	.1700	.1736	.1772	.1808	.1844	.1879
0.5	.1915	.1950	.1985	.2019	.2054	.2088	.2123	.2157	.2190	.2224
0.6	.2257	.2291	.2324	.2357	.2389	.2422	.2454	.2486	.2517	.2549
0.7	.2580	.2612	.2642	.2673	.2704	.2734	.2764	.2794	.2823	.2852
0.8	.2881	.2910	.2939	.2967	.2995	.3023	.3051	.3078	.3106	.3133
0.9	.3159	.3186	.3212	.3238	.3264	.3289	.3315	.3340	.3365	.3389
1.0	.3413	.3438	.3461	.3485	.3508	.3531	.3554	.3577	.3599	.3621
1.1	.3643	.3665	.3686	.3708	.3729	.3749	.3770	.3790	.3810	.3830
1.2	.3849	.3869	.3888	.3907	.3925	.3944	.3962	.3980	.3997	.4015
1.3	.4032	.4049	.4066	.4082	.4099	.4115	.4131	.4147	.4162	.4177
1.4	.4192	.4207	.4222	.4236	.4251	.4265	.4279	.4292	.4306	.4319
1.5	.4332	.4345	.4357	.4370	.4382	.4394	.4406	.4418	.4429	.4441
1.6	.4452	.4463	.4474	.4484	.4495	.4505	.4515	.4525	.4535	.4545
1.7	.4554	.4564	.4573	.4582	.4591	.4599	.4608	.4616	.4625	.4633
1.8	.4641	.4649	.4656	.4664	.4671	.4678	.4686	.4693	.4699	.4706
1.9	.4713	.4719	.4726	.4732	.4738	.4744	.4750	.4756	.4761	.4767
2.0	.4772	.4778	.4783	.4788	.4793	.4798	.4803	.4808	.4812	.4817

z =	0.00	0.01	0.02	0.03	0.04	0.05	0.06	0.07	0.08	0.09
2.1	.4821	.4826	.4830	.4834	.4838	.4842	.4846	.4850	.4854	.4857
2.2	.4861	.4864	.4868	.4871	.4875	.4878	.4881	.4884	.4887	.4890
2.3	.4893	.4896	.4898	.4901	.4904	.4906	.4909	.4911	.4913	.4916
2.4	.4918	.4920	.4922	.4925	.4927	.4929	.4931	.4932	.4934	.4936
2.5	.4938	.4940	.4941	.4943	.4945	.4946	.4948	.4949	.4951	.4952
2.6	.4953	.4955	.4956	.4957	.4959	.4960	.4961	.4962	.4963	.4964
2.7	.4965	.4966	.4967	.4968	.4969	.4970	.4971	.4972	.4973	.4974
2.8	.4974	.4975	.4976	.4977	.4977	.4978	.4979	.4979	.4980	.4981
2.9	.4981	.4982	.4982	.4983	.4984	.4984	.4985	.4985	.4986	.4986
3.0	.4987	.4987	.4987	.4988	.4988	.4989	.4989	.4989	.4990	.4990
3.1	.4990	.4991	.4991	.4991	.4992	.4992	.4992	.4992	.4993	.4993
3.2	.4993	.4993	.4994	.4994	.4994	.4994	.4994	.4995	.4995	.4995
3.3	.4995	.4995	.4995	.4996	.4996	.4996	.4996	.4996	.4996	.4997
3.4	.4997	.4997	.4997	.4997	.4997	.4997	.4997	.4997	.4997	.4998
3.5	.4998	.4998	.4998	.4998	.4998	.4998	.4998	.4998	.4998	.4998

INDEX

About The Author

Howard Dachslager received a Ph.D. in mathematics from the University of California, Berkeley where he specialized in real analysis and probability theory. Prior to beginning his doctoral studies at the University of California, Berkeley, he earned a masters degree in economics from the University of Wisconsin. Since completing his Ph.D. in mathematics, he has taught mathematics to a diverse student population on many levels. As a faculty member of the Department of Mathematics at the University of Toronto he prepared and presented undergraduate level courses in mathematics. For several years he taught undergraduate mathematics courses in the Department of Mathematics, University of California, Berkeley. While working in the State Department's Alliance for Progress program, he taught advanced mathematics courses at a statistics institute in Santiago, Chile. Other teaching experience includes presenting undergraduate and community college mathematics courses.

Throughout his teaching career in mathematics, he has always attempted to find and use the most effective teaching methodologies to communicate an understanding of mathematics. Unable to find an appropriate text for use in his courses in statistics and probability theory, and drawing on his own extensive teaching experience, education and training, he developed a tutorial statistics and probability theory text that has significantly improved the performance of students in those courses. By focusing on problem solving, the student can learn to repeat the methodologies involved, reinforcing on understanding of the concepts clearly explained in the text.

www.ingramcontent.com/pod-product-compliance
Lightning Source LLC
Chambersburg PA
CBHW081428170526
45166CB00008B/2127